Nondestructive Evaluation of Agro-products by Intelligent Sensing Techniques

Edited by

Jiangbo Li

Beijing Research Center of Intelligent Equipment for Agriculture
Beijing Academy of Agriculture and Forestry Sciences
Beijing, China

&

Zhao Zhang

Department of Agricultural and Biosystems Engineering
North Dakota State University
North Dakota, USA

Nondestructive Evaluation of Agro-products by Intelligent Sensing Techniques

Editors: Jiangbo Li and Zhao Zhang

ISBN (Online): 978-981-14-8580-0

ISBN (Print): 978-981-14-8578-7

ISBN (Paperback): 978-981-14-8579-4

CONTENTS

PREFACE

With rapid progress in both theory and practical applications, Artificial Intelligence (AI) is transforming every aspect of life and leading the world to a sustainable future. AI technology is fundamentally and radically affecting agriculture in a positive manner to convert it to be smart – improved efficiency, reduced environmental pollutions, and enhanced productivity.

With such rapid progress in AI transforming the agriculture era, it is appropriate timing to publish a relevant book to update the progress to an academic and industrial domain, which inspires the generation of this book titled *Nondestructive Evaluation of Agro-products by Intelligent Sensing Techniques*. This book focuses on intelligent sensing techniques in the nondestructive evaluation of agro-products and describes existing and innovative techniques that could be or have been applied to agro-products' quality and safety evaluation, processing, harvest, traceability, *etc*.

The book includes 11 individual chapters, with each chapter focusing on a specific topic. Chapter 1 introduces representative techniques and methods for nondestructive evaluation, Chapters 2, 3, 5, 6 and 7 present quality evaluation of agro-products (*e.g.*, fruits, vegetables and meat) based on intelligent sensing technologies, including machine vision, near-infrared spectroscopy, hyperspectral/multispectral imaging, bio-sensing, multi-technology fusion detection, *etc*. Chapter 8 describes intelligent sensing technologies for the processing of agro-products, and Chapters 4 and 9 mainly introduce the grading system and traceability of agricultural products, followed by Chapter 10 on the agricultural products harvest platforms. In addition, Chapter 11 on using unmanned aerial vehicles for crop information extraction expands the topic to field crops, which reflects the future trend.

As a professional book in the subject area, *Nondestructive Evaluation of Agro-products by Intelligent Sensing Techniques.* is written by the most active peers in this field from a number of countries, which significantly highlights the international nature of the work. Through the introduction of methods, systems and applications, this book enables readers to systematically understand the intelligent sensing technologies of nondestructive evaluation of agro-products. This book can also be used as a reference for researchers and managers in the field of nondestructive evaluation of agro-products and food.

Jiangbo Li
Beijing Research Center of Intelligent Equipment for Agriculture
Beijing Academy of Agriculture and Forestry Sciences
Beijing, China

&

Zhao Zhang
Department of Agricultural and Biosystems Engineering
North Dakota State University
North Dakota, USA

List of Contributors

Aichen Wang   School of Agricultural Engineering, Jiangsu University, Zhenjiang 212013, Jiangsu, PR China

Brian J. Steffenson   Department of Plant Pathology, University of Minnesota, Saint Paul, MN 55108, USA

Byoung-Kwan Cho   Department of Biosystems Machinery Engineering, College of Agricultural and Life Science, Chungnam National University, 99 Daehak-ro, Yuseoung-gu, Daejeon 34134, Republic of Korea

Ce Yang   Department of Bioproducts and Biosystems Engineering, University of Minnesota, Saint Paul, MN 55108, USA

Cory D. Hirsch   Department of Plant Pathology, University of Minnesota, Saint Paul, MN 55108, USA

Devrim Ünay   Electrical-Electronics Engineering, Faculty of Engineering, İzmir Demokrasi University, İzmir, Turkey

Dong Hu   School of Engineering, Zhejiang A&F University, Hangzhou 311300, China

Fangfang Gao   College of Mechanical and Electronic Engineering, Northwest A&F University, Yangling 712100, China

Feifei Tao   Geosystems Research Institute, Mississippi State University, Stennis Space Center, Hancock, MS 39529, USA

Haibo Yao   Geosystems Research Institute, Mississippi State University, Stennis Space Center, Hancock, MS 39529, USA

Insuck Baek   Environmental Microbial and Food Safety Laboratory, Agricultural Research Service, U.S. Department of Agriculture, Powder Mill Rd. Bldg. 303, BARC-East, Beltsville, MD 20705, USA

Jiangbo Li   Beijing Research Center of Intelligent Equipment for Agriculture, Beijing Academy of Agriculture and Forestry Sciences, Beijing, China
Key Laboratory of Modern Agricultural Equipment and Technology (Jiangsu University), Ministry of Education, Zhenjiang 212013, Jiangsu, PR China

Jianwei Qin   Environmental Microbial and Food Safety Laboratory, Agricultural Research Service, U.S. Department of Agriculture, Powder Mill Rd. Bldg. 303, BARC-East, Beltsville, MD 20705, USA

Kanniah Rajasekaran   USDA-ARS, Southern Regional Research Center, New Orleans, LA 70124, USA

Lin Zhang   College of Biosystems Engineering and Food Science, Zhejiang University, Hangzhou 310058, China

Longsheng Fu   College of Mechanical and Electronic Engineering, Northwest A&F University, Yangling 712100, China
Key Laboratory of Agricultural Internet of Things, Ministry of Agriculture and Rural Affairs, Yangling 712100, China
Shaanxi Key Laboratory of Agricultural Information Perception and Intelligent Service, Yangling 712100, China
Centre for Precision and Automated Agricultural Systems, Washington State University, Prosser, WA 99350, USA

Moon S. Kim Environmental Microbial and Food Safety Laboratory, Agricultural Research Service, U.S. Department of Agriculture, Powder Mill Rd. Bldg. 303, BARC-East, Beltsville, MD 20705, USA

Paulo Flores Department of Agricultural and Biosystems Engineering, North Dakota State University, North Dakota, USA

Rae Page Department of Plant Pathology, University of Minnesota, Saint Paul, MN 55108, USA

Ryan Johnson Department of Plant Pathology, University of Minnesota, Saint Paul, MN 55108, USA

Tamas Szinyei Department of Plant Pathology, University of Minnesota, Saint Paul, MN 55108, USA

Tong Sun School of Engineering, Zhejiang A&F University, Hangzhou 311300, China

Wen Zhang School of Life Science and Engineering, Southwest University of Science and Technology, Mianyang 621010, Sichuan, PR China

Wen-Hao Su Department of Bioproducts and Biosystems Engineering, University of Minnesota, Saint Paul, MN 55108, USA

Yanhong Dong Department of Plant Pathology, University of Minnesota, Saint Paul, MN 55108, USA

Yingchun Fu College of Biosystems Engineering and Food Science, Zhejiang University, Hangzhou 310058, China

Zhao Zhang Department of Agricultural and Biosystems Engineering, North Dakota State University, North Dakota, USA

Zhiming Guo School of Food and Biological Engineering, Jiangsu University, Zhenjiang 212013, China
International Research Center for Food Nutrition and Safety, Jiangsu University, Zhenjiang 212013, China

Zuzana Hruska Geosystems Research Institute, Mississippi State University, Stennis Space Center, Hancock, MS 39529, USA

CHAPTER 1

Representative Techniques and Methods for Nondestructive Evaluation of Agro-products

Dong Hu[1], Tong Sun[1,*] and Jiangbo Li[2,3,*]

[1] *School of Engineering, Zhejiang A&F University, Hangzhou 311300, China*

[2] *Beijing Research Center of Intelligent Equipment for Agriculture, Beijing 100097, China*

[3] *Key Laboratory of Modern Agricultural Equipment and Technology (Jiangsu University), Ministry of Education, Zhenjiang 212013, Jiangsu, PR China*

Abstract: Property, quality and safety assessment of agro-products are increasingly gaining attention due to the potential human health concern as well as social sustainable development. Emerging techniques and methods have particular advantages in nondestructive evaluation of agro-products due to their simplicity and faster response time, and reliable results, compared with the conventional visual inspection and destructive methods. This chapter briefly elaborates the principles and system components of some representative techniques, in particular, near infrared spectroscopy, infrared spectroscopy, fluorescence spectroscopy, Raman spectroscopy, laser induced breakdown spectroscopy, traditional machine vision, hyperspectral and multispectral imaging, magnetic resonance imaging, X-ray imaging, thermal imaging, light backscattering imaging, electrical nose and acoustics. The recent applications and technical challenges for these representative techniques are also presented.

Keywords: Agro-products, Methods, Nondestructive Evaluation, Techniques.

1. INTRODUCTION

Agro-products, like fruits, vegetables, and meat, are a major category of food products in the human diet. They contain essential nutrients, such as carbohydrates, fats, proteins, vitamins, and minerals. Property, quality and safety evaluation of agro-products, which directly relates to human health and the sustainable development of a country, has received increasing emphasis from government and has attracted great social concern and global attention. A considerable amount of effort has been made in developing techniques and methods to inspect and evaluate the property, quality and safety of agro-products. Conventional evaluation methods are commonly conducted through instrumental

* **Corresponding Authors Tong Sun & Jiangbo Li:** School of Engineering, Zhejiang A&F University, Hangzhou 311300, China; Tel: +86 15170230669; E-mail: suntong980@163.com and Beijing Research Center of Intelligent Equipment for Agriculture, Beijing 100097, China; Tel: +8613683557791; Fax: +86 1051503750; E-mail: jbli2011@163.com

analytical measurements, which can be stationary or hand-held but mostly off-line subjective and destructive in nature [1]. Therefore, there is an increasing demand for nondestructive evaluation of agro-products, because of the importance of determining the optimum time for harvest, monitoring the changes of chemical compositions and structured properties for postharvest, and grading quality and safety of individual pieces of agro-products at the packinghouse.

In recent decades, different nondestructive techniques based on different principles, procedures, and/or instruments, such as vision, spectroscopy, spectral imaging, acoustics, biosensing, and electrical nose/tongue, have been investigated and/or developed for the evaluation of agro-products, including chemical composition, physical structure, mechanical property, and food hazard. Unlike conventional methods, these emerging techniques and methods acquire data without contact with samples, thus providing nondestructive measurements. Generally, nondestructive testing is the evaluation performed on any agro-product, for example, an apple, without changing or altering the sample in any way, in order to determine the absence or presence of conditions that may have an effect on certain characteristics (*e.g.*, quality attributes) [2].

This chapter reviews the representative techniques and methods for nondestructive evaluation of agro-products, including near infrared spectroscopy, infrared spectroscopy, fluorescence spectroscopy, Raman spectroscopy, laser induced breakdown spectroscopy, traditional machine vision, hyperspectral and multispectral imaging, magnetic resonance imaging, X-ray imaging, thermal imaging, light backscattering imaging, electrical nose, acoustics, and other potential techniques. It provides an overview of basic principles, typical system components, and/or popular applications of these nondestructive techniques for evaluating the property, quality and safety of agro-products. A short discussion on the technical challenges and future outlook for these representative nondestructive techniques is also given.

2. EMERGING NONDESTRUCTIVE TECHNIQUES

2.1. Near Infrared Spectroscopy

Near infrared (NIR) spectroscopy is a common and useful nondestructive technique for agricultural product evaluation, which has the advantages of rapid and no sample pretreatment. It has been used for the quality detection of agricultural products such as soluble solid contents in fruit [3], starch in wheat [4], fatty acid in milk [5] and so on. The basic principle of NIR spectroscopy is that when a beam of NIR light illuminates a certain agricultural product, the irradiated agricultural product will selectively absorb light of certain frequencies, thereby generating an NIR absorption spectrum. And the NIR spectrum mainly contains the information of overtone and combination absorption of hydrogen groups (C-H, O-H, N-H), which is related to the quality parameters of agricultural products. Therefore, by establishing the mathematical relationship between the spectral information and the quality of agricultural products, we can detect the quality of agricultural products rapidly and nondestructively. The wavelength range of NIR is 780-2500 nm, which can be divided into short wave NIR (780-1100 nm) and long wave NIR (1100-2500 nm). Sometimes, the visible band is used together with near infrared, and it is called visible/near infrared (Vis/NIR) spectroscopy. Generally, the NIR technique has two modes: reflectance (Fig. **1a**) and transmittance (Fig. **1b**). Liquid samples adopt the transmittance mode; for solid samples, the reflectance mode is usually used in the long wave near infrared region, while the transmittance mode can also be chosen in the short wave near infrared region due to its strong penetration ability.

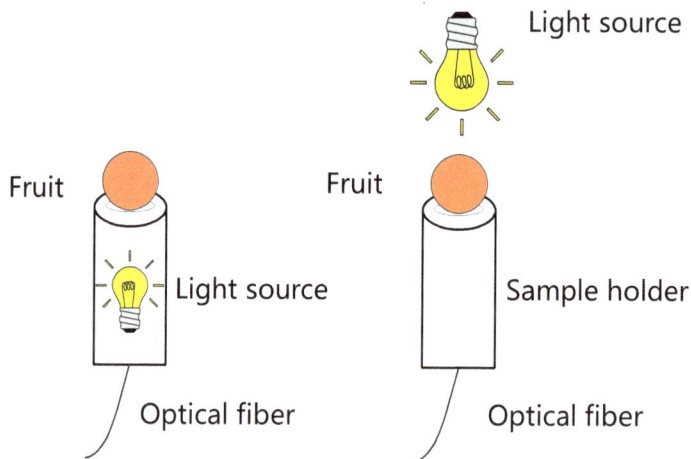

Fig. (1). Two detection modes of Vis/NIR for Nanfeng mandarin fruit: **(a)** reflectance; **(b)** transmittance.

At present, various spectrometers are available and used for NIR spectroscopy. According to different spectroscopic principles, NIR spectrometers can be mainly divided into four types, filter type, dispersion type, Fourier transform and acousto-optic tunable filter. A detector is an important part of the NIR spectrometer, whose function is to transform the optical signal into an electrical signal. In addition, the wavelength range of the NIR spectrometer is also determined by the photosensitive element material used in the detector. The materials of photosensitive elements mainly include Si, Ge, PbS, InSb, InGaAs, *etc.* Halogen tungsten lamps are generally used in NIR spectroscopy as light source, and sometimes light emitting diode (LED) is also used.

Due to the broad absorption bands of the overtone and combination of hydrogen groups, there is a serious band overlap phenomenon in the NIR spectra. Therefore, it is necessary to use chemometrics to process and analyze NIR spectral data, including spectral preprocessing, variable selection, and qualitative/quantitative modeling. The commonly used qualitative modeling methods are discriminant analysis, K-nearest neighbors (KNN), soft independent modelling of class analogy (SIMCA) and cluster analysis, while the quantitative modeling methods mainly include multiple linear regression (MLR), principal component regression (PCR), partial least square (PLS), artificial neural network (ANN) and support vector machine (SVM).

2.2. Infrared Spectroscopy

Generally, infrared (IR) spectroscopy refers to the mid infrared spectroscopy, and its range is 2500-25000 nm. The principle of IR spectroscopy is similar to that of NIR spectroscopy, but it contains different spectral information. The spectra of IR mainly contain the fundamental vibration information of molecules. According to the source of absorption peak, the IR spectra can be divided into four wide regions [6]: X-H stretching region (2500-4000 nm), triple-bond region (4000-5000 nm), double-bond region (5000-6667 nm), and fingerprint region (6667-16667 nm). The X-H stretching region covers the fundamental vibrations of hydrogen groups (C-H, O-H, N-H), the triple-bond region mainly contains vibrations of $C\equiv C$ and $C\equiv N$ bonds, while the double-bond region includes the vibrations of C=C, C=O and C=N. The spectra of the three regions are useful for functional group identification. The fingerprint region contains the abundant fundamental vibrations of key chemical bonds, and is valuable for identifying different molecules. As the fundamental vibration of organic and inorganic substances mostly appears in the IR region, more research and application are conducted on IR spectroscopy.

In IR spectroscopy, there are three main sampling methods named as transmission, transflection and attenuated total reflection (ATR) for spectral acquisition (Fig. **2**). Compared with transmission and transflection methods, ATR has advantages of little or no sample preparations and sample thickness independent, and is used much more in IR spectroscopy by researchers. The spectrometers used in the IR spectroscopy are mainly dispersion type and Fourier transform. Due to the advantages of fast scanning speed and high spectrum quality, Fourier transform spectrometer is used mostly in IR spectroscopy. The core component of the Fourier transform spectrometer is a double beam interferometer.

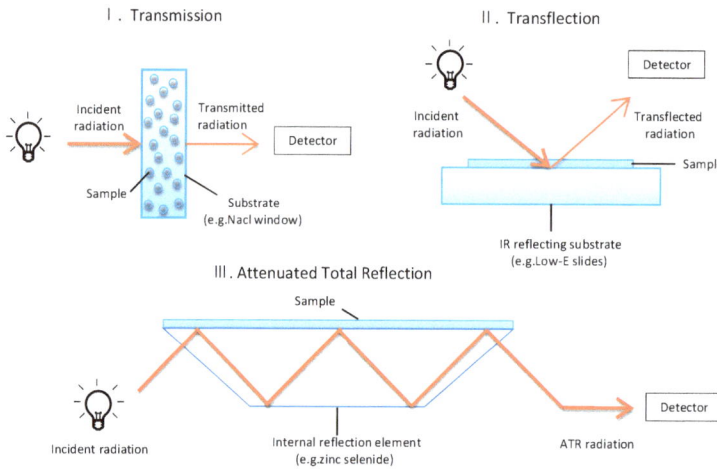

Fig. (2). Three main sampling methods for MIR spectral acquisition: **(a)** transmission; **(b)** transflection; **(c)** attenuated total reflection [6].

Compared with NIR spectra, the IR spectra have distinctive narrow bands without overlap. So comparative law can be used for substance identification and simple methods such as direct calculation method, internal standard method and working curve method can be used for determination in traditional spectral analysis. With the development of IR spectroscopy, chemometrics such as linear discriminant analysis (LDA), ANN and SVM are used for qualitative and quantitative analysis.

2.3. Fluorescence Spectroscopy

Fluorescence spectroscopy is a highly sensitive, rapid and noninvasive method for detecting fluorescence properties of agricultural products. The basic principle of fluorescence generation is that when the fluorescent substance is irradiated with excitation light, the excited molecules absorb energy and then transitions to the electronically excited state. The molecules at electronically excited state are unstable, and will return to the first excited state by releasing part of the energy through a non-radiative transition. Then, emit a longer wavelength of light called fluorescence and return to the ground state. The fluorescence spectrum has two types of characteristic spectra, called the excitation spectrum and emission spectrum. An excitation spectrum is obtained by measuring the fluorescence intensity of a certain wavelength under the excitation light of different wavelengths, while an emission spectrum is acquired by measuring the fluorescence intensity at different wavelengths under the excitation of a certain fixed wavelength. Fluorescence spectroscopy can provide many physical parameters, such as fluorescence intensity, quantum yield, fluorescence lifetime and fluorescence polarization. These parameters reflect the various characteristics of the molecules, and we can get more information about the detected molecules through these parameters.

Fluorescence spectroscopy has several different techniques, such as conventional fluorescence (CF) spectroscopy, three-dimensional fluorescence (TDF) spectroscopy, synchronous fluorescence (SF) spectroscopy and laser-induced fluorescence (LIF) spectroscopy. The CF spectroscopy mainly obtains the excitation spectrum or emission spectrum of substances. While the TDF spectroscopy can be used to get excitation-emission-matrix spectra that are characterized by three-dimensional coordinates of excitation wavelength, emission wavelength and fluorescence intensity. The spectra of TDF are generally expressed in two ways, three-dimensional projection and the contour map. Several methods, such as rank annihilation factor analysis (RAFA), parallel factor analysis (PARAFAC) and alternating trilinear decomposition (ATLD) are used to analyze three-dimensional data. The FS spectroscopy scans the excitation and emission wavelengths simultaneously, and a constant interval between the excitation and emission wavelengths is maintained during the scanning process. And LIF spectroscopy uses laser irradiation to generate fluorescence. The laser has the advantages of high energy, good monochromaticity, and no stray light. Therefore, LIF spectroscopy can obtain lower detection limit and higher sensitivity. Also, the LIF spectroscopy is the best choice for *in-situ* on-line detection of agricultural product quality above these fluorescence spectroscopy techniques. The common schematic of LIF is shown in Fig. (**3**).

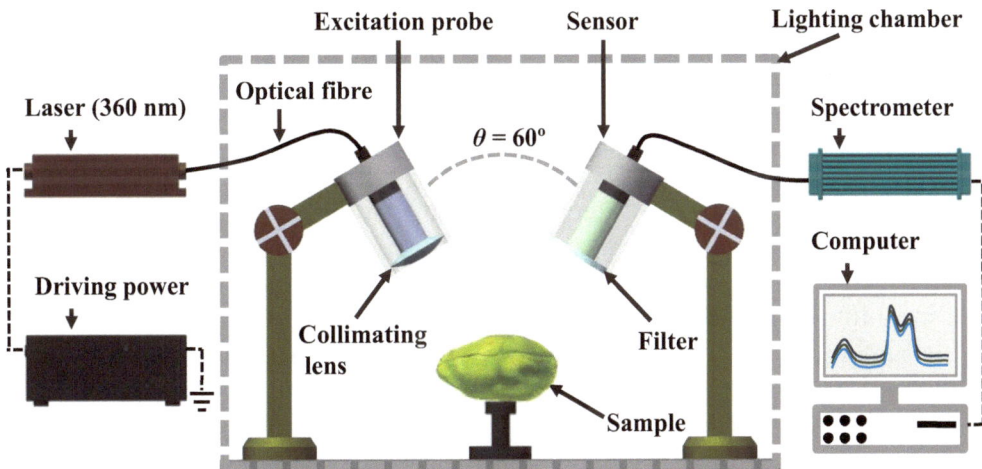

Fig. (3). Schematic of laser induced fluorescence spectroscopy [7].

2.4. Raman Spectroscopy

Raman spectroscopy is a fast and nondestructive analysis method for the quality evaluation of agricultural products. Its principle is based on the Raman scattering effect which is produced by inelastic scattering of light onto matter. In the inelastic scattering process, the molecules in the ground state absorb the incident light hv_0 and are excited to an intermediate virtual state, then return to an excited state, and emit light with a frequency of v_0-Vv, this is called Stokes Raman scattering; while the molecules in the excited state move to the intermediate virtual state after absorbing the incident light hv_0, then return to the ground state, and emit light with a frequency of v_0+Vv, this is called anti-Stokes Raman scattering. Because the intensity of Stokes Raman scattering is much greater than that of anti-Stokes Raman scattering, and Stokes Raman scattering is usually used for Raman spectroscopy analysis. In Raman scattering, the frequency difference Vv between the outgoing light and the incident light is called Raman shift, which is related to the vibrational or rotational energy level of the molecule, but not to the frequency of the incident light. There are differences in the vibrational and rotational energy levels of substance molecules, and this will lead to different Raman shifts, so each substance has its own characteristic Raman spectrum.

The peaks of the Raman spectrum are clear and sharp, basically not affected by moisture, and are suitable for analyzing the molecular structure of substances. However, the signal of the conventional Raman spectrum is weak, and the intensity of its scattered light is about 10^{-6}~10^{-9} of the incident light intensity, which is easily covered up by the fluorescent signal, and this greatly limits the application and development of Raman spectroscopy. With the development of technology, new Raman spectroscopy techniques such as surface-enhanced Raman spectroscopy (SERS), resonance Raman spectroscopy (RRS), confocal Raman micro-spectroscopy (CRM), spatially offset Raman spectroscopy (SORS) are constantly emerging. The SERS technique is to adsorb the tested molecules on some specially treated metal surfaces with nanostructures for Raman signal detection, which can enhance the Raman signal of the molecules by about 6 orders of magnitude, and this technique can be used for detecting trace substances [8, 9]. The RRS technique uses the excitation light with appropriate frequency to make it close to or coincide with an electron absorption peak of the molecule to be measured. Due to the coupling of electronic transition and molecular vibration, the intensity of one or several characteristics Raman bands of the molecule increases abruptly, about 10^4 to 10^6 times of the ordinary Raman spectrum. This technique usually uses a tunable laser in order to select an excitation light with a suitable frequency. The CMR technique is a combination of confocal optical microscopy and Raman spectroscopy. The laser is focused into a tiny spot by the microscope and illuminates the sample, and only the Raman signal in the range of the light spot will be returned to the spectrometer through the microscope. This technique can effectively suppress stray light and reduce fluorescence interference; it can also accurately scan the sample micro area to obtain the spectrum and image information of the sample. The SORS technique is to shift the focal point of the incident laser and the spectrum collection point by a certain distance on the surface space of the sample to be measured, which can obtain the deep-level information of the sample, and it can be used for the detection of the sample with multi-layer opaque or opaque packaging [10].

2.5. Laser Induced Breakdown Spectroscopy

Laser induced breakdown spectroscopy (LIBS) is an elemental analysis technique based on atomic emission spectroscopy, which has the advantages of rapid, *in-situ*, near-destructive and simultaneous detection of multiple elements. It is mainly used for the detection of various heavy metal elements in agricultural products. The basic principle of LIBS is that a high-intensity laser pulse is focused on the sample surface, causing a small amount of sample to burn and instantaneously vaporize to generate a large number of high-temperature plasmas. Then the high temperature plasma will transition from the excited state to the ground state during the process of cooling, and emit the plasma spectrum with sample element information. According to the frequency and intensity of the spectrum, the sample elements can be analyzed qualitatively and quantitatively.

There are two types of LIBS techniques: single pulse laser induced breakdown spectroscopy (SP-LIBS) and double pulse laser induced breakdown spectroscopy (DP-LIBS). The SP-LIBS uses a laser pulse to illuminate the sample. Its typical detection device is shown in Fig. (4), which is mainly composed of laser, spectrometer, delay generator, and optical path system. While the DP-LIBS uses two laser pulses to sequentially illuminate the sample at a certain time interval, which can excite the plasma spectrum better, and obtain a stronger spectral signal. There are two types of structures in DP-LIBS: collinear structure and orthogonal structure. The collinear structure refers to that two parallel laser pulses are focused and incident perpendicularly to the same position on the sample surface at a certain time interval. While orthogonal structure refers to that two laser pulses are orthogonal to each other, one is parallel to the sample surface and the other is perpendicular to the sample surface. There are two working modes for orthogonal structure, namely pre-ablation and reheating. In the pre-ablation mode, a laser pulse parallel to the sample surface is first used to break down the air near the sample surface, then another laser pulse is adopted to ablate the sample surface. In reheating mode, a laser pulse perpendicular to the sample surface is first used to ablate the sample surface to generate plasma, then another laser pulse is adopted to reheat the generated plasma.

The LIBS spectrum has sharp peaks which are related to elements, so the element types can be identified according to the positions of the peaks. For quantitative detection, the calibration curve method can be used to analyze based on the intensity of a characteristic spectral line of the element. In order to make more effective use of multiple characteristic spectral lines or other relevant spectral lines of elements, multiple linear regression and partial least square regression methods are also used for quantitative analysis.

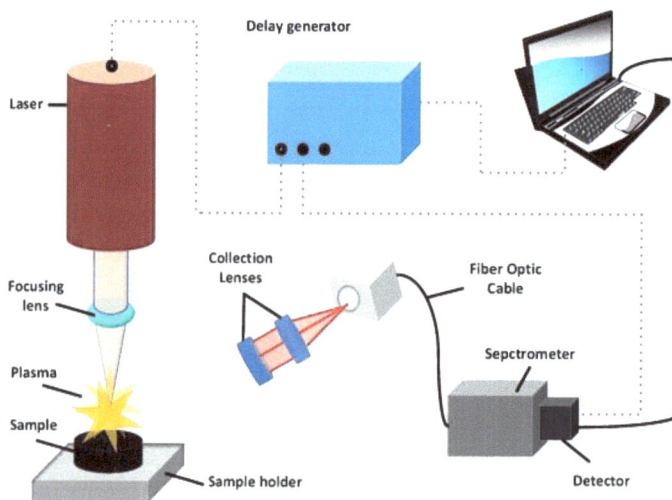

Fig. (4). A typical laser-induced breakdown spectroscopy setup [11].

2.6. Traditional Machine Vision

Traditional machine vision (TMV), also termed as traditional computer vision, is one of the leading optical imaging-based techniques in the nondestructive sensing and inspecting field. The origin of TMV dates back to the 1960s but it had not been exploited in the food and agricultural industry until the 1990s. Owing to the capabilities of high flexibility, accuracy, repeatability and efficiency, the TMV technique has been applied to evaluate the property and quality of agro-products, such as color, texture, shape and size, as well as obvious surface defects of a sample. There are several reasons why TMV has gained popularity in the science and industry during the past three decades. First, it is able to obtain reliable and reproducible data, and thus replacing human vision and perception of images. TMV is also capable of creating accurate descriptive data, which decreases human intervention and speeds up the process. Furthermore, the acquired data proved objective, consistent, effective, nondestructive, un-disturbing and robust, which are suitable for further analysis [12]. However, TMV is restricted to applications in the identification of external quality factors like color, size and surface structure, and it is not able to provide information about the chemical composition and internal quality characteristics (*e.g.*, soluble solid contents, pH, *etc.*) of agro-products.

Generally, a traditional machine vision system consists of the following five basic components: a light source, a camera, an image capture board (frame grabber or digitizer), and computer hardware and software [13]. The light source, which functions like a human eye, is a prerequisite for the success of the imaging analysis by reducing noise, shadow, reflection and enhancing image contrast. The energy distribution of the light source must have a uniform and controlled intensity. The camera, which is used for capturing images, is the key component of the TMV system. The solid-state charged-coupled device (CCD) and complementary metal oxide semiconductor (CMOS) image sensors are two different means used in the cameras for generating the images digitally. The image capture board, or frame grabber, is a vision processor that captures singular still frames from an analog video signal or a digital video stream in the current procedure, and then displayed, stored in raw or compressed digital form. The computer hardware and software is used for imaging processing and analysis, thus determining the quality and resolution of captured images and affecting the entire performance and efficiency of the TMV system. Fig. (**5**) shows the configuration of a typical machine vision system.

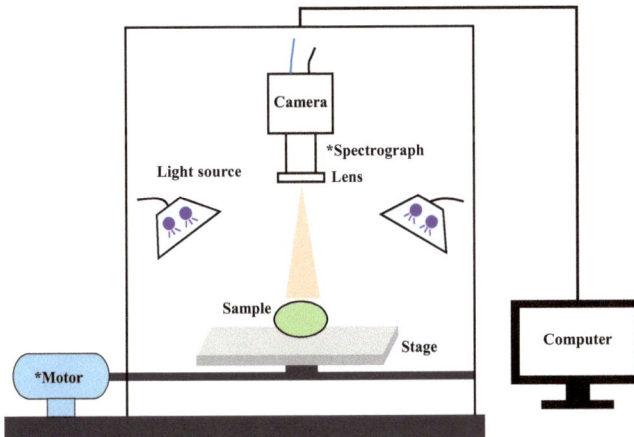

Fig. (5). Schematic of a typical machine vision system. *Just for hyperspectral and multispectral imaging systems [14].

There are many researchers on using TMV for evaluating the agro-products in the past years, including apple, citrus, mango, pear, banana, strawberry, peach, tomato, eggplant, pepper, potato, sweet potato, cheese, beef, carp, salmon, pork, chicken, *etc.* The TMV system is a powerful tool for the inspection of color, texture, size, shape, and some relatively obvious defects, but has less effectivity in detecting defects that are not clearly visible. Furthermore, to realize the defect detection more rapidly and accurately on-line, there are still many challenges to be overcome, such as the uneven distribution of lightness on curvature surface, whole surface inspection, long time consuming of acquisition and processing for the image, and different defects discrimination [15].

2.7. Hyperspectral and Multispectral Imaging

Hyperspectral imaging (HSI), as a spectroscopic imaging analytical tool, integrates conventional imaging and spectroscopy to attain a set of monochromatic images at almost continuous hundreds of thousands of wavelengths. Compared to the traditional machine vision, HSI involves both spatial and spectral information, thus providing the potential for identifying the chemical composition and internal quality characteristics of the sample [16]. The images obtained in the HSI, commonly called hypercubes, are three-dimensional data cubes, which have two spatial dimensions, the same as the TMV, along with spectral information, the same as spectroscopic techniques, for every pixel of the spatial image, as shown in Fig. (**6**). Generally, point scanning, line scanning and area scanning are three commonly used methods to acquire the hypercubes. HSI can be carried out in reflectance, transmittance, or fluorescence modes and scattering, which are selected depending upon specific requirements in practical applications. Like the TMV system, a light source, a camera, an image capture board, and computer hardware and software are basic and essential components for an HSI system, and a wavelength dispersion device (spectrograph) and a transportation stage are additional components (Fig. **5**). Thanks to the extensive information contained in the hyperspectral image, the HSI technique has found applications in diverse fields and exhibited promising results in several research, such as quality inspection of citrus fruits, damage detection in mushrooms, faecal contamination analysis in apples, quality monitoring in stored avocados, analysis of moisture distribution in salmon fish fillet, determination of foreign substances on fresh-cut lettuce, melamine detection in powdery milk, and classification of milk powders [1, 17]. However, the extensive information also brings some drawbacks to the HSI technique, such as long time consuming of image acquisition, as well as the complexity of image processing and analyzing. In consequence, it is always used to acquire images with high spatial and spectral resolutions for some fundamental researches, such as selecting the most efficient wavelengths to develop a multispectral imaging system for real-time quality inspection of agro-products.

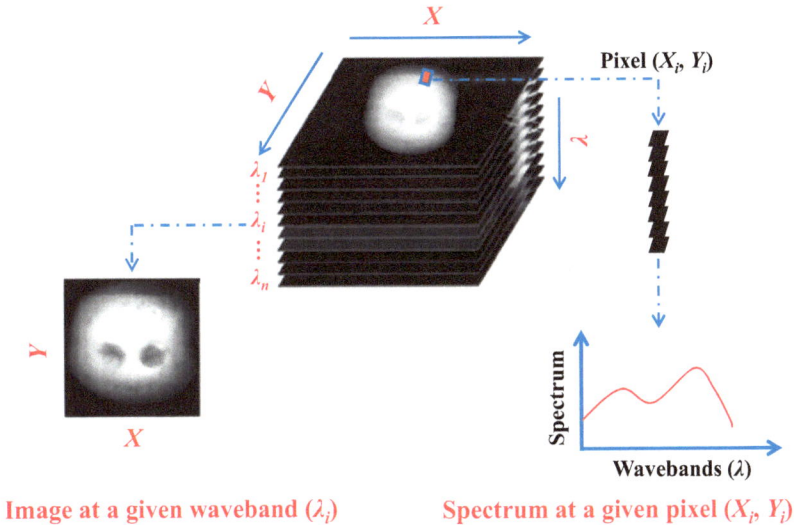

Image at a given waveband (λ_i) **Spectrum at a given pixel (X_i, Y_i)**

Fig. (6). Hypercube showing the relationship between spectral and spatial dimensions [18].

Multispectral imaging (MSI) is different from hyperspectral imaging in the number of the monochromatic images in the spectral domain. In general, MSI is a form of imaging that involves capturing two or more different waveband monochromatic images in the spectrum by employing filters or instruments. As mentioned above, the hypercube acquired by the HSI system is huge, and consumes too much time in further data processing. Therefore, there is an urgent desire to develop the MSI system with the most efficient wavelengths selected based on HSI for improving the efficiency and fulfilling the real-time inspection task. The biggest advantage of the MSI is that the wavelengths of the monochromatic images captured can be chosen freely, while the disadvantage is that the MSI system is always built by ourselves according to the specific imaging task. The constructed MSI system usually needs to be repeatedly checked, calibrated and debugged by the analyst.

2.8. Magnetic Resonance Imaging

Nuclear magnetic resonance (NMR), often referred to as magnetic resonance imaging (MRI), is a unique technique which measures the magnetic properties of spins that can then be related to the physical and chemical properties of a sample. In principle, NMR is a physical process in which the nucleus, whose magnetic moment is not zero, resonantly absorbs radiation of a certain frequency under external magnetic field. The detected NMR signals released as electromagnetic radiation can then be sent to the computer and be converted into the image through data processing. The converted image can be rotated and manipulated to be better able to detect tiny changes of structures within the object. MRI makes use of the fact that food and agricultural product tissues contain lots of water getting aligned in a large magnetic field, and thus working on the principle of resonant magnetic energy absorption by nuclei placed in an alternating magnetic field. MRI shows the image of the object structure, making its physical and chemical information visible.

Generally, an MRI system consists of the following basic components: a magnet and power-supply equipment, a set of gradient magnetic field coil, a controller and power-driven equipment, a radio-frequency system, and a computer system [18]. The magnet and power-supply equipment is used to produce a wide range of uniform, stable and constant magnetic fields. A computer system with a large storage capacity is used for data collection and processing. Some auxiliary equipment are also needed to support the functions of the MRI system. Compared with the conventional imaging techniques, MRI is advantageous in several aspects, such as clear image contrast, in particular between fat and connective tissues, and 3D analysis of samples.

Due to the fact that images are converted from electromagnetic signals that represent internal information of the samples, MRI possesses great merits in the determination of chemical compositions. Moreover, agro-products contain plenty of water, which provides great potential for quality assessment by using the MRI technique. Therefore, MRI has been used for diverse agro-products, such as fruits, vegetables, meat products, and cereals, with the majority concerning the water content [18]. Furthermore, MRI has been used to identify physical structure changes like ageing, infection, microbial detects, and chemical changes. The applications include but are not limited to the identification of decay of postharvest blueberries, evaluation of quality characteristics of Braeburn apples, assessment of maturity states of tomato, monitoring of ripening in persimmon, citrus and oil palm, and monitoring of freezing process [19]. The diverse variety of MRI measurable properties, such as proton density, chemical shifts, relaxation time and diffusion constant, and the spatial distribution of 2D and 3D images, enable the researchers to design a wide variety of assays that can be applied to assess different types of defects, stress and physiological states of agro-products [20].

2.9. X-ray Imaging

X-ray, also called roentgen ray, is one kind of electromagnetic radiation and is an influential tool for nondestructive quality and safety assessment. After being successfully used in medical diagnostics and other industrial implementations, researchers are recently using this technique in the fields of food and agriculture. X-rays have low wavelength range of 0.01-10 nm and high photon energy of 0.1-120 keV, which leads to strong penetrability through numerous materials. X-rays can be divided into soft X-rays and hard X-rays, according to the photon energy and corresponding penetration ability. The photon energy of soft X-rays is up to about 10 keV, while that of hard X-rays is 10-120 keV. Only the soft X-ray imaging (XRI) technique is frequently used in the inspection of agro-products, since the hard XRI pollute the sample.

The principle of the soft XRI technique for inspection is based on the density of the object and the contaminant, as shown in Fig. (7) [18]. Usually, when an X-ray penetrates into an object, the photon energy will be reduced due to the absorption phenomenon. Photons in an X-ray beam, when passing through the object, are either transmitted, scattering, or absorbed. The exited X-ray from the object surface is captured by a sensor, and then the energy signal is converted into an image of the interior of the object. Foreign matter appears as a darker shade of grey that helps to identify foreign contaminants. A typical soft XRI system mainly consists of an X-ray source tube, a line-scanning sensor, conveying belt, stepping motor, image-acquisition card and a computer. In contrast to MRI, the soft XRI is relatively cheaper in the instrument and simpler in accessibility and material restrictions, which expands its application range. Compared with other imaging techniques (*e.g.*, TMV, HSI and MSI), the soft XRI technique is superior in the detection of external contamination and internal tissue distribution of the sample.

As a novel nondestructive method, XRI has witnessed a large number of applications on agro-products. It has been employed for the detection of internal water-core damage in apples, internal defects in sweet onion, fungal infection in wheat, and fish bones in fish fillets. Moreover, the technique was used for apple classification based on the surface bruise, monitoring physiological constituents in peaches, determining the changes in the internal composition of meat products, evaluating frozen products, and monitoring eye formation in cheese [2, 17]. However, XRI cannot detect all kinds of foreign objects, in particular those whose density is similar to that of water, such as hair, paper, and plastics.

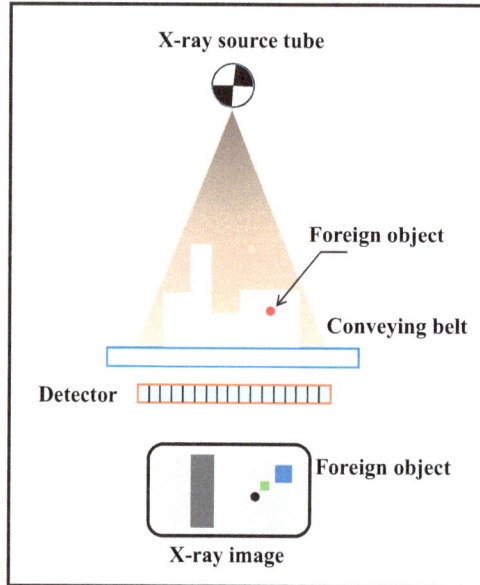

Fig. (7). Principle of soft X-ray imaging [18].

2.10. Thermal Imaging

Thermal imaging (TI) is a non-invasive and non-contact technique, which is based on the fact that all materials emit infrared radiation in the electromagnetic spectrum ranging from 0.75 to 100 µm. The regions can be divided into near (0.75-2.5 µm), short wave (1.4-3 µm), mid (3-8 µm), long wave (>8 µm), and extreme (15-100 µm) infrared regions [21]. TI records thermal distribution by measuring infrared radiation discharged by a body surface to produce a pseudo image of the temperature distribution of the surface. According to the black body radiation law, all objects above the absolute zero temperature (0 K) emit infrared radiation, among which the short-wave to long wave (1.4-15 µm) can be detected by TI systems. Thermography, which measures a large number of point temperatures of the target surface, is a powerful tool for visualizing and analyzing target with thermal gradients. TI technique is able to detect the bodies whose surface temperature is distinguishable from others.

A typical TI system contains a thermal camera, an optical unit (*e.g.*, focusing lens, collimating lenses, and filters), detector array (*e.g.*, micro-bolometers), and a signal processing and image processing unit (Fig. **8**). Thermal images can be captured either by the passive or active imaging modality. The passive system requires no external energy for imaging to the object (*i.e.*, the temperature of the features of interest are naturally higher or lower than the background), while the active system usually uses thermal energy to produce a thermal contrast between the features of interest and the background. There is no need for a light source in the TI system, but the integrated system for active thermography measurements contain a heating or cooling unit, like the halogen lamp, to provide a thermal differential. In the TI cameras, the infrared energy emitted from an object is converted into an electrical signal through infrared detectors and displayed as a monochrome or color thermal image.

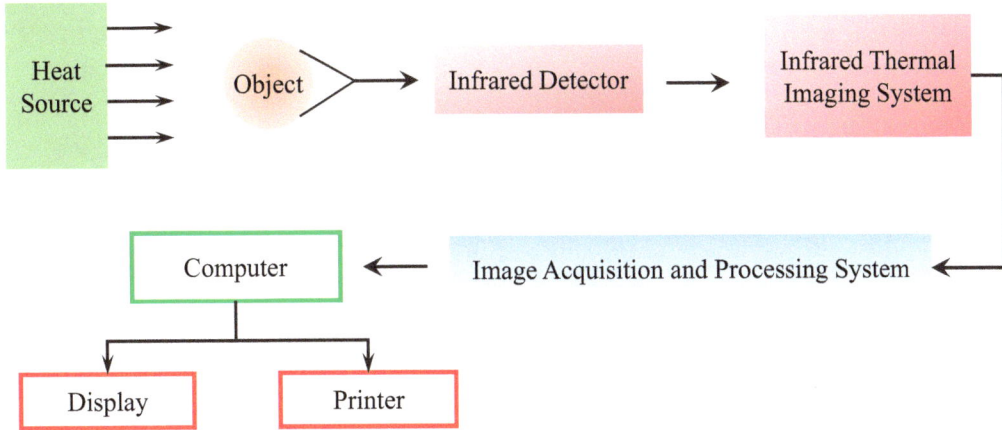

Fig. (8). Schematic of a thermal imaging system [19].

Originally developed for military applications and for surveillance in night vision, TI has attracted increasing attention and witnessed numerous applications in various fields, such as medicine, materials science and fire safety, with technological advances in computer analytical tools. Owing to the capabilities of portability, real-time imaging, and non-contact temperature-measurement, the TI technique is also increasingly used for various agro-products, including tomato, Japanese pear, Japanese persimmon, apple, Satsuma mandarin, natsudaidai fruit, palm fruit, mango, guava, citrus, walnut, grain, cauliflower plant, and chicken meat [18, 19]. The applications include but are not limited to maturity assessment of tomato, Japanese pear, Japanese persimmon, palm fruit, and mango, evaluation of artificial wound of apple, Satsuma mandarin, and natsudaidai fruit, bruise detection of tomato and apple, doneness monitoring of chicken meat, process monitoring of walnut, and detection of the citrus drying process and grain-quality. However, the performance of the TI system is highly affected by environmental condition. Expensive components, like TI camera, hinder the application range to some extends. Moreover, the quality and reliability of the TI system depend largely on the sensitivity and detection speed of the major components, and the spatial resolution of image and intensity resolution. Therefore, thermal cameras with the advantages of easy handling, accurate temperature measuring strategy and lower prices are expected in the future, which could improve the performance of the TI system, make the TI system user friendly and promote the usage of thermal cameras.

2.11. Light Backscattering Imaging

Light scattering is a phenomenon about the change of light traveling direction in a medium. It takes place when light travels in an optically inhomogeneous medium, or when light travels between two optically homogeneous media with different refractive indexes, or when photons encounter scattering particles in the medium. Most agro-products are heterogeneous in structure and composition, and the cellular structures act as scatterers. Light would thus go through multiple scattering events before it exits from the tissue or is being absorbed.

Light backscattering imaging (LBI) technique adopts the principle of capturing the scattered light of photons projected into the agro-products [22]. Light-tissue interaction process carries important information of the structural and chemical characteristics of the object, like an apple, which makes it as a unique feature. Based on the light source and imaging unit used, light backscattering imaging technique can be divided into several categories, which include laser light backscattering imaging (LLBI), hyperspectral backscattering imaging (HBI), and multispectral backscattering imaging (MBI). HBI and MBI belong to HSI and MSI, respectively, which were introduced in Section **1.2.7**. LLBI utilized a laser as the light source, which is the main difference from HBI and MBI. However, the demarcation line between the LBI and spectral imaging techniques is not always clear, as many spectral imaging techniques often capture the scattered signal from the sample, such as spatial-frequency domain imaging (SFDI), Raman scattering, and biospeckle technique. SFDI, also called structured-illumination reflectance imaging, captures diffusely backscattered images from an object subjected to the illumination of sinusoidal pattern with different spatial frequencies (Fig. **9**). Diffuse reflectance can then be demodulated by using proper algorithms, such as conventional three-phase demodulation, spiral phase transform and Gram-Schmidt orthonormalization. Forward and inverse algorithms are often used to deal with the demodulated reflectance for the purposes of optical property estimation, light transfer modeling, and quality assessment (*e.g.*, bruise and defect detection) in agro-products.

As a category of recent rapid imaging technology, LBI has proven its capability in nondestructive quality evaluation of diverse agro-products, such as apple, banana, citrus, mushroom and tomato, papaya, plum, pear, vegetable-based creams, watermelon, peach, and cucumber. The applications include detection of early decay in peaches, chilling injury in cucumbers and bananas, and fresh bruise in apples and pears, and evaluation of ripeness and/or firmness of pears, tomatoes, kiwifruits, apples, and plums [1, 24 - 27]. These are just a few examples of research works done concerning the application of this technology. However, the optical property estimation, which can be realized by using the light backscattering imaging techniques (*e.g.*, SFDI), is prone to errors due to the complicated mathematical model and parameter estimation algorithm. Applications of the estimated absorption and scattering coefficients, which are related with the food property and quality, are thus always hindered or limited. Future research is expected to focus on the algorithm simplification for optical property estimation, as well as accuracy and efficiency improvement.

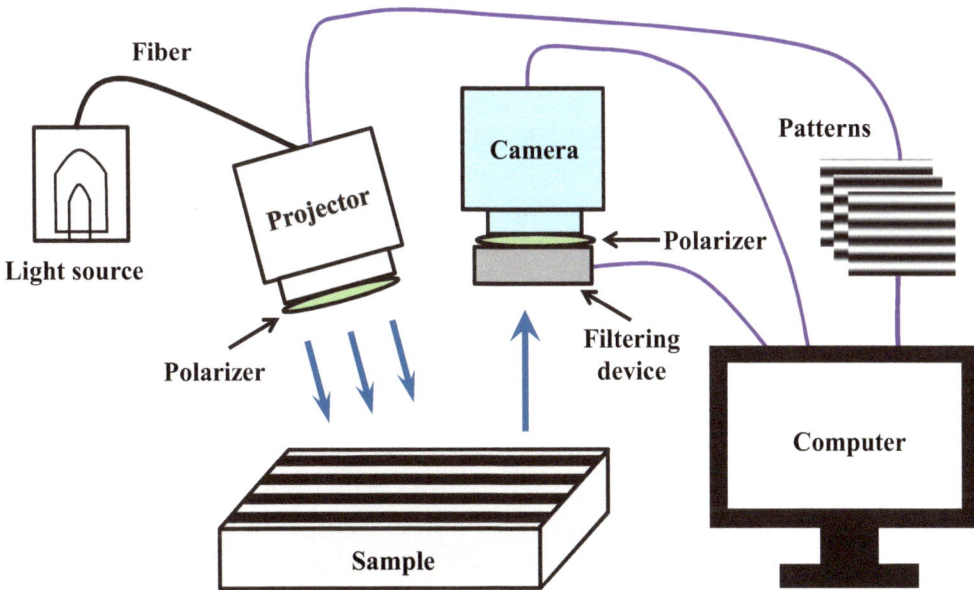

Fig. (9). Schematic of a typical spatial-frequency domain imaging system [23].

2.12. Electrical Nose Technique

Electronic nose is an intelligent sensory instrument that imitates the human olfactory system, and can perform qualitative or quantitative detection of gas or volatile components through odor fingerprint information. Due to the advantages of rapid detection, high sensitivity, no sample pretreatment, simple operation and low cost, the electronic nose technique has become one of the main means for nondestructive detection of agricultural product quality, such as damage in fruit [28], freshness in meat [29] and disease in vegetables [30]. Electronic nose is usually composed of three parts: gas sensor array, signal processing system and pattern recognition system. The gas sensor array is composed of different gas sensors, which can generate different responses to odors and convert the response into a group of measurable physical signals. The signal processing system is used to filter, exchange and extract the physical signal from the gas sensor, the most important of which is the signal feature extraction. At present, the commonly used feature extraction methods are a relative method, difference method, logarithm method and normalization method. The pattern recognition system uses a certain algorithm to process the extracted characteristic signals of odor, and gives the qualitative or quantitative analysis results of the odor. The main pattern recognition methods include principal component analysis (PCA), DA, cluster analysis, and ANN.

The gas sensor is the core component of the electronic nose, which is crucial to the overall performance of the electronic nose detection system. The common gas sensors mainly include metal oxide sensors (MOS), conductive polymer (CP) sensors, quartz crystal microbalance (QCM), surface acoustic wave (SAW), and fiber optic sensors. The principle of MOS is to adsorb the measured gas on the surface of the sensor and change the conductivity of the sensor, thereby generating different signal values. Due to the advantages of simple preparation, cheap price, high sensitivity and wide application range, it has become the most extensive sensor used in the electronic nose system. But it has several shortcomings such as poor selectivity to odor and gas, high working temperature, and "poisoning" reaction to sulfide in the mixture. The common metal oxide sensors are SnO_2, ZnO, WO_3, Fe_2O_3, Co_3O_4, *etc.* The conductive polymer sensors are generally composed of active materials such as thiophene, indole, and furan. Its biggest advantage is that it can work at a normal temperature, does not need heating, and has a small size, which can be used for portable equipment; however, it has the disadvantages of the time-consuming manufacturing process, drift phenomenon and susceptibility to humidity.

2.13. Acoustics Techniques

Acoustics deals with the generation and reception of mechanical waves and propagation. When an acoustic wave reaches an object, the reflected or transmitted acoustic wave depends on the acoustic characteristics of the object, which can provide information on the interaction between acoustic wave and the object. Acoustic vibration, as one of the acoustics techniques, is a common and efficient way for nondestructive quality inspection of agro-products by capturing, processing and analyzing the vibration response under some kind of excitations. The acoustic vibration waves include reflection, scattering, transmission, and absorption when they interact with the agro-products, which depend on the acoustic vibration characteristics of agro-products. The acoustic vibration characteristics are related to mechanical and structural properties of agro-products, such as modulus of elasticity, Poisson's ratio, density, mass, and shape.

In the acoustic vibration method, a test sample is assumed to be a simple elastic body, which is composed of two masses connected with a spring, as shown in Fig. (10). The natural frequency (*f*) of the system can be calculated by Eq. (1).

$$f = \frac{1}{2\pi}\sqrt{\frac{4k}{m}} \tag{1}$$

where *m* is the sample mass, and *k* is the spring constant of the system. The equation can be converted into:

$$k = \pi^2 f^2 m \tag{2}$$

It can be observed that f^2m is proportional to *k*, which is related to the elastic properties of the tested sample. Therefore, f^2m can be used as an index for elasticity or texture evaluation.

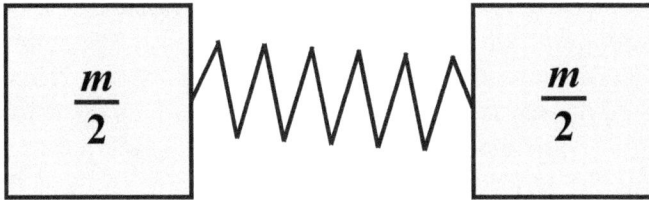

Fig. (10). Mass-spring model for an elastic body of mass *m* [31].

The general process of the acoustic vibration method for quality evaluation of agro-products consists of the excitation module, signal acquisition module, and signal processing module, as shown in Fig. (11). The tested sample was excited by an excitation signal from the excitation module, and the response signal was collected using a signal acquisition module and analyzed in a signal processing module. Impact method (*e.g.*, hammer, stick, and pendulum) and forced method (*e.g.*, vibrator, speaker, and piezoelectric vibration generator) are two excitation methods in the excitation module, while contact (*e.g.*, acceleration pickup and piezoelectric sensor) and noncontact sensors (*e.g.*, microphone, and laser Doppler vibrometer) are used for vibration measurement in the signal acquisition module.

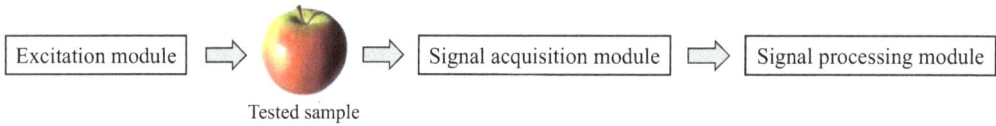

Excitation module ⟹ Tested sample ⟹ Signal acquisition module ⟹ Signal processing module

Fig. (11). The general process of acoustic vibration method for quality evaluation of agro-products [32].

Due to the advantages of high repeatability and sensitivity, the acoustic vibration technique was widely used for texture evaluation, ripeness classification and optimum eating ripeness determination, and defect detection of agro-products, including apple, pear, kiwifruit, watermelon, melon, tomato, peach, pitaya, dates, avocado, grape, persimmon, guava, preserved egg, potato, and meat [32 - 35]. Since the contact measurement in the acoustic vibration can prevent the free vibration and damage the surface of agro-products, and is also not suitable for on-line detection, future research may be conducted to increase detection accuracy and speed, as well as reduce cost in the noncontact measurements.

2.14. Other Techniques

Other techniques including odor imaging, ultrasound imaging, and bio-sensing were also used for property, quality and safety evaluation of agro-products. Odor imaging (OI) is based on the color change induced by the reaction between volatile material and an array of chemically-responsive dyes upon ligand binding. A low-cost, sensitive colorimetric sensor array is the key component in the OI technique. During the past years, the OI technique has been used for evaluating tea quality and classifying tea variety, discriminating different brands of bottled water, analyzing alcohol, and assessing meat freshness [18]. However, only odorous food samples are suitable for this inspection tool, because the OI technique deals with signal according to the aroma of materials. Some problems related with this newly developing technology, such as the selection of responsive dyes for colorimetric sensor array, are also needed to be solved. Ultrasound imaging is easy to use, cheap in the instrument, and without complicated post-imaging processing procedures in food and agricultural product quality assessment, but is dependent on the amount of energy reflected through materials. It should be noted that many techniques mentioned above contain both spectral and imaging information of the samples, such as Raman spectroscopy/imaging, fluorescence spectroscopy/imaging, hyperspectral/multispectral imaging, and spatial-frequency domain imaging, which can always be categorized as spectral imaging techniques.

3. FUTURE PERSPECTIVES

Over the past years, we have seen significant research efforts in the development and application of nondestructive techniques and methods for characterizing food and agricultural products. While these emerging techniques have become valuable tools for inspecting property, quality and safety of agro-products, there still exist considerable issues and challenges in using these techniques. First, agro-product property, quality and safety contain various aspects involving external factors (*e.g.*, size, shape, color, *etc.*) and internal factors (*e.g.*, chemical, physical, microbial, *etc.*). Most of emerging techniques cannot simultaneously solve all aspects. Thus, cost-effective solutions to challenging problems of the future will make use of the integration of multiple representative techniques. For example, the HSI technique can be integrated with an electronic nose or electronic tongue to evaluate food quality and safety broadly, and machine/deep learning can be combined with TMV to enhance evaluation ability and capability. Second, budget constrain is the major barrier to applications of some emerging techniques in agro-products, in particular for techniques traditionally used in medical diagnostics such as MRI and XRI. Even an image-processing system is still unviable in many potential applications because the cost is unacceptable. Thus, simpler instrumentation with lower cost and higher efficiency should be developed in order to satisfy the requirements of the agro-product industry. Furthermore, performance (*e.g.*, accuracy, sensitivity, real-time capability, robustness, *etc.*) is rather critical for the evaluation of agro-products. More researches should be conducted to improve the evaluation performance of the representative techniques, especially for recently developed techniques such as fluorescence spectroscopy, Raman spectroscopy, spatial-frequency domain imaging, and laser biospeckle. Finally, most of the nondestructive techniques, integrated with the remarkable trends of computer science, have the potential use for on-line grading and sorting of agro-products in future days. Automation will continue to increase by being applied to more applications that are not currently automated. Further integration is possible between automated processing units, which is a transition from automation of a single processing unit to a more coherent, potentially more flexible automation of the entire processing chain.

CONCLUSION

This chapter summarized the representative nondestructive techniques (*i.e.*, NIR spectroscopy, IR spectroscopy, fluorescence spectroscopy, Raman spectroscopy, LIBS, TMV, HSI, MSI, MRI, XRI, TI, LBI, electrical nose and acoustics) that have been employed for the evaluation of agro-products. The basic principles, typical system components, and/or popular applications, by using the nondestructive techniques, were reviewed. The techniques have the ability to replace conventional techniques like visual inspection and other destructive methods, with higher accuracy and efficiency to a greater extent. Various qualitative evaluations like assessing maturity, ripeness, firmness, defects, and physiochemical properties of agro-products, and quantitative modeling of quality attributes can be analyzed more efficiently. The need to change from the conventional way for evaluation to advanced nondestructive techniques is a must for this time that the quality and safety criteria of all agro-products are highly important than the cost. Given the potential applications discussed, increased adoption of these representative techniques by the food and agricultural engineers for improved efficiency is likely. Furthermore, the integration of multiple nondestructive techniques could broaden the applicability and improve the evaluation performance. Grading and sorting agro-products in real-time based on the inspected properties or quality attributes by using the nondestructive techniques are worth researching sustainably in the future.

CONSENT FOR PUBLICATION

Not applicable.

CONFLICT OF INTEREST

The authors confirm that this chapter contents have no conflict of interest.

ACKNOWLEDGEMENTS

The authors gratefully acknowledge the financial support provided the National Natural Science Foundation of China (No. 32001414 and 31972152), Beijing Talents foundation (No. 2018000021223ZK06).

REFERENCES

[1]   P.D.C. Sanchez, N. Hashim, R. Shamsudin, and M.Z.M. Nor, "Applications of imaging and spectroscopy techniques for non-destructive quality evaluation of potatoes and sweet potatoes: A review", *Trends Food Sci. Technol.,* vol. 96, pp. 208-221, 2020. [http://dx.doi.org/10.1016/j.tifs.2019.12.027]

[2]   K. Narsaiah, A.K. Biswas, and P.K. Mandal, "Nondestructive methods for carcass and meat quality evaluation", In: *Meat Quality Analysis: Advanced Evaluation Methods, Techniques, and Technologies.,* A.K. Biswas, P.K. Mandal, Eds., Academic Press: UK, 2020, pp. 37-49. [http://dx.doi.org/10.1016/B978-0-12-819233-7.00003-3]

[3]   S. Escribano, W.V. Biasi, R. Lerud, D.C. Slaughter, and E.J. Mitcham, "Non-destructive prediction of soluble solids and dry matter content using NIR spectroscopy and its relationship with sensory quality in sweet cherries", *Postharvest Biol. Technol.,* vol. 128, pp. 112-120, 2017. [http://dx.doi.org/10.1016/j.postharvbio.2017.01.016]

[4]   H. Jiang, and J. Lu, "Using an optimal CC-PLSR-RBFNN model and NIR spectroscopy for the starch content determination in corn", *Spectrochim. Acta A Mol. Biomol. Spectrosc.,* vol. 196, pp. 131-140,

2018.
[http://dx.doi.org/10.1016/j.saa.2018.02.017] [PMID: 29444495]

[5]   P.L. Suarez, A. Soldado, A. Gonzalez-Arrojo, F. Vicente, and B. de la Roza-Delgado, "Rapid on-site monitoring of fatty acid profile in raw milk using a handheld near infrared sensor", *J. Food Compos. Anal.,* vol. 70, pp. 1-8, 2018.
[http://dx.doi.org/10.1016/j.jfca.2018.03.003]

[6]   S. De Bruyne, M.M. Speeckaert, and J.R. Delanghe, "Applications of mid-infrared spectroscopy in the clinical laboratory setting", *Crit. Rev. Clin. Lab. Sci.,* vol. 55, no. 1, pp. 1-20, 2018.
[http://dx.doi.org/10.1080/10408363.2017.1414142] [PMID: 29239240]

[7]   Q.F. Wu, J. Xu, and H.R. Xu, "Discrimination of aflatoxin B1 contaminated pistachio kernels using laser induced fluorescence spectroscopy", *Biosyst. Eng.,* vol. 179, pp. 22-34, 2019.
[http://dx.doi.org/10.1016/j.biosystemseng.2018.12.009]

[8]   S. Dhakal, K. Chao, Q. Huang, M. Kim, W. Schmidt, J. Qin, and C.L. Broadhurst, "A simple surface-enhanced Raman spectroscopic method for on-site screening of tetracycline residue in whole milk", *Sensors (Basel),* vol. 18, no. 2, p. 424, 2018.
[http://dx.doi.org/10.3390/s18020424] [PMID: 29389871]

[9]   L.L. Sun, and C.S. Wang, "Highly sensitive and rapid surface enhanced Raman spectroscopic (SERS) determination of thiram on the epidermis of fruits and vegetables using a silver nanoparticle-modified fibrous swab", *Anal. Lett.,* vol. 53, no. 6, pp. 973-983, 2019.
[http://dx.doi.org/10.1080/00032719.2019.1687509]

[10]   K.M. Khan, S.B. Dutta, H. Krishna, and S.K. Majumder, "Inverse SORS for detecting a low Raman-active turbid sample placed inside a highly Raman-active diffusely scattering matrix - A feasibility study", *J. Biophotonics,* vol. 9, no. 9, pp. 879-887, 2016.
[http://dx.doi.org/10.1002/jbio.201600075] [PMID: 27433790]

[11]   P.R. Villas-Boas, M.A. Franco, L. Martin-Neto, H.T. Gollany, and D.M.B.P. Milori, "Applications of laser-induced breakdown spectroscopy for soil analysis, part I: Review of fundamentals and chemical and physical properties", *Eur. J. Soil Sci.,* vol. 7, no. 5, pp. 1-16, 2019.
[http://dx.doi.org/10.1111/ejss.12888]

[12]   A. Taheri-Garavand, S. Fatahi, M. Omid, and Y. Makino, "Meat quality evaluation based on computer vision technique: A review", *Meat Sci.,* vol. 156, pp. 183-195, 2019.
[http://dx.doi.org/10.1016/j.meatsci.2019.06.002] [PMID: 31202093]

[13]   H.H. Wang, and D.W. Sun, "Correlation between cheese meltability determined with a computer vision method and with Arnott and Schreiber tests", *J. Food Sci.,* vol. 67, no. 2, pp. 745-749, 2002.
[http://dx.doi.org/10.1111/j.1365-2621.2002.tb10670.x]

[14]   B.H. Zhang, W.Q. Huang, J.B. Li, C.J. Zhao, S.X. Fan, J.T. Wu, and C.L. Liu, "Principles, developments and applications of computer vision for external quality inspection of fruits and vegetables: A review", *Food Res. Int.,* vol. 62, pp. 326-343, 2014.
[http://dx.doi.org/10.1016/j.foodres.2014.03.012]

[15]   H.K. Tian, T.H. Wang, Y.D. Liu, X. Qiao, and Y.Z. Li, "Computer vision technology in agricultural Automation-A review", *Inf. Process. Agric.,* vol. 7, pp. 1-19, 2020.
[http://dx.doi.org/10.1016/j.inpa.2019.09.006]

[16]   D. Wu, and D.W. Sun, "Advanced applications of hyperspectral imaging technology for food quality and safety analysis and assessment: A review-Part I: Fundamentals", *Innov. Food Sci. Emerg. Technol.,* vol. 19, pp. 1-14, 2013.
[http://dx.doi.org/10.1016/j.ifset.2013.04.014]

[17]   T. Lei, and D.W. Sun, "Developments of nondestructive techniques for evaluating quality attributes of cheeses: A review", *Trends Food Sci. Technol.,* vol. 88, pp. 527-542, 2019.
[http://dx.doi.org/10.1016/j.tifs.2019.04.013]

[18]   Q.S. Chen, C.J. Zhang, J.W. Zhao, and Q. Ouyang, "Recent advances in emerging imaging techniques for non-destructive detection of food quality and safety", *Trends Analyt. Chem.,* vol. 52, pp. 261-274, 2013.
[http://dx.doi.org/10.1016/j.trac.2013.09.007]

[19]   P. Pathmanaban, B.K. Gnanavel, and S.S. Anandan, "Recent application of imaging techniques for fruit quality assessment", *Trends Food Sci. Technol.,* vol. 94, pp. 32-42, 2019.
[http://dx.doi.org/10.1016/j.tifs.2019.10.004]

[20]   R.K. Srivastava, S. Talluri, S.K. Beebi, and B.R. Kumar, "Magnetic resonance imaging for quality evaluation of fruits: a review", *Food Anal. Methods,* no. 11, pp. 2943-2960, 2018.
[http://dx.doi.org/10.1007/s12161-018-1262-6]

[21]   J.M. Lloyd, *Thermal Imaging Systems.* Springer Science & Business Media: Berlin, 2013.

[22]   S.E. Adebayo, N. Hashim, K. Abdan, and M. Hanafi, "Application and potential of backscattering imaging techniques in agricultural and food processing-A review", *J. Food Eng.,* vol. 169, pp. 155-164, 2016.
[http://dx.doi.org/10.1016/j.jfoodeng.2015.08.006]

[23]   D. Hu, R.F. Lu, and Y.B. Ying, "Spatial-frequency domain imaging coupled with frequency optimization for estimating optical properties of two-layered food and agricultural products", *J. Food Eng.,* vol. 277, p. 109909, 2020.
[http://dx.doi.org/10.1016/j.jfoodeng.2020.109909]

[24]   Y. Sun, R.F. Lu, Y.Z. Lu, K. Tu, and L.Q. Pan, "Detection of early decay in peaches by structured-illumination reflectance imaging", *Postharvest Biol. Technol.,* vol. 151, pp. 68-78, 2019.
[http://dx.doi.org/10.1016/j.postharvbio.2019.01.011]

[25]   Y.Z. Lu, and R.F. Lu, "Enhancing chlorophyll fluorescence imaging under structured illumination with automatic vignetting correction for detection of chilling injury in cucumbers", *Comput. Electron. Agric.,* vol. 168, p. 105145, 2020.
[http://dx.doi.org/10.1016/j.compag.2019.105145]

[26]   D. Hu, X. Fu, X. He, and Y. Ying, "Noncontact and wide-field characterization of the absorption and scattering properties of apple fruit using spatial-frequency domain imaging", *Sci. Rep.,* vol. 6, p. 37920, 2016.
[http://dx.doi.org/10.1038/srep37920] [PMID: 27910871]

[27]   R.F. Lu, *Light Scattering Technology for Food Property, Quality and Safety Assessment.* CRC Press: Boca Raton, 2016.

[28]   Y.M. Ren, H.S. Ramaswamy, Y. Li, C.L. Yuan, and X.L. Ren, "Classification of impact injury of apples using electronic nose coupled with multivariate statistical analyses", *J. Food Process Eng.,* vol. 41, no. 5, p. e12698, 2018.
[http://dx.doi.org/10.1111/jfpe.12698]

[29]   Z. Haddi, N. El Barbri, K. Tahri, M. Bougrini, N. El Bari, E. Llobet, and B. Bouchikhi, "Instrumental assessment of red meat origins and their storage time using electronic sensing systems", *Anal. Methods,* vol. 7, pp. 5193-5203, 2015.
[http://dx.doi.org/10.1039/C5AY00572H]

[30]   T. Konduru, G.C. Rains, and C.Y. Li, "Detecting sour skin infected onions using a customized gas sensor array", *J. Food Eng.,* vol. 160, pp. 19-27, 2015.
[http://dx.doi.org/10.1016/j.jfoodeng.2015.03.025]

[31]   M. Taniwaki, and N. Sakurai, "Evaluation of the internal quality of agricultural products using acoustic vibration techniques", *J. Jpn. Soc. Hortic. Sci.,* vol. 79, no. 2, pp. 113-128, 2010.
[http://dx.doi.org/10.2503/jjshs1.79.113]

[32]   W. Zhang, Z. Lv, and S. Xiong, "Nondestructive quality evaluation of agro-products using acoustic vibration methods-A review", *Crit. Rev. Food Sci. Nutr.,* vol. 58, no. 14, pp. 2386-2397, 2018.

[http://dx.doi.org/10.1080/10408398.2017.1324830] [PMID: 28613932]

[33]   T. Sun, K. Huang, H.R. Xu, and Y.B. Ying, "Research advances in nondestructive determination of internal quality in watermelon/melon: A review", *J. Food Eng.,* vol. 100, pp. 569-577, 2010.
[http://dx.doi.org/10.1016/j.jfoodeng.2010.05.019]

[34]   Z. Fathizadeh, M. Aboonajmi, and S.R.H. Beygi, "Nondestructive firmness prediction of apple fruit using acoustic vibration response", *Sci. Hortic. (Amsterdam),* vol. 262, p. 109073, 2020.
[http://dx.doi.org/10.1016/j.scienta.2019.109073]

[35]   C.Q. Ding, H.L. Wu, Z. Feng, D.C. Wang, W.H. Li, and D. Cui, "Online assessment of pear firmness by acoustic vibration analysis", *Postharvest Biol. Technol.,* vol. 160, p. 111042, 2020.
[http://dx.doi.org/10.1016/j.postharvbio.2019.111042]

CHAPTER 2

Evaluation of Quality of Agro-Products by Imaging and Spectroscopy

Insuck Baek[1], Jianwei Qin[1], Byoung-Kwan Cho[2] and Moon S. Kim[1,*]

[1] *Environmental Microbial and Food Safety Laboratory, Agricultural Research Service, U.S. Department of Agriculture, Powder Mill Rd. Bldg. 303, BARC-East, Beltsville, MD 20705, USA*

[2] *Department of Biosystems Machinery Engineering, College of Agricultural and Life Science, Chungnam National University, 99 Daehak-ro, Yuseoung-gu, Daejeon 34134, Republic of Korea*

Abstract: The quality of agro-products is the foremost current issue for the food industry and consumers. Healthful agro-products such as fruits and vegetables, meat, grains, and dairy products are essential for human life, and reliable quality evaluation is important for product safety and consumer appeal. As a result, rapid and precise evaluation methods for the quality of agro-products are required. In this regard, optical sensing techniques such as imaging and spectroscopy are among the most promising techniques currently investigated for quality assessment purposes in agricultural fields. This chapter aims to present the basic concepts, components and principles of imaging and spectroscopy techniques in a comparative manner for agriculture application. Moreover, this chapter also elaborates upon the partiality of the optical sensing techniques by highlighting previous studies in agricultural applications. The insights in this chapter will help a novice to understand and encourage further knowledge about optical sensing techniques.

Keywords: Agro-product, Hyperspectral imaging, Imaging, Spectroscopy.

1. INTRODUCTION

The quality of agro-products is affected by a variety of factors which, aside from those factors that are also related to food safety problems, may be weighted or flexible depending on customer awareness and current market conditions. Furthermore, agro-products have sample variations among batches or individual units, even when assessing the same product or cultivar type. Thus, evaluation of the quality of agro-products is more difficult than that for industrial products and needs a more sophisticated sensing technique. New developments in sensing

---

* **Corresponding Author Moon S. Kim:** Environmental Microbial and Food Safety Laboratory, Agricultural Research Service, U.S. Department of Agriculture, Powder Mill Rd. Bldg. 303, BARC-East, Beltsville, MD 20705, USA; Tel: +1-3015048462; Fax: +1-3015049466; E-mail: moon.kim@usda.gov

Jiangbo Li & Zhao Zhang (Eds.)

devices have allowed us to open a world of novel inspection methods. Optical instruments are a prominent example of advanced technologies used for food quality, including techniques such as imaging, spectroscopy, and hyperspectral imaging. In contrast with traditional methods for evaluating food and agro-products, these rapid techniques can deal with high throughput inspection. It is very important to quickly assess the external and internal attributes of agro-products in the product pipeline since those attributes are directly associated with both manufacturer profits and customer safety [1]. As a result, a wide range of evaluation techniques for agro-products has been suggested for assessing appearance, texture and chemical components on agro-products. This chapter targets understanding of the basic principles and concepts of optical techniques, especially spectroscopy and imaging, and also broadly presents their application for the assessment of agro-products in different parts of the world. The discussion presents information about the application of optical technologies for the quality evaluation of agro-products categorized according to major agricultural products.

2. SPECTROSCOPY TECHNIQUES

A good way to work with agro-products, determine them or define their attributes is to see how light interacts with them. Spectroscopy techniques are based on the interaction of light with matter and examine how light behaves in the target. Seeing how light interacts with agro-products is a good way to characterize or define some of their various qualities attributes. A spectrum is a measure of the amount of light detected at different wavelengths, showing how much light is reflected, absorbed, or transmitted from the target. Spectral data acquired through spectroscopy techniques can be interpreted as fingerprints for recognizing different materials based on their different spectral signatures. Conversely, these spectral signatures can be identified from the spectrum of the target. Therefore, spectroscopy techniques enable us to measure properties that are invisible to the human eye. In the agricultural field, the spectroscopy techniques are generally used to evaluate qualities that cannot be visually determined by the eye, such as the pigment content and soluble solids content of apples, mango maturity, or water content perdition of grape leaves [2 - 5]. Techniques using ultraviolet-visible (UV-VIS), fluorescence, infrared (IR), and Raman spectroscopy are usually adopted in the agriculture field. For evaluating the quality of agro-products, the technique selected is dependent on the target material and the attributes of interest, since spectroscopy techniques differ in their principles and effectiveness with various chemical substances.

2.1. Types of Spectroscopy

Ultraviolet-Visible (UV-VIS) spectroscopy is called absorption spectroscopy or reflectance spectroscopy and can measure wavelengths of light ranging from 100 nm to 380 nm (UV) and from 380 nm to 750 nm (VIS). The light absorption or reflectance in the visible range is related to the color of the chemicals involved. Fluorescence spectroscopy is complementary to absorption spectroscopy. Fluorescence deals with transitions from the excited state of a system to the ground state of the system, while absorption deals with transitions from the ground state to the excited state. Fluorescence spectroscopy is distinguished from other spectroscopy by the emission of light from the targeted substance material. When some energy from the incident light (excitation) is absorbed by the substance, the substance radiates light (emission) at typically lower energy. This emission of light is called fluorescence. Fluorescence spectroscopy is commonly used for food analysis due to the high sensitivity and selectivity of fluorescence measurements in food materials. Infrared (IR) spectroscopy is performed at IR wavelengths from 780 nm to 1 mm. In general, subdivisions of this spectral region are often described using the following scheme: near-infrared (NIR), short-wavelength infrared (SWIR), mid-wavelength infrared (MWIR), long-wavelength infrared (LWIR) and far infrared. Sometimes, NIR and SWIR are known as reflected infrared, while MWIR and LWIR are called thermal infrared. As the full scope of IR spectroscopy and its applications are so voluminous as to merit its own book, this chapter will only discuss IR spectroscopy and imaging in terms of NIR and SWIR techniques, which are popularly used for applications in agriculture. Raman spectroscopy is a light scattering technique, whereby molecules in a substance scatter incident light from a high intensity laser light source. Most of the scattered light-called Rayleigh scattering-occurs at the same wavelength as the laser source, but does not provide meaningful information. On the other hand, a small amount of the scattered light-called Raman scattering-occurs at various wavelengths different from the laser source, depending on the chemical structure of the analyte.

2.2. Spectroscopy Measurement

UV-VIS spectroscopy can use single-beam or double-beam instruments. In a double-beam instrument, the light is split into two beams, of which one is used for the sample and the other is used for a reference. Photodetectors measure the intensity of both beams, with the reference beam's intensity used to provide a value for 100% transmission, relative to which the sample's absorbance can be calculated. A single-beam instrument performs in a similar way using a single beam to measure two separate times-once for the empty sample holder (reference) and once for the sample holder containing the sample material.

Fluorescence spectroscopy uses light selected by a monochromator at a very narrow waveband for excitation of the sample substance. The resulting fluorescence emission from the sample is measured after passing through another monochromator to eliminate second order scattering effects from both excitation and emission. The concentration of the fluorescent sample substance is directly proportional to the intensity of the emission. There are many different types of excitation light sources, such as xenon arc lamps and mercury-vapor lamps, that can be used for fluorescence spectroscopy.

For NIR and SWIR measurements in agricultural applications, there are typically three types of data acquisition configurations. Those methods are reflectance, transmittance and interactance. Each method leads to a different effect on data acquisition, even for the same target. Reflectance uses a light source to illuminate the sample surfaced and measure the fraction of incident electromagnetic power which is reflected at the interface, providing good data for surface attributes but little information for internal or subsurface quality attributes [6]. The transmittance method gauges light passing through a sample target by placing the measurement probe and light source opposite each other with the target between the two. Transmittance is useful for measuring internal sample attributes but demands a stronger light source than reflectance since the light must penetrate the depth of the sample. The interactance method is sometimes called diffuse reflectance. Interactance combines aspects of the other two methods, typically using probes with a concentric outer ring for illumination and an inner ring for measurement of both surface and internal sample attributes as a two-in-one method.

Four techniques are commonly used for obtaining Raman spectral measurements in the agricultural field: dispersive Raman spectroscopy, Fourier transform Raman spectroscopy, surface-enhanced Raman spectroscopy, and spatially offset Raman spectroscopy. The most common and basic technique is dispersive Raman spectroscopy [7], in which the laser creates Raman scattering from the sample and notch filters block the overlapping of the Raman signal by stray light from the far more intense Rayleigh scattering. The role of the grating in a Raman system is to improve the resolution of the spectrum.

The basic concepts and schemes for these four general types of spectroscopy are presented in Fig. (1).

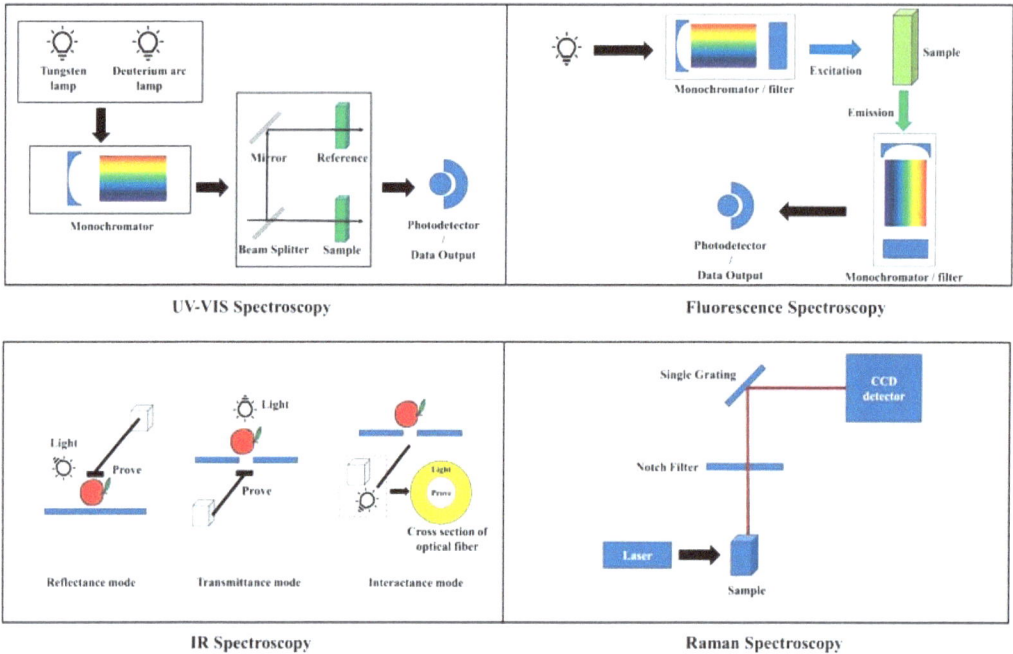

Fig. (1). Schemes for the four types of spectroscopy.

2.3. Data Preprocessing and Analysis

Data preprocessing or pre-treatment is often performed on raw spectral data to eliminate noise in the spectral data that is the result of the acquisition environment and baseline variations. Common preprocessing techniques can be divided into two groups: scatter correction methods and spectral derivatives. Scatter correction methods include multiplicative scattering correction (MSC), Inverse MSC, Extended MSC, de-trending and standard normal variate (SNV) and normalization (max, mean and range). Spectral derivative techniques include the Norris-Williams and Savitzky-Golay polynomial derivative filters. The purpose of preprocessing is to eliminate extraneous spectral noise in order to enhance a subsequent exploratory analysis, enhance a subsequent bi-linear calibration model, and enhance a subsequent classification model [8].

Calibration models for qualitative and quantitative evaluation of agro-products are commonly built with the use of modeling methods such as multiple linear regression (MLR), step multiple linear regression (SMLR), principle component regression (PCR), partial least squares regression (PLSR), artificial neural networks (ANN), support vector machine regression (SVM), K-nearest neighbors (KNN), soft independent modeling of class analogy (SIMCA) and discriminating PLS (PLS-DA). Those methods play an important role in system accuracy because the calibration model directly relates to the precision of property prediction and the accuracy of discrimination.

The ability of the final model is evaluated by using calibration and prediction (validation) sets. The performance of the final model is determined by the root mean squared error of calibration and the coefficient of determination (R squared) between real and prediction value for both data sets. If the number of data samples is lacking, cross-validation (rotation estimation) is used for making a calibration model to prevent over-fitting or selection bias problems. Cross-validation methods include the holdout method, K-fold cross validation method, and leave-one-out method. Ideally, an effective model should have a low root mean square error and a high coefficient of determination, with a minimum difference between those corresponding values of the calibration and prediction sets. This is important for determining the trade-off value between calibration and prediction accuracy since this value is directly related to the generalization ability of the model to new data.

3. IMAGING TECHNIQUES

Imaging techniques are very well suited for quality evaluation in agriculture due to reliable performance for many agro-products in automatic or rapid processing operations. There are many imaging applications in agricultural sectors, such as land-based and aerial-based remote sensing for crop resources assessments, digital farming, and sorting machines for food quality and safety [9, 10]. Imaging techniques obtain physical or meaningful information from images or scenes. This section reviews how imaging techniques are used through machine vision systems, including necessary system components and principles for using imaging techniques. The machine vision systems analyze image data to produce output that is used to guide physical manufacturing processes. For example, machine vision is frequently employed to automatically assess attributes for product inspection, quality control, and robot guidance. A machine vision system generally consists of several components: illumination, a frame grabber or digitizer, an imaging processing and analysis software, and a camera. The basic concept of the machine vision system is shown in Fig. (2). However, the mechanical design of a machine vision system should be built to suit the particular details involved in the processing operations for any specific agro-product. For instance, the implementation of imaging by machine vision on a modern apple packing line that moves the fruit on roller conveyors made of dark colored rubber

[11] will be different from implementation for inspection of raw chickens suspended from steel shackles on poultry processing lines [12].

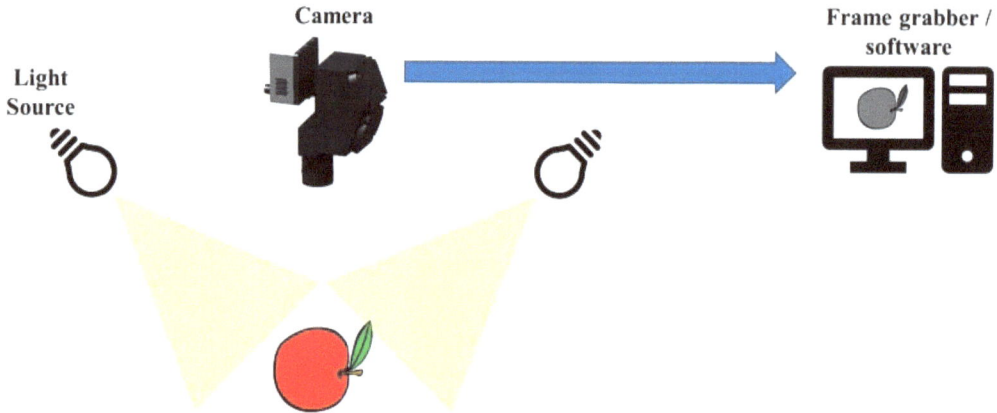

Fig. (2). Basic concept of the machine vision system.

3.1. Illumination

Lighting is a crucial factor in any imaging system. The lighting system should provide continuous and uniform illumination of the target as it is important to eliminate variations in appearance that are not due to the sample target. The type of light sources (*e.g.*, incandescent, fluorescent, halogen, Xenon, and LED), their placement (back, front, top), their shape (ring lighting, structured lighting), and their illumination geometry (point lighting, collimated lighting, diffuse lighting) can greatly influence image quality, enhance image contrast, and affect the overall efficiency and accuracy of the imaging system [13]. Good lighting sources can help to solve problems related to reflection, shadow, and some types of noise, thereby reducing image processing time. Therefore, the selection of proper lighting sources for a machine vision system is essential.

3.2. Digitizer or Frame Grabber

A machine vision system requires a frame grabber, which captures a digital form from an analog video signal or a digital video stream. This is called digitization. The functions of a frame grabber include image acquisition, camera handling, and image data pre-processing. For camera handling, the minimum requirements are accurate A/D circuitry and camera timing. In addition, the ability to control value related to gain and offset is also important for reducing effects from camera variability and fluctuations of the lighting source [14]. A traditional frame grabber needs a peripheral card slot (a peripheral component interconnect, commonly termed PCI) and thus cannot be implemented for machine vision on systems such as laptop computers, tablets, and embedded computer systems that lack space for traditional PCI frame grabbers. However, in the current market, external frame grabbers can easily be integrated any for the camera into any type of system. A camera with an integrated frame grabber can deliver video to computing platforms through GigE, USB 3.0 and Camera Link, *etc.*

3.3. Camera

To create a digital image, most digital cameras use a charge-coupled device (CCD) and complementary metal oxide semiconductor (CMOS) sensor. Both CCD and CMOS sensors convert light from the target surface into electrons. The electrical signals are made proportional to the intensity of the light by using an analog-to-digital converter (ADC). The CCD and CMOS sensors mainly differ in how each reads the value (accumulated charge) of each cell in the image. A CCD transports the charge across the chip and reads it at one readout circuit, and then the ADC turns each pixel's value into a digital value such as 8- or 16-bit data by measuring the amount of charge. A CMOS image sensor uses several transistors at each pixel to amplify and move the charge using wires. The different characteristics of these image sensors are explained in Table **1**. Fig. (**3**) presents the conceptual diagrams illustrating the difference between CCD and CMOS imaging chip architectures.

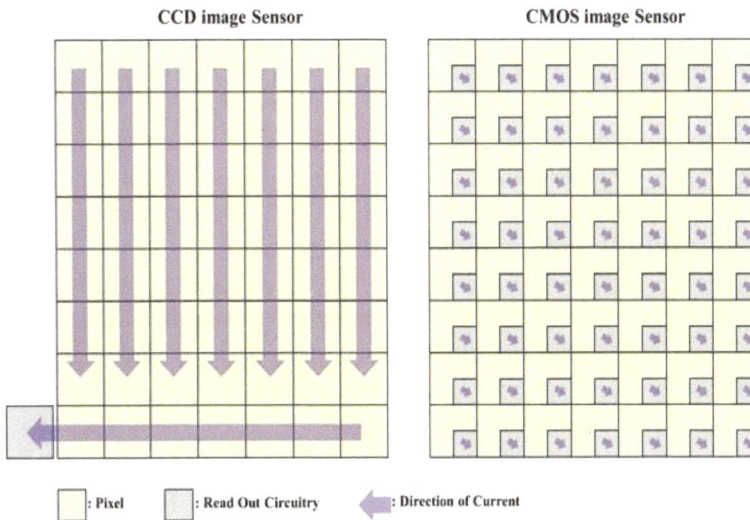

CCD image Sensor   CMOS image Sensor

☐ : Pixel   ☐ : Read Out Circuitry   ◀ : Direction of Current

Fig. (3). Conceptual diagrams illustrating the difference between CCD and CMOS imaging chip architectures.

Table 1. The Comparisons between CCD and CMOS image sensors.

Feature	CCD Sensor	CMOS Sensor
Susceptibility	Less (creates high quality and low noise image)	More
Sensitivity	Greater (Work in low light condition)	Lower
Power consumption	More	Less
Cost	More	Less
Architecture used	Passive Pixel sensor	Active Pixel sensor
Readout mechanism	Serial	Parallel
Speed	Lower	Higher
Dynamic Range	High	Low

3.4. Types of Imaging

Image processing and analysis software have a prominent role in any machine vision system for evaluating the quality of agro-products. Image processing is performed with a computer to create useful information such as color, morphological features, or texture features which are closely related to product quality. After extracting these features, the target is identified by using classification methods such as chemometrics, machine learning, and deep learning. The image processing method should be chosen based on the type of imaging being used, which typically falls into one of these four imaging categories: monochrome, color, multispectral, or hyperspectral [14]. Each type of imaging has specific advantages for extracting features from an image. Monochrome (grayscale) imaging provides images with better resolution, smaller output file size, and greater sensitivity to light. Traditional machine vision systems have been coupled with color cameras to acquire color images. The benefits of color imaging include high efficiency, ease in the extraction of true color and size, robust separation of target images and background removal. Both monochromatic and color imaging provide basic spatial and color information of the target, and thus represent the most basic concept for classifying physical differences in agro-products [14, 15]. In contrast, multispectral and hyperspectral imaging both combine spectral and spatial information about the target. Therefore, multispectral and hyperspectral image data are three-dimensional (3-D) in nature and usually used for creating chemical images or prediction images for visually non-obvious attributes such as sugar and acidity. Multispectral and hyperspectral imaging have advantages of powerful detection, flexible setup, and application versatility due to image association with chemical information in the form of spectral data. The main difference between multispectral imaging and hyperspectral imaging is the number of wavelengths presented in the image data and the resolution of those wavelengths. Hyperspectral imaging typically consists of numerous (tens or hundreds) continuous wavelength images, while multispectral imaging contains only a few (generally less than 10) discrete wavelength images [1]. Thus, multispectral imaging is frequently used for real-time applications in agriculture, while hyperspectral imaging is usually employed as a fundamental tool for finding key imaging wavelengths that can be implemented by a multispectral imaging system for practical use.

4. APPLICATIONS

As optical sensing methods, spectroscopy and imaging techniques greatly improve detection and evaluation capabilities and expand the scope of application. Thus, these optical sensing techniques have been explored for evaluating the physical, chemical, and biological attributes of a wide range of agro-products for many different applications.

4.1. Fruits & Vegetables

Fruits and vegetables are important foodstuffs that are rich in nutrients for the human body. When selecting their fresh produce, consumers desire trustworthy indicators of quality and safety in order to eat healthful fruits and vegetables without contamination [6]. The quality of fruits can be evaluated by physical and chemical properties [16]. Usually, external and internal quality factors indicate physical and chemical properties. External quality attributes include size (weight, volume, dimension), shape (diameter and depth ratio), color (uniformity intensity) and outwardly visible defects (bruise, stab, spot), while internal quality factors include flavor (sweetness, sourness, astringency, aroma), texture (firmness, crispness, juiciness), nutrition (carbohydrates, proteins, vitamins, functional properties), and outwardly invisible defects (internal cavity, water core, frost damage, rotten) [17]. Image data can be used to determine morphology, texture, and types of disease present, while spectroscopy can be used for the abovementioned internal attributes related to chemical composition. Hyper-spectral and multispectral imaging, which combine both spectroscopy and imaging techniques, have been used for simultaneously assessing both types of information for fruits and vegetables [18]. The results of previous studies in fruits and vegetables show that these techniques enable early sorting of fruits and vegetables and thereby can enhance total quality control. Table **2** summarizes the different applications and methods for quality evaluation of fruit and vegetables that have been employed by numerous researchers.

Many researchers have endeavored to apply spectroscopy and imaging techniques in the classification and quantification of fruits and vegetables such as apple, orange, peach, strawberry, carrot and onion. For example, because the 678 nm wavelength is associated with low chlorophyll content, reflectance around 678 nm was suggested for monitoring of apple ripeness [2]. In addition, for crucial attributes affecting the taste of apples such as firmness and soluble solids content, an SSC model was built by using SWNIR and LWIR, and this prediction model showed adequate coefficients of determination [3]. Another study about using SSC for apple was reported to have excellent accuracy [19]. In addition, imaging techniques have been used for detecting defects or disease on the surface [20 - 22].

As shown in Table **2**, spectroscopy techniques mainly were used for quality evaluation and prediction, while imaging techniques dealt with texture, defect, size, and color.

Table 2. Optical sensing technique applications for fruits and vegetables.

Product	Method	Application	Ref.
Apple	UV-VIS-NIR Spectroscopy	Pigment content	[2]
	VIS-NIR Spectroscopy	SSC prediction	[3]
	NIR Spectroscopy	SSC prediction	[19]
	Raman Spectroscopy	Pesticide residue detection	[23]
	NIR Spectroscopy	Firmness	[24]
	Color Imaging	Defects Detection	[25]
	Color Imaging	Quality grading	[26]
	Hyperspectral imaging	Bruise detection	[22]
	Hyperspectral imaging	Defect and feces detection	[21]
	Multispectral imaging	Faces Detection	[20]
Orange	VIS-NIR Spectroscopy	SSC, pH prediction	[27]
	NIR Spectroscopy	Brix prediction	[28]
	Color Imaging	Surface defect detection	[29]
	Hyperspectral imaging	Detecting rottenness	[30]
Peach	NIR Spectroscopy	Variety identification	[31]
	Color Imaging	Color and Size	[32]
	Hyperspectral imaging	Firmness	[33]

(Table 2) cont.....

Product	Method	Application	Ref.
Strawberry	VIS/NIR Spectroscopy	Acidity	[34]
	NIR spectroscopy	SSC prediction	[35]
	Color Image	Color Grading	[36]
	Hyperspectral Imaging	pH, Brix, Moisture content	[37]
Carrot	NIR spectroscopy	Carotenoid, fructose, glucose, sucrose and sugar content detection	[38]
Potato	NIR spectroscopy	Protein, fructose, glucose, starch and sucrose content	[39]
	Color Imaging	Irregular potatoes detecting	[40]
	Hyperspectral Imaging	Cooking time	[41]
Onion	Hyperspectral Imaging	Sour skin disease detection	[42]

4.2. Meats

As with fruits and vegetables, pork is a highly nutritious food and can be an important constituent of the human diet [43, 44]. Moreover, it is served as a major source of animal protein [45], with over 100 million tons of pork produced worldwide during the decade [46]. Consumers have been highly attentive to the quality and safety of the last meat [47]. Meat quality can be determined by measuring attributes or characters related to the suitability of meat to be consumed fresh or to be stored for appropriate periods without deterioration. Moreover, meat quality attributes can include chemical, microbiological, sensory, and functional attributes [45].

Many research studies have confirmed the ability of optical sensing technique to predict the chemical composition of beef such as protein, fat and moisture, dry matter, and other constituents. Meat quality can change by a series of reactions among different chemical constituents. The main example is water in meat since water is the main factor for the evaluation of quality of meat [48]. In the NIR spectra of meat, the absorbance of O-H bonds at 1450 and 1940 nm specifically enables the prediction of moisture, a result that many studies include is listed in Table 3. Sensory characteristics are also important such as appearance, juiciness, tenderness, and firmness, which can determine the acceptability of raw meat as perceived by consumers. Consumers tend to prefer bright red beef and lamb, pink pork, and uniformly and finely distributed fat marbling in red meats. These attributes are easily assessed by the imaging technique. In addition, pH and water holding capacity (WHC) are crucial meat quality factors. pH affects the storage and quality of meat by influencing WHC and color. WHC is related to protein, intramuscular fat, and water. During the meat aging process, a decrease in pH leads to electrostatic repulsion between water and proteins, resulting in a closed structure with a high WHC. The previous studies about meats are summarized in Table **3**.

Table 3. Optical sensing technique applications for meats.

Product	Method	Application	Ref.
Beef	NIR spectroscopy	Prediction of protein, intramuscular fat, ash	[49]
	NIR spectroscopy	Prediction of protein, intramuscular fat, moisture	[50]
	NIR spectroscopy	Prediction of intramuscular fat	[51]
	Fluorescence spectroscopy	Thermal change inspection	[52]
	Hyperspectral imaging	Prediction of tenderness	[53]
	Hyperspectral imaging	Detection of microbial spoilage	[54]
	Hyperspectral imaging	Prediction of tenderness, color and pH	[55]
Pork	NIR spectroscopy	Prediction of WHC, intramuscular fat	[56]
	NIR spectroscopy	Prediction of intramuscular fat, moisture	[57]
	NIR spectroscopy	Prediction of WHC, dip loss in fresh pork	[58]
	Hyperspectral imaging	Determination of tenderness and *E. coli*	[59]
	Hyperspectral imaging	Grading and classification	[60]
	Hyperspectral imaging	Prediction of protein, moisture, fat content	[61]
	Hyperspectral imaging	Prediction of pH content	[62]
Lamb	VIS-NIR spectroscopy	Prediction of protein, intramuscular fat, moisture	[63]
	Hyperspectral imaging	Discrimination of lamb muscles	[64]
	Hyperspectral imaging	Prediction of tenderness	[65]
Poultry	NIR spectroscopy	Prediction of protein, intramuscular fat, ash	[66]
	Hyperspectral and multispectral imaging	Skin tumor detection	[67]
	Multispectral imaging	Detection of fecal contamination	[68]
	Multispectral imaging	Chicken heart disease detection	[69]

4.3. Miscellaneous Applications

There are many successful applications of optical technology in other agro-products. The grains are among the most important of the agro-products since they are used to prepare common staple foods worldwide, including bread, cake, biscuits and so on [70]. Research studies regarding grains have focused mainly on classification and identification of grain types, disease or viability detection, and constituent prediction. For differentiating between diseased and sound grains, imaging techniques have advantages over spectroscopy because of the small size of grains and the need to inspect bulk samples. Specific examples include the color camera measurement of rice milling [71] and hyperspectral imaging for the classification of bulk samples for different classes of wheat [72].

Spectroscopy and imaging techniques have also been tested for quality and safety evaluation of dairy products (*e.g.*, cheese, yoghurt, cream, and butter). For example, Liu *et al.* conducted experiments to detect yolk contamination in egg white by using spectroscopy [73]. The fatty acid content in goat milk was predicted *via* VIS-NIR spectroscopy [74]. Hyperspectral imaging technique was demonstrated to be capable of detecting melamine in milk powders [75, 76].

The evaluation of oil and beverage quality is another example of the application of optical technique. Specific examples include origin identification of coffee [77] and quality control of coffee [78], tomato juice [79], apple wine [80] and rice wine [81]. Liquid beverage samples were mainly tested by spectroscopy techniques since spatial information is generally not needed for the homogenous composition presented by these products.

Table 4. Optical sensing technique applications for miscellaneous products.

Product	Method	Application	Ref.
Grain	NIR and Raman spectroscopy	Viability assessment	[82]
	NIR spectroscopy	Detecting damaged almonds	[83]
	Color imaging	Cultivar identification	[84]
	Color imaging	Measuring the degree of milling for rice	[71]
	Hyperspectral imaging	Classification	[85]
	Hyperspectral imaging	Classification	[72]
	Hyperspectral imaging	Detecting toxigenic fungi	[86]
	Hyperspectral imaging	Variety discrimination of maize seeds	[87]
Dairy	UV-VIS-NIR spectroscopy	Contamination detection of egg	[73]
	VIS-NIR spectroscopy	Fatty acid content detection in goat milk	[74]
	Raman and Hyperspectral imaging	Detecting melamine in milk powders	[75, 76]

(Table 4) cont.....

Product	Method	Application	Ref.
Beverage	NIR spectroscopy	Identification of origin coffee	[77]
	NIR spectroscopy	Determination of roasting degree and robusta ratio in coffee	[78]
	NIR-MIR spectroscopy	Monitoring multicomponent quality traits in tomato juice	[79]
	NIR spectroscopy	Monitoring volatile compounds in apple wines	[80]
	NIR-MIR spectroscopy	Quality control of rice wines	[81]
	Hyperspectral imaging	Classification of quality of green tea	[88]
	Hyperspectral imaging	Prediction of moisture content in tealeaf	[89]

CONCLUSION

In conclusion, optical sensing techniques can produce great benefits for producers and processors when integrated into agricultural operations. Complete knowledge of the properties and characteristic interactions between light and food materials enables quantitative and qualitative assessment of many attributes of agro-products for quality and safety analysis. Many optical techniques exist but mostly used for agricultural purposes can be categorized primarily as either imaging-based or spectroscopy-based techniques. The spectroscopy techniques are effective in obtaining spectral information, whereas the imaging techniques are essential for collecting spatial and color imaging data. In addition, hyperspectral and multispectral imaging are combinations of both imaging and spectroscopy, allowing simultaneous collection of both spectral and spatial data sets. These optical sensing techniques have been used for a wide range of food and agricultural products such as fruits, vegetable, meats, grains, dairy products, and beverages. Optical sensing techniques can be directly implemented in both fundamental research tools and industrial applications to inspect and evaluate a great range of agro-products. The development of reliable, fast, and efficient systems using optical sensing techniques is very promising. Therefore, the industry and researchers could choose these techniques to overcome the limitations of traditional evaluation methods.

CONSENT FOR PUBLICATION

Not applicable.

CONFLICT OF INTEREST

The authors confirm that this chapter contents have no conflict of interest.

ACKNOWLEDGEMENTS

None declared.

REFERENCES

[1]   J. Qin, K. Chao, M.S. Kim, R. Lu, and T.F. Burks, "Hyperspectral and multispectral imaging for evaluating food safety and quality", *J. Food Eng.,* vol. 118, no. 2, pp. 157-171, 2013.
[http://dx.doi.org/10.1016/j.jfoodeng.2013.04.001]

[2]   A. Nagy, P. Riczu, and J. Tamás, "Spectral evaluation of apple fruit ripening and pigment content alteration", *Sci. Hortic. (Amsterdam),* vol. 201, pp. 256-264, 2016.
[http://dx.doi.org/10.1016/j.scienta.2016.02.016]

[3]   Z. Guo, W. Huang, Y. Peng, Q. Chen, Q. Ouyang, and J. Zhao, "Color compensation and comparison of shortwave near infrared and long wave near infrared spectroscopy for determination of soluble solids content of 'Fuji' apple", *Postharvest Biol. Technol.,* vol. 115, pp. 81-90, 2016.
[http://dx.doi.org/10.1016/j.postharvbio.2015.12.027]

[4]   S.N. Jha, "Nondestructive prediction of maturity of mango using near infrared spectroscopy", *J. Food Eng.,* vol. 124, pp. 152-157, 2014.
[http://dx.doi.org/10.1016/j.jfoodeng.2013.10.012]

[5]   A.B. González-Fernández, J.R. Rodríguez-Pérez, M. Marabel, and F. Álvarez-Taboada, "Spectroscopic estimation of leaf water content in commercial vineyards using continuum removal and partial least squares regression", *Sci. Hortic. (Amsterdam),* vol. 188, pp. 15-22, 2015.
[http://dx.doi.org/10.1016/j.scienta.2015.03.012]

[6]   H. Lin, and Y. Ying, "Theory and application of near infrared spectroscopy in assessment of fruit quality: A review", *Sens. Instrum. Food Qual. Saf.,* vol. 3, no. 2, pp. 130-141, 2009.
[http://dx.doi.org/10.1007/s11694-009-9079-z]

[7]   D. Yang, and Y. Ying, "Applications of raman spectroscopy in agricultural products and food analysis: A review", *Appl. Spectrosc. Rev.,* vol. 46, no. 7, pp. 539-560, 2011.
[http://dx.doi.org/10.1080/05704928.2011.593216]

[8]   Å. Rinnan, F. van den Berg, and S.B. Engelsen, "Review of the most common pre-processing techniques for near-infrared spectra, TrAC", *Trends Analyt. Chem.,* vol. 28, no. 10, pp. 1201-1222, 2009.
[http://dx.doi.org/10.1016/j.trac.2009.07.007]

[9]   T. Brosnan, and D-W. Sun, "Improving quality inspection of food products by computer vision—a review", *J. Food Eng.,* vol. 61, no. 1, pp. 3-16, 2004.
[http://dx.doi.org/10.1016/S0260-8774(03)00183-3]

[10]   K.K. Patel, A. Kar, S.N. Jha, and M.A. Khan, "Machine vision system: a tool for quality inspection of food and agricultural products", *J. Food Sci. Technol.,* vol. 49, no. 2, pp. 123-141, 2012.
[http://dx.doi.org/10.1007/s13197-011-0321-4] [PMID: 23572836]

[11]   Y. Tao, "Spherical transform of fruit images for on-line defect extraction of mass objects", *Opt. Eng.,* vol. 35, no. 2, p. 344, 1996.
[http://dx.doi.org/10.1117/1.600902]

[12]   K. Chao, C-C. Yang, M.S. Kim, and D.E. Chan, "High throughput spectral imaging system for wholesomeness inspection of chicken", *Appl. Eng. Agric.,* vol. 24, no. 4, pp. 475-485, 2008.
[http://dx.doi.org/10.13031/2013.25135]

[13]   S. Gunasekaran, "Computer vision technology for food quality assurance", *Trends Food Sci. Technol.,* vol. 7, no. 8, pp. 245-256, 1996.
[http://dx.doi.org/10.1016/0924-2244(96)10028-5]

[14]   Y.R. Chen, K. Chao, and M.S. Kim, "Machine vision technology for agricultural applications", *Comput. Electron. Agric.,* vol. 36, no. 2–3, pp. 173-191, 2002.
[http://dx.doi.org/10.1016/S0168-1699(02)00100-X]

[15]   M. Aboonajmi, and H. Faridi, Nondestructive quality assessment of Agro-food products. *IRNDT 2016-A13105, Proceedings of the 3ʳᵈ Iranian International NDT Conference.* Olympic Hotel: Tehran, Iran, 2016.

[16]   D. Liu, X.A. Zeng, and D.W. Sun, "Recent developments and applications of hyperspectral imaging for quality evaluation of agricultural products: a review", *Crit. Rev. Food Sci. Nutr.,* vol. 55, no. 12, pp. 1744-1757, 2015.
[http://dx.doi.org/10.1080/10408398.2013.777020] [PMID: 24915395]

[17]   H.S. El-Mesery, H. Mao, and A.E.F. Abomohra, "Applications of non-destructive technologies for agricultural and food products quality inspection", *Sensors (Basel),* vol. 19, no. 4, pp. 1-23, 2019.
[http://dx.doi.org/10.3390/s19040846] [PMID: 30781709]

[18]   A. Bhargava, and A. Bansal, "Fruits and vegetables quality evaluation using computer vision: A review", In: *J. King Saud Univ. Comput. Inf. Sci.* In press, 2018.
[http://dx.doi.org/10.1016/j.jksuci.2018.06.002]

[19]   B. Park, J.A. Abbott, K.J. Lee, C.H. Choi, and K.H. Choi, "Near-infrared diffuse reflectance for quantitative and qualitative measurement of soluble solids and firmness of delicious and gala apples", *Trans. ASAE,* vol. 46, no. 6, pp. 1721-1731, 2003.
[http://dx.doi.org/10.13031/2013.15628]

[20]   M.S. Kim, A.M. Lefcourt, and Y-R. Chen, "Multispectral laser-induced fluorescence imaging system for large biological samples", *Appl. Opt.,* vol. 42, no. 19, pp. 3927-3934, 2003.
[http://dx.doi.org/10.1364/AO.42.003927] [PMID: 12868832]

[21]   M.S. Kim, "Hyperspectral reflectance and fluorescence line-scan imaging for online defect and fecal contamination inspection of apples", *Sens. Instrum. Food Qual. Saf.,* vol. 1, no. 3, pp. 151-159, 2007.
[http://dx.doi.org/10.1007/s11694-007-9017-x]

[22]   R. Lu, "Detection of bruises on apples using near-infrared hyperspectral imaging", *Trans. ASAE,* vol. 46, no. 2, pp. 523-530, 2003.

[23]   H. Luo, Y. Huang, K. Lai, B.A. Rasco, and Y. Fan, "Surface-enhanced Raman spectroscopy coupled with gold nanoparticles for rapid detection of phosmet and thiabendazole residues in apples", *Food Control,* vol. 68, pp. 229-235, 2016.
[http://dx.doi.org/10.1016/j.foodcont.2016.04.003]

[24]   B. Shi, L. Zhao, H. Wang, and D. Zhu, "Signal optimization approaches on the prediction of apples firmness by near infrared spectroscopy", *Sens. Lett.,* vol. 9, no. 3, pp. 1062-1068, 2011.
[http://dx.doi.org/10.1166/sl.2011.1381]

[25]   Z. Xiao-bo, Z. Jie-wen, L. Yanxiao, and M. Holmes, "In-line detection of apple defects using three color cameras system", *Comput. Electron. Agric.,* vol. 70, no. 1, pp. 129-134, 2010.
[http://dx.doi.org/10.1016/j.compag.2009.09.014]

[26]   J. Blasco, N. Aleixos, and E. Moltó, "Machine vision system for automatic quality grading of fruit", *Biosyst. Eng.,* vol. 85, no. 4, pp. 415-423, 2003.
[http://dx.doi.org/10.1016/S1537-5110(03)00088-6]

[27]   H. Lu, H. Jiang, X. Fu, H. Yu, H. Xu, and Y. Ying, "Non-invasive measurements of the internal quality of intact 'Gannan' navel orange by vis/nir spectroscopy", *Trans. ASABE,* vol. 51, no. 3, pp. 1009-1014, 2008.
[http://dx.doi.org/10.13031/2013.24505]

[28]   Y. Liu, R. Gao, Y. Hao, X. Sun, and A. Ouyang, "Improvement of near-infrared spectral calibration models for brix prediction in 'Gannan' navel oranges by a portable near-infrared device", *Food Bioprocess Technol.,* vol. 5, no. 3, pp. 1106-1112, 2012.

[http://dx.doi.org/10.1007/s11947-010-0449-7]

[29]   D. Rong, X. Rao, and Y. Ying, "Computer vision detection of surface defect on oranges by means of a sliding comparison window local segmentation algorithm", *Comput. Electron. Agric.*, vol. 137, pp. 59-68, 2017.
[http://dx.doi.org/10.1016/j.compag.2017.02.027]

[30]   J. Gómez-Sanchis, J.D. Martín-Guerrero, E. Soria-Olivas, M. Martínez-Sober, R. Magdalena-Benedito, and J. Blasco, "Detecting rottenness caused by Penicillium genus fungi in citrus fruits using machine learning techniques", *Expert Syst. Appl.*, vol. 39, no. 1, pp. 780-785, 2012.
[http://dx.doi.org/10.1016/j.eswa.2011.07.073]

[31]   W. Guo, J. Gu, D. Liu, and L. Shang, "Peach variety identification using near-infrared diffuse reflectance spectroscopy", *Comput. Electron. Agric.*, vol. 123, pp. 297-303, 2016.
[http://dx.doi.org/10.1016/j.compag.2016.03.005]

[32]   A. Esehaghbeygi, M. Ardforoushan, S.A.H. Monajemf, and A.A. Masoumi, "Digital image processing for quality ranking of saffron peach", *Int. Agrophys.*, vol. 24, no. 2, pp. 115-120, 2010.

[33]   R. Lu, and Y. Peng, "Hyperspectral scattering for assessing peach fruit firmness", *Biosyst. Eng.*, vol. 93, no. 2, pp. 161-171, 2006.
[http://dx.doi.org/10.1016/j.biosystemseng.2005.11.004]

[34]   Y. Shao, and Y. He, "Nondestructive measurement of acidity of strawberry using Vis/NIR spectroscopy", *Int. J. Food Prop.*, vol. 11, no. 1, pp. 102-111, 2008.
[http://dx.doi.org/10.1080/10942910701257057]

[35]   Z. Guo, W. Huang, L. Chen, X. Wang, and Y. Peng, ""Nondestructive evaluation of soluble solid content in strawberry by near infrared spectroscopy," PIAGENG 2013 Image Process", *Photonics Agric. Eng.*, vol. 8761, 2013.87610O

[36]   X. Liming, and Z. Yanchao, "Automated strawberry grading system based on image processing", *Comput. Electron. Agric*, vol. 71, no. SUPPL. 1, pp. 32-39, 2010.
[http://dx.doi.org/10.1016/j.compag.2009.09.013]

[37]   G. ElMasry, N. Wang, A. ElSayed, and M. Ngadi, "Hyperspectral imaging for nondestructive determination of some quality attributes for strawberry", *J. Food Eng.*, vol. 81, no. 1, pp. 98-107, 2007.
[http://dx.doi.org/10.1016/j.jfoodeng.2006.10.016]

[38]   H. Schulz, H.H. Drews, R. Quilitzsch, and H. Krüger, "Application of near infrared spectroscopy for the quantification of quality parameters in selected vegetables and essential oil plants", *J. Near Infrared Spectrosc.*, vol. 6, no. 1–4, pp. 125-130, 1998.
[http://dx.doi.org/10.1255/jnirs.179]

[39]   R. Hartmann, and H. Büning-Pfaue, "NIR determination of potato constituents", *Potato Res.*, vol. 41, no. 4, pp. 327-334, 1998.
[http://dx.doi.org/10.1007/BF02358965]

[40]   G. Elmasry, S. Cubero, E. Moltó, and J. Blasco, "In-line sorting of irregular potatoes by using automated computer-based machine vision system", *J. Food Eng.*, vol. 112, no. 1–2, pp. 60-68, 2012.
[http://dx.doi.org/10.1016/j.jfoodeng.2012.03.027]

[41]   N. Nguyen Do Trong, M. Tsuta, B.M. Nicolaï, J. De Baerdemaeker, and W. Saeys, "Prediction of optimal cooking time for boiled potatoes by hyperspectral imaging", *J. Food Eng.*, vol. 105, no. 4, pp. 617-624, 2011.
[http://dx.doi.org/10.1016/j.jfoodeng.2011.03.031]

[42]   W. Wang, C. Li, E.W. Tollner, R.D. Gitaitis, and G.C. Rains, "Shortwave infrared hyperspectral imaging for detecting sour skin (*Burkholderia cepacia*)-infected onions", *J. Food Eng.*, vol. 109, no. 1, pp. 38-48, 2012.
[http://dx.doi.org/10.1016/j.jfoodeng.2011.10.001]

[43]   F. Qu, D. Ren, Y. He, P. Nie, L. Lin, C. Cai, and T. Dong, "Predicting pork freshness using multi-index statistical information fusion method based on near infrared spectroscopy", *Meat Sci.,* vol. 146, no. February, pp. 59-67, 2018.
[http://dx.doi.org/10.1016/j.meatsci.2018.07.023] [PMID: 30099231]

[44]   M. Kamruzzaman, Y. Makino, and S. Oshita, "Non-invasive analytical technology for the detection of contamination, adulteration, and authenticity of meat, poultry, and fish: a review", *Anal. Chim. Acta,* vol. 853, no. 1, pp. 19-29, 2015.
[http://dx.doi.org/10.1016/j.aca.2014.08.043] [PMID: 25467447]

[45]   Z. Xiong, D-W. Sun, X-A. Zeng, and A. Xie, "Recent developments of hyperspectral imaging systems and their applications in detecting quality attributes of red meats: A review", *J. Food Eng.,* vol. 132, pp. 1-13, 2014.
[http://dx.doi.org/10.1016/j.jfoodeng.2014.02.004]

[46]   FAOSTAT, "FAOSTAT statistics database", http://www.fao.org/home/en

[47]   H. Lee, M.S. Kim, W.H. Lee, and B.K. Cho, "Determination of the total volatile basic nitrogen (TVB-N) content in pork meat using hyperspectral fluorescence imaging", *Sens. Actuators B Chem.,* vol. 259, pp. 532-539, 2018.
[http://dx.doi.org/10.1016/j.snb.2017.12.102]

[48]   N. Prieto, R. Roehe, P. Lavín, G. Batten, and S. Andrés, "Application of near infrared reflectance spectroscopy to predict meat and meat products quality: A review", *Meat Sci.,* vol. 83, no. 2, pp. 175-186, 2009.
[http://dx.doi.org/10.1016/j.meatsci.2009.04.016] [PMID: 20416766]

[49]   R. Sanderson, S.J. Lister, M.S. Dhanoa, R.J. Barnes, and C. Thomas, "Use of near infrared reflectance spectroscopy to predict and compare the composition of carcass samples from young steers", *Anim. Sci.,* vol. 65, no. 1, pp. 45-54, 1997.
[http://dx.doi.org/10.1017/S1357729800016283]

[50]   G. Tøgersen, T. Isaksson, B.N. Nilsen, E.A. Bakker, and K.I. Hildrum, "On-line NIR analysis of fat, water and protein in industrial scale ground meat batches", *Meat Sci.,* vol. 51, no. 1, pp. 97-102, 1999.
[http://dx.doi.org/10.1016/S0309-1740(98)00106-5] [PMID: 22061541]

[51]   R. Rødbotten, "B. Nilsen, and K. Hildrum, "Prediction of beef quality attributes from early post mortem near infrared reflectance spectra", *Food Chem.,* vol. 69, no. 4, pp. 427-436, 2000.
[http://dx.doi.org/10.1016/S0308-8146(00)00059-5]

[52]   A. Sahar, "U. ur Rahman, A. Kondjoyan, S. Portanguen, and E. Dufour, "Monitoring of thermal changes in meat by synchronous fluorescence spectroscopy", *J. Food Eng.,* vol. 168, pp. 160-165, 2016.
[http://dx.doi.org/10.1016/j.jfoodeng.2015.07.038]

[53]   G.K. Naganathan, L.M. Grimes, J. Subbiah, C.R. Calkins, A. Samal, and G.E. Meyer, "Visible/near-infrared hyperspectral imaging for beef tenderness prediction", *Comput. Electron. Agric.,* vol. 64, no. 2, pp. 225-233, 2008.
[http://dx.doi.org/10.1016/j.compag.2008.05.020]

[54]   Y. Peng, "Potential prediction of the microbial spoilage of beef using spatially resolved hyperspectral scattering profiles", *J. Food Eng.,* vol. 102, no. 2, pp. 163-169, 2011.
[http://dx.doi.org/10.1016/j.jfoodeng.2010.08.014]

[55]   G. ElMasry, D-W. Sun, and P. Allen, "Near-infrared hyperspectral imaging for predicting colour, pH and tenderness of fresh beef", *J. Food Eng.,* vol. 110, no. 1, pp. 127-140, 2012.
[http://dx.doi.org/10.1016/j.jfoodeng.2011.11.028]

[56]   J. Brøndum, L. Munck, P. Henckel, A. Karlsson, E. Tornberg, and S.B. Engelsen, "Prediction of water-holding capacity and composition of porcine meat by comparative spectroscopy", *Meat Sci.,* vol. 55, no. 2, pp. 177-185, 2000.

[http://dx.doi.org/10.1016/S0309-1740(99)00141-2] [PMID: 22061083]

[57] N. Barlocco, A. Vadell, F. Ballesteros, G. Galietta, and D. Cozzolino, "Predicting intramuscular fat, moisture and Warner-Bratzler shear force in pork muscle using near infrared reflectance spectroscopy", *Anim. Sci.,* vol. 82, no. 1, pp. 111-116, 2006.
[http://dx.doi.org/10.1079/ASC20055]

[58] J.C. Forrest, M.T. Morgan, C. Borggaard, A.J. Rasmussen, B.L. Jespersen, and J.R. Andersen, "Development of technology for the early post mortem prediction of water holding capacity and drip loss in fresh pork", *Meat Sci.,* vol. 55, no. 1, pp. 115-122, 2000.
[http://dx.doi.org/10.1016/S0309-1740(99)00133-3] [PMID: 22060911]

[59] F. Tao, Y. Peng, Y. Li, K. Chao, and S. Dhakal, "Simultaneous determination of tenderness and *Escherichia coli* contamination of pork using hyperspectral scattering technique", *Meat Sci.,* vol. 90, no. 3, pp. 851-857, 2012.
[http://dx.doi.org/10.1016/j.meatsci.2011.11.028] [PMID: 22169338]

[60] D. Barbin, G. Elmasry, D-W.W. Sun, and P. Allen, "Near-infrared hyperspectral imaging for grading and classification of pork", *Meat Sci.,* vol. 90, no. 1, pp. 259-268, 2012.
[http://dx.doi.org/10.1016/j.meatsci.2011.07.011] [PMID: 21821367]

[61] D.F. Barbin, G. ElMasry, D-W. Sun, and P. Allen, "Non-destructive determination of chemical composition in intact and minced pork using near-infrared hyperspectral imaging", *Food Chem.,* vol. 138, no. 2-3, pp. 1162-1171, 2013.
[http://dx.doi.org/10.1016/j.foodchem.2012.11.120] [PMID: 23411227]

[62] D.F. Barbin, G. ElMasry, D-W.W. Sun, and P. Allen, "Predicting quality and sensory attributes of pork using near-infrared hyperspectral imaging", *Anal. Chim. Acta,* vol. 719, pp. 30-42, 2012.
[http://dx.doi.org/10.1016/j.aca.2012.01.004] [PMID: 22340528]

[63] D. Cozzolino, I. Murray, J.R. Scaife, and R. Paterson, "Study of dissected lamb muscles by visible and near infrared reflectance spectroscopy for composition assessment", *Anim. Sci.,* vol. 70, no. 3, pp. 417-423, 2000.
[http://dx.doi.org/10.1017/S1357729800051766]

[64] M. Kamruzzaman, G. ElMasry, D-W. Sun, and P. Allen, "Application of NIR hyperspectral imaging for discrimination of lamb muscles", *J. Food Eng.,* vol. 104, no. 3, pp. 332-340, 2011.
[http://dx.doi.org/10.1016/j.jfoodeng.2010.12.024]

[65] M. Kamruzzaman, G. Elmasry, D-W. Sun, and P. Allen, "Non-destructive assessment of instrumental and sensory tenderness of lamb meat using NIR hyperspectral imaging", *Food Chem.,* vol. 141, no. 1, pp. 389-396, 2013.
[http://dx.doi.org/10.1016/j.foodchem.2013.02.094] [PMID: 23768372]

[66] F. Abeni, and G. Bergoglio, "Characterization of different strains of broiler chicken by carcass measurements, chemical and physical parameters and NIRS on breast muscle", *Meat Sci.,* vol. 57, no. 2, pp. 133-137, 2001.
[http://dx.doi.org/10.1016/S0309-1740(00)00084-X] [PMID: 22061355]

[67] K. Chao, P.M. Mehl, and Y.R. Chen, "Use of hyper and multi spectral imaging for detection of chicken skin tumors", *Appl. Eng. Agric.,* vol. 18, no. 1, 2002.
[http://dx.doi.org/10.13031/2013.7700]

[68] B. Park, M. Kise, K.C. Lawrence, W.R. Windham, D.P. Smith, and C.N. Thai, "Real-time multispectral imaging system for online poultry fecal inspection using unified modeling language", *Sens. Instrum. Food Qual. Saf.,* vol. 1, no. 2, pp. 45-54, 2007.
[http://dx.doi.org/10.1007/s11694-007-9006-0]

[69] K. Chao, Y.R. Chen, W.R. Hruschka, and B. Park, "Chicken heart disease characterization by multi-spectral imaging", *Appl. Eng. Agric.,* vol. 17, no. 1, pp. 99-106, 2001.
[http://dx.doi.org/10.13031/2013.1926]

[70] D. Cozzolino, "An overview of the use of infrared spectroscopy and chemometrics in authenticity and traceability of cereals", *Food Res. Int.,* vol. 60, pp. 262-265, 2014.
[http://dx.doi.org/10.1016/j.foodres.2013.08.034]

[71] W. Liu, Y. Tao, T.J. Siebenmorgen, and H. Chen, "Digital image analysis method for rapid measurement of rice degree of milling", *Cereal Chem.,* vol. 75, no. 3, pp. 380-385, 1998.
[http://dx.doi.org/10.1094/CCHEM.1998.75.3.380]

[72] R. Choudhary, S. Mahesh, J. Paliwal, and D.S. Jayas, "Identification of wheat classes using wavelet features from near infrared hyperspectral images of bulk samples", *Biosyst. Eng.,* vol. 102, no. 2, pp. 115-127, 2009.
[http://dx.doi.org/10.1016/j.biosystemseng.2008.09.028]

[73] M. Liu, L. Yao, T. Wang, J. Li, and C. Yu, "Rapid determination of egg yolk contamination in egg white by VIS spectroscopy", *J. Food Eng.,* vol. 124, pp. 117-121, 2014.
[http://dx.doi.org/10.1016/j.jfoodeng.2013.10.004]

[74] N. Núñez-Sánchez, A.L. Martínez-Marín, O. Polvillo, V.M. Fernández-Cabanás, J. Carrizosa, B. Urrutia, and J.M. Serradilla, "Near Infrared Spectroscopy (NIRS) for the determination of the milk fat fatty acid profile of goats", *Food Chem.,* vol. 190, pp. 244-252, 2016.
[http://dx.doi.org/10.1016/j.foodchem.2015.05.083] [PMID: 26212967]

[75] J. Qin, K. Chao, and M.S. Kim, "Raman chemical imaging system for food safety and quality inspection", *Trans. ASABE,* vol. 53, no. 6, pp. 1873-1882, 2010.
[http://dx.doi.org/10.13031/2013.35796]

[76] J. Lim, G. Kim, C. Mo, M.S. Kim, K. Chao, J. Qin, X. Fu, I. Baek, and B.K. Cho, "Detection of melamine in milk powders using near-infrared hyperspectral imaging combined with regression coefficient of partial least square regression model", *Talanta,* vol. 151, pp. 183-191, 2016.
[http://dx.doi.org/10.1016/j.talanta.2016.01.035] [PMID: 26946026]

[77] I. Marquetti, J.V. Link, and A.L.G. Lemes, "M. B. dos S. Scholz, P. Valderrama, and E. Bona, "Partial least square with discriminant analysis and near infrared spectroscopy for evaluation of geographic and genotypic origin of arabica coffee", *Comput. Electron. Agric.,* vol. 121, pp. 313-319, 2016.
[http://dx.doi.org/10.1016/j.compag.2015.12.018]

[78] E. Bertone, A. Venturello, A. Giraudo, G. Pellegrino, and F. Geobaldo, "Simultaneous determination by NIR spectroscopy of the roasting degree and Arabica/Robusta ratio in roasted and ground coffee", *Food Control,* vol. 59, pp. 683-689, 2016.
[http://dx.doi.org/10.1016/j.foodcont.2015.06.055]

[79] H. Ayvaz, A. Sierra-Cadavid, D.P. Aykas, B. Mulqueeney, S. Sullivan, and L.E. Rodriguez-Saona, "Monitoring multicomponent quality traits in tomato juice using portable mid-infrared (MIR) spectroscopy and multivariate analysis", *Food Control,* vol. 66, pp. 79-86, 2016.
[http://dx.doi.org/10.1016/j.foodcont.2016.01.031]

[80] M. Ye, Z. Gao, Z. Li, Y. Yuan, and T. Yue, "Rapid detection of volatile compounds in apple wines using FT-NIR spectroscopy", *Food Chem.,* vol. 190, pp. 701-708, 2016.
[http://dx.doi.org/10.1016/j.foodchem.2015.05.112] [PMID: 26213028]

[81] D-Y. Kim, B-K. Cho, S.H. Lee, K. Kwon, E.S. Park, and W-H. Lee, "Application of Fourier transform-mid infrared reflectance spectroscopy for monitoring Korean traditional rice wine 'Makgeolli' fermentation", *Sens. Actuators B Chem.,* vol. 230, pp. 753-760, 2016.
[http://dx.doi.org/10.1016/j.snb.2016.02.076]

[82] A. Ambrose, S. Lohumi, W.H. Lee, and B.K. Cho, "Comparative nondestructive measurement of corn seed viability using Fourier transform near-infrared (FT-NIR) and Raman spectroscopy", *Sens. Actuators B Chem.,* vol. 224, pp. 500-506, 2016.
[http://dx.doi.org/10.1016/j.snb.2015.10.082]

[83] T.C. Pearson, "Use of near infrared transmittance to automatically detect almonds with concealed

damage", *Lebensm. Wiss. Technol.,* vol. 32, no. 2, pp. 73-78, 1999.
[http://dx.doi.org/10.1006/fstl.1998.0489]

[84] H. Lu, Z. Cao, Y. Xiao, Z. Fang, and Y. Zhu, "Toward good practices for fine-grained maize cultivar identification with filter-specific convolutional activations", *IEEE Trans. Autom. Sci. Eng.,* vol. 15, no. 2, pp. 430-442, 2018.
[http://dx.doi.org/10.1109/TASE.2016.2616485]

[85] S. Mahesh, A. Manickavasagan, D.S. Jayas, J. Paliwal, and N.D.G. White, "Feasibility of near-infrared hyperspectral imaging to differentiate Canadian wheat classes", *Biosyst. Eng.,* vol. 101, no. 1, pp. 50-57, 2008.
[http://dx.doi.org/10.1016/j.biosystemseng.2008.05.017]

[86] A. Del Fiore, M. Reverberi, A. Ricelli, F. Pinzari, S. Serranti, A.A. Fabbri, G. Bonifazi, and C. Fanelli, "Early detection of toxigenic fungi on maize by hyperspectral imaging analysis", *Int. J. Food Microbiol.,* vol. 144, no. 1, pp. 64-71, 2010.
[http://dx.doi.org/10.1016/j.ijfoodmicro.2010.08.001] [PMID: 20869132]

[87] X. Zhang, F. Liu, Y. He, and X. Li, "Application of hyperspectral imaging and chemometric calibrations for variety discrimination of maize seeds", *Sensors (Basel),* vol. 12, no. 12, pp. 17234-17246, 2012.
[http://dx.doi.org/10.3390/s121217234] [PMID: 23235456]

[88] J. Zhao, Q. Chen, J. Cai, and Q. Ouyang, "Automated tea quality classification by hyperspectral imaging", *Appl. Opt.,* vol. 48, no. 19, pp. 3557-3564, 2009.
[http://dx.doi.org/10.1364/AO.48.003557] [PMID: 19571909]

[89] S. Deng, Y. Xu, X. Li, and Y. He, "Moisture content prediction in tealeaf with near infrared hyperspectral imaging", *Comput. Electron. Agric.,* vol. 118, pp. 38-46, 2015.
[http://dx.doi.org/10.1016/j.compag.2015.08.014]

CHAPTER 3

Evaluation of Quality and Safety of Agro-products Based on Bio-sensing Technique

Lin Zhang and **Yingchun Fu**[*]

College of Biosystems Engineering and Food Science, Zhejiang University, Hangzhou 310058, China

Abstract: The quality and safety of agro-products are a global concern due to their significant role in human health and economy, and the detection of hazards or ingredients in agro-products is thus essential to ensure safety. Biosensor, as a newly-emerging but promising detection tool, has contributed a lot in this field. On the one hand, based on the high sensitivity and specificity of bio-receptors for target capture and the diversity of transducers for signal transduction, biosensors exhibit capabilities for highly sensitive, specific, accurate and rapid detection. On the other hand, the combination/integration with miniaturized and portable platforms/devices endows biosensors with unrivaled advantages in low-cost, in-field and nondestructive detection. This chapter gives a systematical introduction of biosensors for the evaluation of quality and safety of agro-products, emphasizing on new biosensing principles and the advantages of exceptional analytical performance for rapid and in-field evaluation. Recent advances in biosensors for the detection of pesticide residues, antibiotic residues, pathogenic bacteria and mycotoxins, heavy metal ions, food allergens, and ingredients in agro-products are surveyed (mainly in 2018-2020).

Keywords: Agro-product, Allergen, Antibiotic, Biosensor, Food, Heavy metal ion, Mycotoxin, Nanotechnology, Nondestructive detection, Pathogenic bacteria, Pesticide.

1. BRIEF INTRODUCTION OF BIO-SENSING TECHNIQUE FOR THE EVALUATION OF QUALITY AND SAFETY OF AGRO-PRODUCTS

Agro-product is one of the most important necessities for human survival and health. However, the safety of agro-product has been broadly threatened by microbial contamination, pesticides and antibiotic residues, heavy metal ions, spoilage and adulteration. Therefore, strict evaluation of the quality and safety of agro-product has long been regarded as one of the most important issues to ensure the safety of the whole production and supply processes. For effective evaluation,

[*] **Corresponding author Yingchun Fu:** College of Biosystems Engineering and Food Science, Zhejiang University, Hangzhou 310058, China; Tel: +86 13588384722; Fax: +86 571 88982534; E-mail: ycfu@zju.edu.cn

Jiangbo Li & Zhao Zhang (Eds.)

the following concerns have to be addressed. (1) The low concentration but high toxicity of hazards in agro-product requires the evaluation to be highly sensitive. (2) Complex ingredients and a variety of homologous hazards in agro-products highlight the significance of specific and accurate identification and quantitation of target against diverse interferences. (3) Low-cost, user-friendly and portable detection devices are preferred since agricultural products are relatively cheap while the detection is generally completed by workers without professional skills in the field. (4) Smart detection has attracted increasing attention due to the strong trend to integrate the detection results with the informatics system. (5) Additionally, non-destructive detection is also in demand since it not only benefits fast analysis without time-consuming and complex sample pretreatment but also maintains the intact state of agro-products for follow-up growth or sale.

The above concerns have flourished the development of a wide range of evaluation techniques for the quality and safety of agro-products, such as traditional plate counting, high-performance liquid chromatography (HPLC), gas chromatography, mass spectrometer, *etc.* They have contributed to high sensitivity, accuracy and sample throughput but still suffer from some disadvantages, including complicated pretreatment and tedious analytical procedures, the requirement of sophisticated and expensive instruments, highly trained personnel, as well as controlled lab atmosphere. Therefore, their practical and broader applications are rather limited. It always remains a strong impetus to develop rapid, sensitive and portable detection tools.

Biosensors are defined as analytical devices incorporating a biological material, a biologically derived or a biomimetic material (termed as a bio-receptor), intimately associated with or integrated within a physicochemical transducer or transducing microsystem (termed as a transducer) [1, 2]. The bio-receptor can be an enzyme, antibody (Ab), nucleic acid (both DNA and RNA), tissue, cell, molecularly imprinted polymer (MIP), *etc.* General signals (transducers) include electrochemical (EC), optical, gassy, piezoelectric, magnetic, thermal signal and so on [2]. The bio-receptor recognizes the target and the transducer converts the recognition event into a readable signal that is proportional to the amount/concentration of the target (Fig. **1**). Since the birth of the first biosensor (enzyme transducer/electrode) in 1962 [3], biosensors have progressed intensively with interdisciplinary efforts, including biology, engineering, chemistry, electronics, informatics, materials science and nanotechnology. Beyond the well-known glucose biosensor, nowadays, biosensors have been widely applied as powerful analytical tools in a wide range of fields such as biomedical diagnosis, environment monitoring, food safety surveillance, and agricultural applications of growth monitoring, quality analysis and safety detection [1, 4, 5].

Biosensors present the features of high accuracy, sensitivity, specificity, and rapidity. Therefore, they have garnered substantial attention in fulfilling the performance requirement for agro-product evaluation. Meanwhile, due to the characteristics of low cost, ease of fabrication, miniaturization and operation, biosensors have shown great promise in rapid, in-field and nondestructive detection. To date, considerable types of hazards or ingredients in agro-products have been successfully detected *via* various biosensors, ranging from microorganisms (*e.g.* bacteria and viruses) to molecules such as proteins, organic acids, pesticides, antibiotics and toxins. For specific and rapid recognition of targets in complex agro-product samples, a variety of biological or biomimetic materials/entities with remarkable affinities and specificities have been selected or synthesized as bio-receptors for the construction of biosensors, such as Ab, aptamer, enzyme and cell. On the other hand, different transducers have been developed to improve the sensitivity, response speed and portability. EC and optical transducers are two kinds of the most attractive and widely applied transducing techniques for agro-product evaluation by virtue of high sensitivity, rapid response, and significantly, facile demands for instruments and plentiful strategies for signal amplification. EC biosensors are based on electro-analytical chemistry techniques, such as amperometric, voltammetric, impedimetric, and photoelectrochemical measurements. Quantitative sensing is made by varying the electric field and measuring the resulting changes of electrical signals (current, potential, impedance, and *etc.*) as the signal reporter (target/substrate/label) reacts electrochemically on the surface of the working electrode (the transducer) [6]. Optical transduction mechanisms include the change in color, absorbance, fluorescence, luminescence, Raman scattering, plasmon resonance, and so on. These techniques utilize light as the delivery/collection medium to obtain intrinsic information of the physicochemical properties of the optical signal reporter to detect changes/induced changes by the target [7]. Besides, other measurement techniques, such as piezoelectric, magnetic and calorimetric measurements, have also been applied as transducers in biosensors, enriching the evaluation methods for agro-products.

In addition to the performance improvement of biosensors, the demands of in-field evaluation for agro-products have galvanized the development of a series of portable and miniaturized biosensors [8]. The simple constitutions make biosensors powerful candidates for in-field evaluation: biosensors can not only be miniaturized *via* integrating both bio-receptors and transducers on miniaturized systems/devices (*e.g.* chips and flexible polymer substrates), but also be portable, smart and user-friendly *via* applying hand-held readers (*e.g.* smartphone, gas detector and glucose meter) for signal transduction and readout.

In this chapter, recent advances of biosensors for the evaluation of the quality and safety of agro-products are overviewed with the categorization of different targets in agro-products (mainly in 2018-2020), aiming to provide a systematical introduction to the principles and performance characteristics of biosensors. Three kinds of critical factors to evaluate an analytical tool for agro-product analysis are concerned: 1) Analytical performance such as sensitivity, limit of detection (LOD), linear detection range (LDR) and specificity; 2) Applicability in agro-product samples; 3) The miniaturization and portability of biosensors for in-field detection. Biosensor-based nondestructive evaluation is especially delivered. Finally, conclusions, current challenges and future perspectives of this rapidly-developing field are discussed.

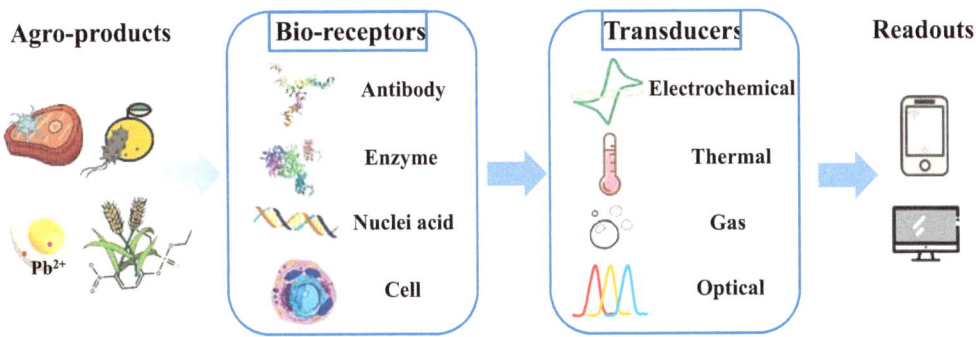

Fig. (1). Constitutions and principles of biosensors for the evaluation of quality and safety of agro-products.

2. ADVANCES IN BIOSENSORS FOR THE EVALUATION OF QUALITY AND SAFETY OF AGRO-PRODUCTS

As mentioned above, there are a series of species that need to be evaluated to indicate and guarantee the quality and safety of agro-products. A great progress has been achieved on biosensors for the detection of pesticide residues, antibiotic residues, pathogenic bacteria and mycotoxins, heavy metals, allergens, and ingredient analysis of agro-products (Table 1). In this part, recent advances of biosensors in terms of targets were surveyed and discussed.

2.1. Biosensors for Pesticide Residues

Pesticides, *e.g.* organophosphorus (OPs), carbamates, organochlorines, pyrethrin and pyrethroids compounds, have been widely used to protect crops against insects and pests in modern agriculture production. Nevertheless, due to their inherent toxicity and misuse-caused accumulation in the food chain, pesticide residues in agro-products bring horrific effects on human health and the environment. For example, OPs can irreversibly inhibit the catalytic activity of acetylcholinesterase (AChE, an important central-nervous enzyme), thereby leading to the accumulation of the neurotransmitter acetylcholine and detrimental effects or even death. Therefore, the evaluation of pesticide residues is of great importance to food and public safety. The urgent demands of sensitive, reliable and fast quantification have been fulfilled by the rapidly-developing biosensors with various well-developed transducers, such as EC [9], fluorescent [10 - 12], colorimetric [13], phosphorescent [14], magnetic [15], piezoelectric [16] and microcantilever-based ones [17]. Among them, EC biosensors with prominent sensitivity and quick response have been widely explored for pesticide residues' detection. On the other hand, plentiful bio-receptors have been selected and applied to construct biosensors for pesticide residues, principally enzymes, aptamers, antibodies and MIPs [9]. These bio-receptors not only benefit rapid and sensitive analysis based on their high affinity, but also improve the specificity for the identification of pesticides in complex agro-product samples.

Enzymes, a kind of the most significant biomolecules with high catalytic ability and specificity, have been immobilized on the surface of working electrodes (the transducer) to construct EC biosensors for pesticide residues detection *via* the inhibition or catalysis mechanism [18]. For the inhibition-type EC enzymatic biosensors, the pesticide (*e.g.* OPs and carbamates) inhibits the activity of an enzyme (*e.g.* AChE, tyrosinase and alkaline phosphatase) and thus results in the decrease of EC response (inhibition effect). For example, a voltammetric biosensor based on reduced graphene oxide (rGO) and AChE was successfully applied to detect carbaryl with a LOD of 1.9 nM [19]. The performance of EC enzymatic biosensors, particularly sensitivity, stability and reproducibility, depends largely on the efficient immobilization of enzymes on the surface of electrode. Hence, some efforts have been devoted to exploring reliable immobilization strategies and matrices for enzymes [20]. Recently, a bio-inspired fibrin-bone@polydopamine-shell adhesive composite matrix has been developed for AChE immobilization (fibrin@PDA-AChE) in an amperometric biosensor for paraoxon detection [21]. Due to the porosity and stability of the matrix, the biosensor exhibited a low LOD of 4 ng L^{-1} and a recovery of 96.7% in spiked apple sample. Inhibition-type biosensors provide high sensitivity and low LOD, however, the necessary incubation step (recognition process) increases additional analysis time and procedures, limiting their application for rapid and in-field evaluation.

The limitations associated with enzyme inhibition have been easily overcome by furnishing the direct catalysis of enzymes. Based on a typical organophosphorus hydrolase (OPH)-catalyzed mechanism, a flexible and wearable glove EC biosensor can offer rapid, nondestructive sampling and in-field detection of OPs in agro-products (Fig. **2**) [22]. The biosensor consists of a sampling finger (collector), a sensing finger containing the immobilized OPH layer on the working electrode, a portable potentiostat with wireless communication to a smartphone for rapid collection of the voltammetric results. For the detection procedure, OP on tomato surface was firstly collected by the sampling finger *via* swipe sampling (Fig. **2a**). And then, the EC sensing was performed by joining the sampling and sensing fingers, in which OPs were hydrolyzed by the catalysis of OPH to produce electrochemically active *p*-nitrophenol (Fig. **2b**). Finally, the EC detection of the *p*-nitrophenol was recorded and displayed using the potentiostat and smartphone, which indicated the concentration of OPs on tomato surface (Fig. **2c**). The proposed biosensor can accurately detect the absence/ presence of methyl parathion (MP) and methyl paraoxon on the skin surfaces of four types of fruit samples with high inter-glove reproducibility of relative standard deviation (RSD) of 3.1% and 2.1%.

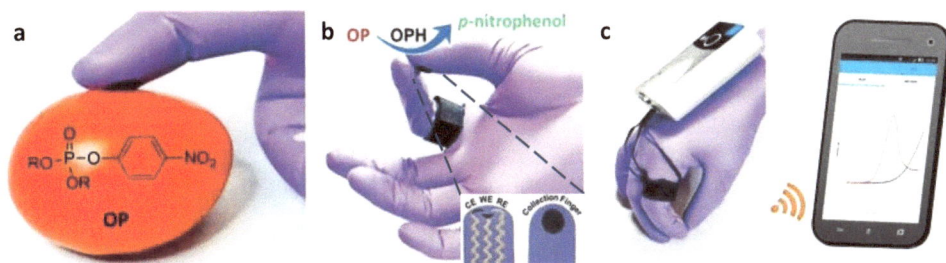

Fig. (2). Illustration of the fabrication and principle of the catalysis-type EC enzymatic biosensors for methyl parathion detection [22].

Specific evaluation of the targeted pesticide is essential because there are a series of interfering substances in agro-products that may cause false detection, such as heavy metal ions with inhibition effects and other analogues with similar structures. To further improve the specificity of EC biosensors for pesticide residues beyond enzymatic biosensors, Ab-based immunosensors and aptamer-based aptasensors have aroused considerable interest in recent years [23, 24]. In a typical EC immunosensor for chlorpyrifos detection, anti-chlorpyrifos Ab was immobilized on the surface of gold nanoparticles (AuNPs)-modified electrode [25]. With the binding of chlorpyrifos to the immobilized Ab, the peak current of ferro/ferricyanide redox probe changed for signal output. The immunosensor can specifically detect chlorpyrifos with concentrations of 10 nM for apple and cabbage, and 50 nM for pomegranate.

Similar to Ab, an aptamer can also capture the target *via* non-covalent bonds and form a complex to arouse the change of EC signal. Furthermore, aptamer provides additional advantages such as low cost, desirable chemical stability, long shelf-life, and *in vitro* synthesis by "systematic evolution of ligands by exponential enrichment" (SELEX) technique [9]. Hence, various aptasensors have been employed for pesticide detection. Due to the synergistic effect between aptamer-based specific recognition and nanomaterial-based signal amplification, an AuNPs/nickel hexacyanoferrate nanoparticles/rGO-based electrochemical aptasensor exhibited satisfying performance for atrazine detection with a wide LDR of 0.25-250 pM, a LOD of 0.1 pM, and remarkable specificity towards other pesticides, such as simazine, propanil and malathion [26].

Although biomolecules-based biosensors are the most commonly used biosensors for pesticide detection, they still suffer from unsatisfactory stability and reusability due to the fragile nature of biomolecules. The usage of biomolecules in biosensors increases the storage costs; more importantly, limits the practical application in many scenarios, for example, outside in-field detection near farms. MIP exhibits great potential in this field due to its facile synthesis procedure, remarkable stability and adaptability in normal and even harsh conditions. *In situ* electro-polymerization is a common route for the fabrication of MIP-based EC biosensor, which includes the polymerization of functional monomers in the presence of the template and the subsequent removal of the template from MIP film. The leaving imprinted cavities are complementary in size, shape and functionality to the template. In the subsequent detection procedure, sorption of the target (the template or analogue) onto the MIP film (*i.e.* the recognition process) leads to a change in EC response for signal output. Following this strategy, a novel EC biosensor based on MIP-modified nitrogen and sulfur-doped hollow Mo_2C spheres (N, S-Mo_2C) has been fabricated for carbendazim detection with a LDR of 1 pM - 8 nM and a LOD of 0.67 pM (Fig. **3**) [27]. Moreover, the biosensor exhibited robust stability, with 94.3% of the initial response value after 15-day storage at room temperature.

2.2. Biosensors for Antibiotic Residues

Antibiotics have emerged as revolutionary antimicrobials in animal husbandry to improve the production, health and welfare of animals. Unfortunately, extensive use of antibiotics not only leads to the emergence of drug-resistant microbes, but also produces antibiotic residues in animal-derived agro-products, posing serious threats to human health. Hence, there is an urgent need to develop effective analytical techniques to detect antibiotics. Due to low content of antibiotic residues in complex agro-products, the accuracy, sensitivity and specificity are three concerned indexes to guide the development of evaluation techniques. Meanwhile, simplicity for detection procedures is also influential. Satisfying above requirements, a series of substantial progress has been achieved by biosensors, such as optical, EC, piezoelectric and calorimetric ones [28 - 33]. In particular, optical biosensors have garnered considerable attention in this field owing to high sensitivity, simple operation, cost effectiveness, as well as the diversity of signal-output techniques.

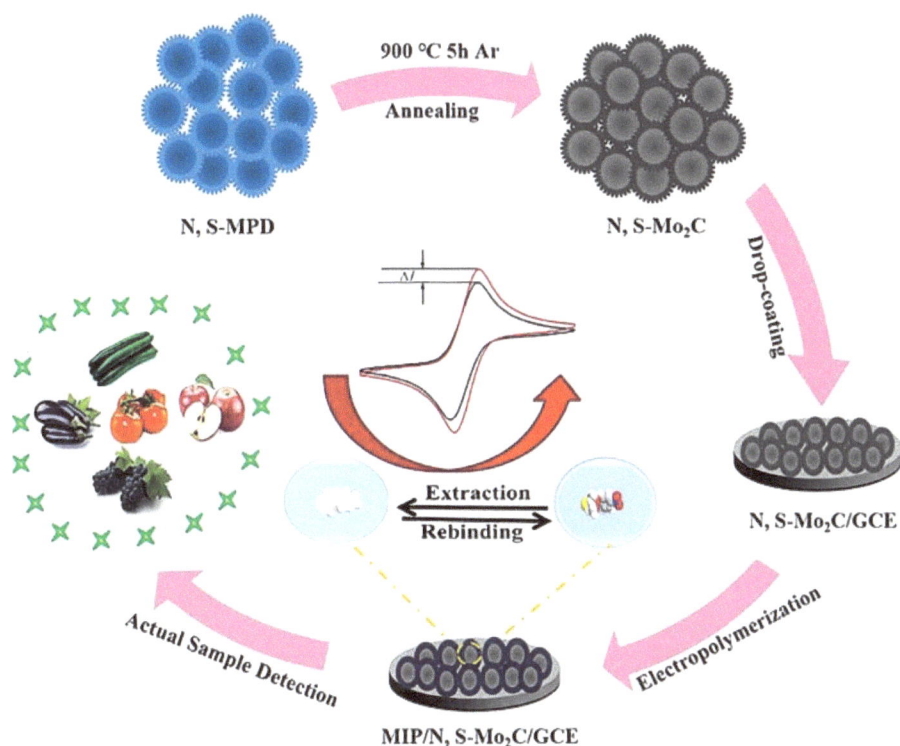

Fig. (3). Illustration of fabrication of EC biosensor based on MIP-modified N, S-Mo$_2$C for carbendazim detection [27].

Colorimetric biosensors with visible signals have been proven to be capable of providing low-cost, rapid, simple and in-field detection towards antibiotic residues in agro-products [34, 35]. AuNPs-based colorimetric biosensor is the most popular and commercially available ones. Generally, the capture of antibiotic by AuNPs-bio-receptor complex is related to the change of dispersion state of AuNPs, which triggers the color conversion between red and violet-red/blue. For instance, a colorimetric aptasensor based on lanthanum ion-assisted AuNPs aggregation has been fabricated for chloramphenicol (CAP) detection with a LOD of 1.9 ng mL^{-1} (Fig. **4**) [36]. Using a smartphone to read color changes, the biosensor exhibited satisfactory detection results with recovery of 93.0%-104.5% and RSD of 1.94%-3.57% for CAP detection in milk and chicken samples.

Although colorimetric biosensors are characterized by simplicity, rapidity and visualization, the LODs of some colorimetric biosensors are still above the maximum residue limits of antibiotic residues. Meanwhile, the color observation may be disturbed by colored agro-products, limiting their broader application. Hence, other optical biosensors with sensitive qualification ability have also been developed [37, 38]. Among them, fluorescent biosensor has become a hotspot for the detection of antibiotic residues in agro-products on account of high sensitivity and multiplexed detection capacity [39]. For instance, in a fluorescent aptasensor for CAP detection, the fluorescence of the dye that labeled on the aptamer can be quenched by PCN-222 (a zirconium-porphyrin metal-organic framework material) [40]. When CAP was introduced, a dye-labeled aptamer was released from PCN-222 surface to bind CAP, resulting in the recovery of fluorescence for signal output. The total detection time for CAP was shortened to be 26 min. More importantly, the biosensor possessed a wide LDR of 0.0001-10 ng mL^{-1} and a very low LOD of 0.08 pg mL^{-1}. The LOD was orders of magnitude lower than colorimetric methods. Furthermore, the detection results of real milk and shrimp samples were also consistent with those of the commercial enzyme-linked immunosorbent assay kits, demonstrating its potential for practical applications.

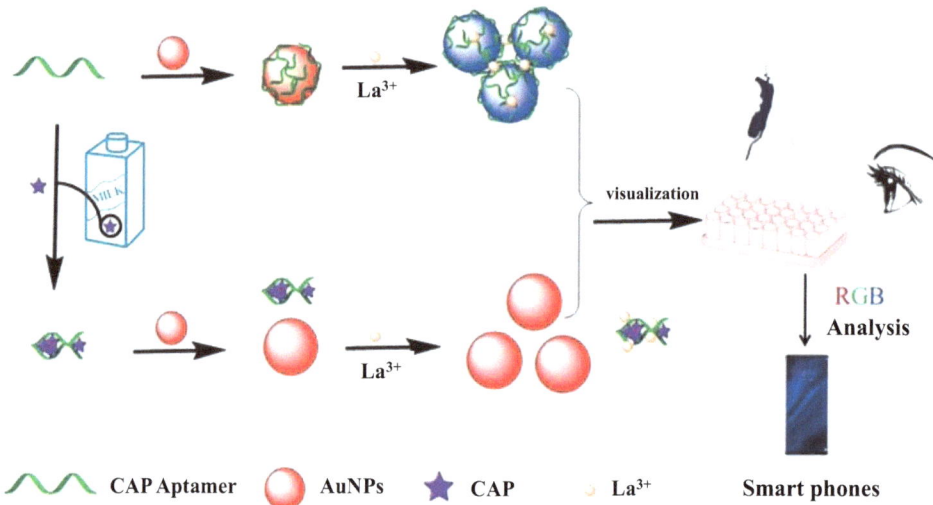

Fig. (4). Illustration of the AuNPs-based colorimetric aptasensor for CAP detection [36].

Simultaneous detection of multiplexed antibiotic residues is also in demand because different types of antibiotics may be applied and co-exist in some agro-products. In this case, high sensitivity and high throughput analysis are favorable. Optical biosensors have attracted considerable attention in this field [41]. For instance, a surface-enhanced Raman scattering (SERS)-based immunosensor with two test line-based lateral flow assays has been reported for ultrasensitive simultaneous detection of neomycin (NEO) and norfloxacin (NOR) with low LODs of 0.37 ng L^{-1} and 0.55 ng L^{-1} [42].

2.3. Biosensors for Pathogenic Bacteria and Mycotoxins

Foodborne pathogenic bacteria are ubiquitous and are capable of infecting humans, leading to illness and even death *via* contaminated agro-products or water. Therefore, rapid and accurate detection of pathogenic bacteria and mycotoxins are urgent. Biosensors have been widely applied to the detection of pathogenic bacteria. A majority of recent advances have been focused on the improvement of analytical performance. For highly-specific identification and quantification, aptamers and antibodies (common immunoglobulin G and novel immunoglobulin Y) have been widely employed as effective bio-receptors toward varied pathogenic bacteria, such as *Escherichia coli* O157:H7 (*E. coli*), *Salmonella enterica* (*S. enterica*), *Salmonella typhimurium*, *Staphylococcus aureus* (*S. aureus*), *Listeria monocytogenes* (*L. monocytogenes*), *Campylobacter jejuni*, *Pseudomonas aeruginosa* (*P. aeruginosa*) and *Shigella flexner* [43 - 51]. On the other hand, to achieve ultra-sensitive detection, versatile sensing principles with sensitive signal reporters and signal amplification strategies have been explored to improve the performance dramatically. In particular, nanomaterial-based biosensors have attracted tremendous attention in this field. Nanomaterials can serve as not only direct reporters for sensitive signal generation, but also important signal promoters *via* diversified amplification routes, including highly-conductive substrates for EC biosensors, robust immobilization matrices for bio-receptors and reporters, and highly-active catalysts for signal generation reactions [8].

In addition to ongoing efforts to meet the performance requirements for the detection of bacteria in agro-products, in-field and real-time detection is another urgent demand in this field. In this case, users can obtain immediate information. Biosensors show promising potentials in this field *via* integrating or combining with miniaturized and portable platforms/devices, such as paper strip [52, 53], cotton swab [54], flexible polymer [55], microfluidic chip [56] and silicon chip [57]. For instance, a fluorescent biosensor has been fabricated on a flexible and transparent cyclo-olefin polymer (COP) film to achieve non-destructive real-time monitoring of live *E. coli* (Fig. **5**) [55]. The RNA-cleaving fluorescent DNAzyme (RFD) was selected as the specific bio-receptor, while its fluorescence signal can be turned on through its cleavage triggered by bacterium-related protein. By monitoring the fluorescence intensity of the RFD-COP film surface, the biosensor was capable of real-time probing the presence of bacteria on the packages of real raw beef and sliced apple samples. The total amount of bacteriain each food sample was down to 1 cfu mg^{-1}. Moreover, the biosensor was specific and stable for at least the shelf-life of perishable packaged products (ca. 14 days).

Besides, portable hand-held readers have also been employed for signal readout, such as smartphone [58], portable fluorescent biochemical rapid analyzer [59], gas detector [60] and glucose meter [61]. Recently, an aptasensor coupled with volumetric bar-chart spin-chip for signal readout was fabricated for visually quantitative detection of multiple pathogenic bacteria (Fig. **6**) [62]. For the construction of the biosensor, the DNA probe consisting of magnetic beads-DNA and aptamer-DNA-platinum nanoparticles (aptamer-PtNPs) was pre-immobilized in the sample recognition microwell in the chip using magnetic field (Fig. **6a**). H_2O_2 and food dyes were also pre-injected into the amplification and the indicator microwell for signal readout, respectively. After the injection of sample solution, targeted pathogenic bacteria would interact with aptamer-PtNPs to form free binding complexes in solution. The binding complexes with catalytic-active PtNPs would catalyze the oxidation of H_2O_2 to produce O_2, causing a dramatic pressure increase to drive the move of co-existing dyes in the sealed chambers. Hence, the concentration of bacteria was proportional to the moving distance of the dyes (*i.e.* bar-chart signal). This biosensor achieved simultaneous visual quantitation of three types of pathogenic bacteria in apple juice (*S. enterica*, *E. coli*, and *L. monocytogenes*) with low LODs of 10 cfu mL^{-1}.

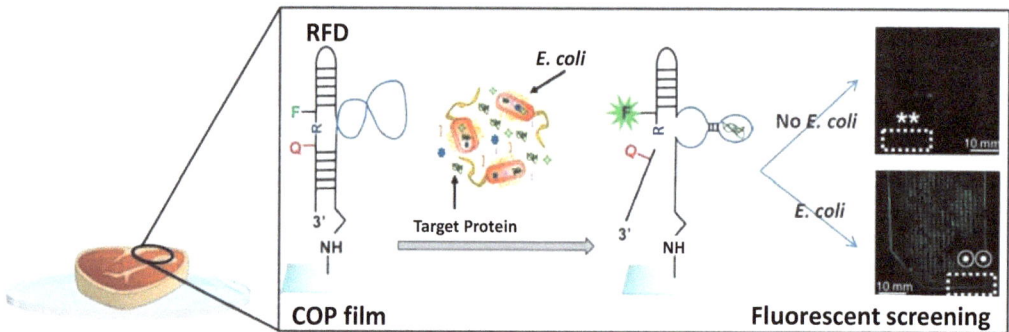

Fig. (5). Illustration of highly sensitive RFD-based fluorescent biosensor for non-destructive detection of live *E. coli* cells in packaged agro-products [55].

In addition to bacteria, mycotoxins produced by fungi and mold can also result in not only severe health threats to humans and animals, but also huge industrial and agricultural losses to food crops [63]. Timely evaluation of mycotoxins in agro-products is crucial to reduce economic loss and mitigate the negative public-health-related impacts of foodborne illnesses. In the past decades, biosensors appeared to be one of the most powerful techniques for quantification of mycotoxins in agro-products [63]. A variety of EC and optical biosensors have arisen to detect common mycotoxins, such as voltammetric aptasensor for AFB1 in peanut and corn samples [64], label-free impedimetric aptasensor for OTA in coffee [65], label-free microfluidic surface plasmon resonance (SPR) biosensor for aflatoxin B1 (AFB1) in wheat [66], fluorescent DNA hydrogel aptasensor for ochratoxin A (OTA) [67], chemiluminescent aptasensor for AFB1 in peanut and milk [68], and colorimetric biosensors for AFB1 in peanut [69] and zearalenone

(ZEN) in cereals and feed [70]. Not limited to single-target detection, simultaneous detection of multiple mycotoxins has also been achieved using biosensors. A colorimetric immunosensor has been fabricated for simultaneous detection of OTA, AFB1 and deoxynivalenol in a corn-based feed sample [71]. With the combination of a microfluidic capillary chip for automation, specific antigen-Ab interactions, horseradish peroxidase (HRP)-catalytic TMB-H_2O_2 chromogenic reaction, and fast grayscale quantification procedure in smartphones, the multiplexed assay was completed within only 10 min and achieved LODs of < 40, 0.1-0.2 and < 10 ng mL^{-1} for OTA, AFB1 and deoxynivalenol, respectively. Those LODs all fell in the majority of currently enforced regulatory and/or recommended limits.

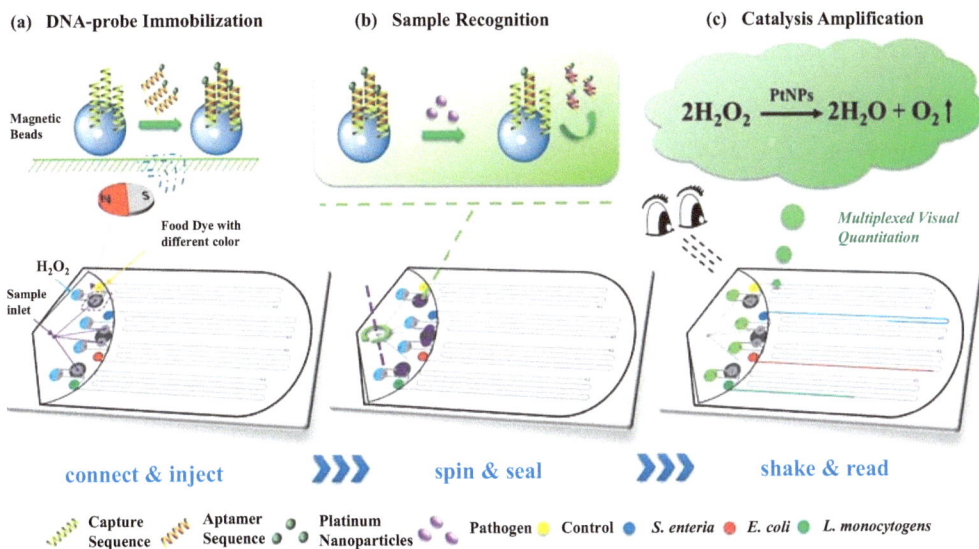

Fig. (6). Illustration of the aptasensor coupled with volumetric bar-chart spin-chip for visual simultaneous quantitative detection of *S. enterica*, *E. coli*, and *L. monocytogenes* [62].

2.4. Biosensors for Heavy Metal Ions

Heavy metal ions, *e.g.* mercury (Hg^{2+}), lead (Pb^{2+}), cadmium (Cd^{2+}), chromium (Cr^{6+}), arsenic (As^{3+} and As^{5+}) and copper (Cu^{2+}), can pose serious risks to human health and the ecosystem by accumulating in the soil, groundwater and agro-products. Identification and quantification of heavy metals is, therefore a global concern.

Comparing with traditional detection methods for heavy metal ions, *e.g.* atomic absorption spectrometry, atomic fluorescence spectrometry and inductively coupled plasma atomic emission spectrometry, biosensors exhibit exceptional advantages in quick response, simple operation, good sensitivity as well as anti-interference ability. Hence, various biosensors have been developed using various bio-receptors, such as aptamers, enzymes and cells [72 - 77].

Due to the high affinity and specificity of aptamers toward heavy metal ions, aptasensors provide sensitive, accurate, specific and rapid detection for heavy metal ions in agro-products. Commonly, aptamers can form special structures that are triggered by heavy metal ions to achieve target recognition, such as G-quadruplex for Pb^{2+} and As^{3+}, $T-Hg^{2+}-T$ complex formed by T-T mismatches for Hg^{2+}, and complex of thymine and guanine-rich nonrepeating single-strand DNA with Cd^{2+} [72, 78]. For example, a turn-on fluorescence aptasensor was fabricated for the detection of Hg^{2+} with a LOD of 60 nM [79]. Meanwhile, it can detect Hg^{2+} pollution in milk and tap water with recoveries of 95.18-108.22%.

Additionally, DNAzyme with excellent specificity in target recognition and catalytic capacity, also attracts increasing attention for *in situ* detection and imaging of heavy metal ions [74, 80]. For example, an ultrasensitive DNAzyme-based fluorescent biosensor was fabricated for Pb^{2+} detection with a LOD of 0.22 nM [81]. It was also successfully applied to detect Pb^{2+} in eggs with recoveries of 87.1%-101.2%.

2.5. Biosensors for Food Allergens

Food allergy, an overactive immune response causing by given food, is becoming increasingly recognized as a major healthcare concern around the world. According to Food and Agriculture Organization (FAO), major food allergens are peanut, wheat, egg, milk, soybean, tree nuts, fish, and shellfish [82], all of which are essential agro-products. Accordingly, timely detection of allergens in agro-products is an efficient route to protect food allergic consumers from acute and potentially life-threatening allergic reactions. A variety of biosensors have provided rapid and accurate detection for many allergens in agro-products, such as tropomyosin in shellfish [83], lectin in kidney bean [84], gluten in grain and corn [85, 86], non-specific lipid transfer protein Sola l 7 in tomato seeds [87], and even simultaneous detection of the four allergens (bovine k-casein, peanut protein, soy protein and gliadin) [88].

Rapid, low-cost and in-field detection of allergens in agro-products using portable devices has attracted increasing attention [89, 90]. An EC immunosensor with a keychain-size reader (termed as integrated exogenous antigen testing (*i*EAT) system) has been elaborately designed to detect five major allergens in peanuts, hazelnuts, wheat, milk, and eggs (Fig. **7**) [91]. The portable *i*EAT system comprises a keychain reader, a disposable extraction kit, and a smartphone app (Fig. **7a**). For detection (Fig. **7b**), a common immunomagnetic separation/ enrichment technique was firstly applied for allergen extraction from agro-products, in which allergen was captured on immunomagnetic beads and labeled with Ab$_2$ conjugated with HRP. When mixed with TMB, HRP catalyzed the oxidation of TMB and the generated TMB oxide was then reduced by the electrode. The corresponding current signal was applied to quantitative sensing of allergen. The mini-reader not only displayed results, but also wirelessly communicated with smartphone *via* Bluetooth. Benefiting from highly-sensitive EC biosensing technique, mini-reader and smartphone-based cloud server for web-based data collection and sharing, users can obtain quantitative information of allergens in 10 min with low costs of the system (< $40) and assay (< $4 per antigen).

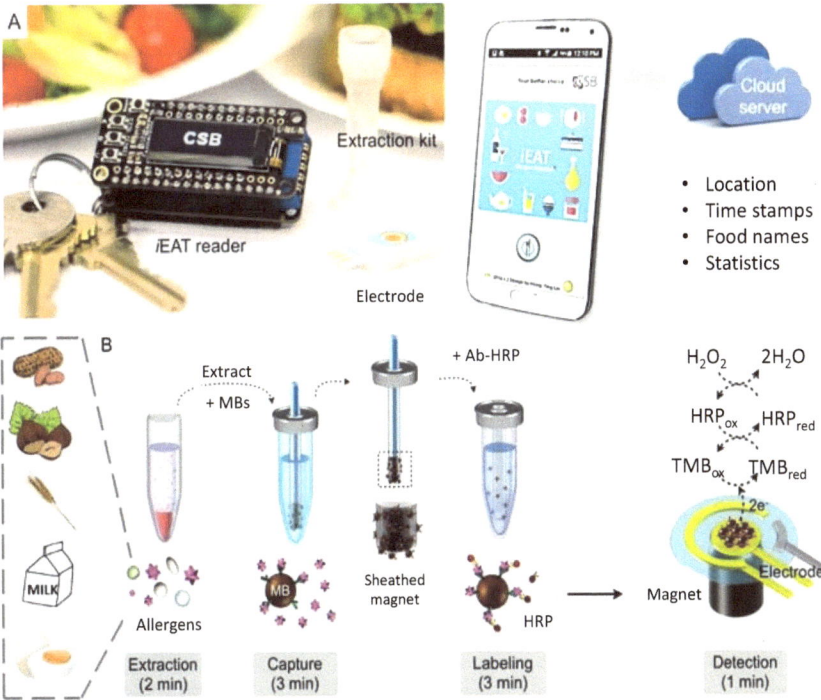

Fig. (7). *i*EAT system for portable detection of allergens. **A)** Constitution of *i*EAT system. **B)** Detection principle and procedure of the EC biosensor for allergens [91].

2.6. Biosensors for Ingredients

Besides the above hazards, biosensors have also been developed as powerful tools for the analysis of active ingredients of agro-products. Recently, many active ingredients of agro-products, such as sugars [92, 93], phenols [94], amino acids [95], coenzyme [96], and organic acids [97, 98], have been detected by biosensors. A voltammetric biosensor detected chlorogenic acid (5-O-caffeoyl quinic acid) with a LDR of 0.56-7.3 μM and a LOD of 0.18 μM, which can be applied to discriminate the quality of specialty and traditional coffee beverages

[97]. Sugars are common constituents of food, and their contents are important indicators of the quality of agro-products. An amperometric multi-enzymatic biosensor provided rapid sucrose monitoring in 50 s with a LDR of 10-1200 µM and low LOD of 8.4 µM [92]. The biosensor exhibited an extraordinary operational (3 days) and storage (1 year) stability. Moreover, the results of the proposed biosensor for sucrose monitoring in 17 samples of green coffee were comparable to those obtained by the standard HPLC method, demonstrating its potential for real agro-product evaluation.

Table 1. Typical biosensors for evaluation of quality and safety of agro-products.

Category	Targets	Biosensors	Performance	Recoveries for Real Agro-Product Samples	References
Pesticide residues	MP, methyl paraoxon	OPH-based voltammetric biosensor	200 µM-contaminated skin surfaces of apple, tomato, grape, green pepper		[22]
	Paraoxon	Fibrin@PDA-AChE-based amperometric biosensor	LOD of 4 ng L^{-1}	96.7% for apple	[21]
	Carbendazim	N, S-Mo$_2$C-MIP-based voltammetric biosensor	LDR of 1 pM - 8 nM, LOD of 0.67 pM	(98.4-100.1)% for grape, (99.6-99.8)% for eggplant	[27]
	[1]Dichlorvos, [2]MP	Fluorescent biosensor based on [a]N-CDs and AChE	LODs of [1]3.2 µg L^{-1} and [2]13 µg L^{-1}	[1](93.4-97.6)% and [2](95-106)% for rice	[10]
	Chlorpyrifos	Colorimetric/chemiluminescent immunosensor based on MnO$_2$ nanoflower	LDR of 0.1-50 µg L^{-1}, LOD of 0.033 µg L^{-1}	(103-114)% for *Astragalus*, (90-120)% for *Poria cocos*	[13]
	Isocarbophos	Phosphorescent aptasensor based on AuNPs and [b]PLNRs	LDR of 5-160µg L^{-1}, LOD of 0.54 µg L^{-1}	(96.7-104.5)% for cabbage, (97.8-103.2)% for lettuce	[14]
	Chlorpyrifos	[c]CuAAC-mediated magnetic immunosensor	LOD of 0.022 µg L^{-1}	(92-120)% for drinking water	[15]

(Table 1) cont.....

Category	Targets	Biosensors	Performance	Recoveries for Real Agro-Product Samples	References
Antibiotic residues	Chloramphenicol	Voltammetric aptasensor based on [d]ECC	LDR of 1-1000µg L^{-1} LOD of 1µg L^{-1}	(82-113)% for skim milk	[33]
	[1]NEO, [2]NOR	SERS-based immunosensor	LODs of [1]0.37 ng L^{-1} and [2]0.55 ng L^{-1}	(86-121)% for milk	[42]
	Captan	Luminescent bacteria-based biosensor in combination with [e]HPTLC	LDR of 10-80 ng zone^{-1}	(75-96)% for apple, pear, plum, apricot, cherry, peach	[37]
	Tobramycin	Aptasensor based on resonance scattering spectra	LDR of 0.50-17 nM, LOD of 0.19 nM	(95.3-103)% for milk	[31]
	Amantadine	Label-free piezoelectric immunosensor	LOD of 1.3 µg L^{-1}	82.6-91.5% for chicken, duck, pork	[29]
	Enrofloxacin	Photothermal-sensing immunochromatographic biosensor based on [f]PB-Au	LDR of 0.03-10 µg L^{-1}, LOD of 0.023 µg L^{-1}	(72.6-126.2)% for chicken, egg, pork, beef, crucian	[30]

(Table 1) cont.....

Category	Targets	Biosensors	Performance	Recoveries for Real Agro-Product Samples	References
Pathogenic bacteria and mycotoxins	*E. coli* O157:H7	Disposable impedance immunosensor based on [g]SPCE/ERGNO/AuNPs	LDR of 1.5-15000 cfu μL^{-1}, LOD of 1.5 cfu μL^{-1}	/	[48]
	S. aureus	IgY-based colorimetric biosensor	LDR of 0.5-50 cfu μL^{-1}, LOD of 0.11 cfu μL^{-1}	(101.4-102.8)% for apple	[43]
	P. aeruginosa	[h]GOQDs-based fluorescent aptasensor	LDR of 1.28×10^3-2×10^7 cfu mL^{-1}, LOD of 100 cfu mL^{-1}	(95.2-108%) for juice	[44]
	[i]3 bacteria cells	Multichannel SPR immunosensor	14, 6, and 28 CFU/25 g chicken for 3 bacteria cells		[57]
	E. coli O157:H7	Au nanoflower-enhanced dynamic light scattering immunosensor	LDR of 6-60000 cfu mL^{-1}, LOD of 2.7 cfu mL^{-1}	(96.7-105)% for milk	[46]
	S. enterica	Magnetic immunosensor with membrane filtration	LOD of 10^4 cfu mL^{-1}	/	[47]
	AFB_1	[j]Voltammetric aptasensor	LDR of 1 ng L^{-1} - 200 μg L^{-1}, LOD of 0.33 ng L^{-1}	(93.6-107)% for peanut, (88.5-110.2)% for corn	[64]
	OTA	Fluorescent DNA hydrogel aptasensor	LDR of 0.05-100 ng mL^{-1}, LOD of 0.01 ng mL^{-1}	(94.2-105)% for beer	[67]
	ZEN	Colorimetric immunochromatographic biosensor with smartphone	LODs of 0.08 μg kg^{-1} in cereals and 0.18 μg kg^{-1} in feed	(92-105)% for corn, (86-107.5)% for feed	[70]

(Table 1) cont.....

Category	Targets	Biosensors	Performance	Recoveries for Real Agro-Product Samples	References
Heavy metals	As^{3+}	Impedimetric aptasensor	LOD of 74 pM	(92.4-118.5)% for drinking water	[99]
	Hg^{2+}	Fluoresence aptasensor based on upconversion nanoparticles and AuNPs	LDR of 0.2-20 μM, LOD of 60 nM	(99.6-103.8)% for milk	[79]
	Pb^{2+}	DNAzyme-based fluorescence biosensor	LOD of 0.22 nM,	(87-114)% for egg	[81]
Allergens	Tropomyosin in shellfish	AuNP trimer-based immunosensor with Circular dichroism signal	LDR of 0.1-15 ng mL^{-1}, LOD of 21 pg mL^{-1}	(84.9-108.1)% for *Penaeus, Oratosquilla, Procambarus clarkia*	[83]
	Lectin in kidney bean	^{k}AuNPs-PEI-MWCNTs-based voltammetric immunosensor	LDR of 0.05-100 μg mL^{-1}, LOD of 0.023 μg mL^{-1}	(91-97.1)% for raw kidney bean milk	[84]
	Gluten in grain	Immunosensor based on floating-gate transistor	detect gluten at/below 10^{-3} mg mL^{-1}	/	[85]
	^{1}4 allergens	Immunosensor on monolithic interferometric silicon chip	LODs of 0.04, 1.0, 0.8, 0.1 μg mL^{-1} for 4 allergens	(88-118)% for 4 allergens	[88]

(Table 1) cont.....

Category	Targets	Biosensors	Performance	Recoveries for Real Agro-Product Samples	References
Ingredient analysis	Sucrose	Amperometric biosensor based on FAD-dependent glucose dehydrogenase	LDR of 10-1200 μM, LOD of 8.4 μM	Monitoring in green coffee	[92]
	Catechol	Amperometric biosensor based on bacterial laccase on engineered *E. coli*	LDR of 0.5-300 μM, LOD of 0.1μM	(98.2-102.1)% for tea	[94]
	Tyrosine	[m]Voltammetric biosensor	LDR of 0.01-8 nM and 8-160 nM, LOD of 9 pM	101.7% for egg	[95]
	Alkaline phosphatase	Paper-based immunosensor with smartphone	LDR of 10-1000 U mL^{-1}, LOD of 0.87 U mL^{-1}	(91-100)% for milk	[100]

[a]N-CDs: Nitrogen-doped carbon dot;
[b]PLNRs: persistent luminescence nanorods;
[c]CuAAC: the coordination chemistry and Cu(I)-catalyzed 1,3-dipolar cycloaddition of azide andalkyne;
[d]ECC: electrochemical conversion (ECC) of magnetic nanoparticles to electrochemically-active Prussian blue;
[e]HPTLC: high performance thin-layer chromatography;
[f]PB-Au: Au nanoparticle-enhanced two-dimensional black phosphorus;
[g]SPCE/ERGNO/AuNPs: screen-printed carbon electrode/electrochemically reduced graphene oxide/AuNPs;
[h]GOQDs: graphene oxide quantum dots;
[i]Simultaneous detection of 3 bacteria cells: *E. coli* O157:H7, *S. enterica* and L. monocytogenes;
[j]Aptasensor based on DNA-AuNPs-HRP and exonuclease-assisted signal amplification;
[k]AuNPs-PEI-MWCNTs: gold nanoparticles-polyethyleneimine-multiwalled carbon nanotubes nanocomposite;
[l]Simultaneous detection of 4 allergens: bovine k-casein, peanut protein, soy protein, and gliadin;
[m]Biosensor based on tyrosine hydroxylase/PdPt NPs/chitosan-1-ethyl-3-methylimida-zolium bis (trifluoromethylsulfonyl) imide/graphene-multiwalled carbon nanotubes-IL/glassy carbon electrode

CONCLUSIONS AND PERSPECTIVES

In conclusion, this chapter reviews recent advances in biosensors for the evaluation of the quality and safety of agro-products. Evaluation targets range from hazards of pesticides, antibiotics, pathogenic bacteria and mycotoxins, and heavy metal ions, to food allergens, as well as active ingredients. By taking full advantage of bio-receptors with high specificity and affinity to capture targets, as well as various currently available transducing techniques, biosensors provide accurate, sensitive, specific, and rapid evaluation for complex agro-products.

Meanwhile, in combination with miniaturized and portable platforms/devices for integration and/or signal readout, biosensors exhibit distinguishing advantages in satisfying the important and urgent demands of in-field and even nondestructive evaluation for agro-products, *e.g.* low cost, simplicity, portability and flexibility.

Despite advances in biosensors, there are still some challenges to be addressed to realize their deeper and broader application in the evaluation of the quality and safety of agro-products. 1) High throughput analysis and simultaneous detection of multiple-targets in agro-products are favorable but relatively rare. Delicate design in biosensing principles with different signal reporters or multiple signal output routes is a contributing factor. Meanwhile, the integration of biosensors in multichannel chips can also help. 2) Complex ingredients of agro-products may disturb the evaluation results of specific targets (*i.e.* matrix effects). Accordingly, the anti-interference capacity of biosensors against analogues of the target, food matrix, as well as solution environment (*e.g.* pH and ions) needs to be improved. High specificity towards the target can be realized *via* elaborate selection/fabrication of appropriate bio-receptors, and specific conformation of bio-receptors on transducer surfaces (*e.g.* antibody orientation). Besides, the development of effective sample pretreatment methods to withstand matrix effects or to concentrate the target is important. 3) The reliability, stability and reproducibility of biosensors under a normal evaluation environment need to be concerned, considering the application scenarios of biosensors for agro-product evaluation, commonly in farms, supermarkets, or at home. The poor stability of bio-receptors under these environments is always a key limitation of this part. In this case, fundamental knowledge about the structure and properties of bio-receptors and advanced stabilization/preservation methods would make a significant contribution. On the other hand, unsatisfactory reproducibility also comes from manual operation. The automation of biosensors with the efforts of engineering is meaningful. 4) Non-destructive evaluation is one of the most significant trends in the development of biosensors for agro-product safety. Attempts in this field need to be highlighted. 5) Although the costs of biosensors are generally lower than the conventional bulky analytical instrument, it is still highly desirable to sharply decrease the costs of fabrication and assay of biosensors in terms of practical application. 6) Intelligentization and informatization will help biosensors be grounded to users. Cutting-edge technology such as artificial intelligence and the internet of things may provide novel solutions to minimize the gap between laboratory researches and real-world applications. They can serve for signal transduction, output and sharing for users, as well as the collection of requirements and feedback from consumers to developers.

CONSENT FOR PUBLICATION

Not applicable

CONFLICT OF INTEREST

The authors confirm that this chapter contents have no conflict of interest.

ACKNOWLEDGEMENTS

None declared.

REFERENCES

[1]   M. Lv, Y. Liu, J. Geng, X. Kou, Z. Xin, and D. Yang, "Engineering nanomaterials-based biosensors for food safety detection", *Biosens. Bioelectron.,* vol. 106, pp. 122-128, 2018.
        [http://dx.doi.org/10.1016/j.bios.2018.01.049] [PMID: 29414078]

[2]   N. Wongkaew, M. Simsek, C. Griesche, and A.J. Baeumner, "Functional nanomaterials and nanostructures enhancing electrochemical biosensors and lab-on-a-chip performances: recent progress, applications, and future perspective", *Chem. Rev.,* vol. 119, no. 1, pp. 120-194, 2019.
        [http://dx.doi.org/10.1021/acs.chemrev.8b00172] [PMID: 30247026]

[3]   L.C. Clark Jr, and C. Lyons, "Electrode systems for continuous monitoring in cardiovascular surgery", *Ann. N. Y. Acad. Sci.,* vol. 102, pp. 29-45, 1962.
        [http://dx.doi.org/10.1111/j.1749-6632.1962.tb13623.x] [PMID: 14021529]

[4]   Y. Wu, R.D. Tilley, and J.J. Gooding, "Challenges and solutions in developing ultrasensitive biosensors", *J. Am. Chem. Soc.,* vol. 141, no. 3, pp. 1162-1170, 2019.
        [http://dx.doi.org/10.1021/jacs.8b09397] [PMID: 30463401]

[5]   X. Huang, Y. Liu, B. Yung, Y. Xiong, and X. Chen, "Nanotechnology-enhanced no-wash biosensors for *in vitro* diagnostics of cancer", *ACS Nano,* vol. 11, no. 6, pp. 5238-5292, 2017.
        [http://dx.doi.org/10.1021/acsnano.7b02618] [PMID: 28590117]

[6]   J.N. Tiwari, V. Vij, K.C. Kemp, and K.S. Kim, "Engineered carbon-nanomaterial-based electrochemical sensors for biomolecules", *ACS Nano,* vol. 10, no. 1, pp. 46-80, 2016.
        [http://dx.doi.org/10.1021/acsnano.5b05690] [PMID: 26579616]

[7]   P. Devi, A. Thakur, R.Y. Lai, S. Saini, R. Jain, and P. Kumar, "Progress in the materials for optical detection of arsenic in water", *Trends Analyt. Chem.,* vol. 110, pp. 97-115, 2019.
        [http://dx.doi.org/10.1016/j.trac.2018.10.008]

[8]   L. Zhang, Y. Ying, Y. Li, and Y. Fu, "Integration and synergy in protein-nanomaterial hybrids for biosensing: Strategies and in-field detection applications", *Biosens. Bioelectron.,* vol. 154, pp. 112036-112045, 2020.
        [http://dx.doi.org/10.1016/j.bios.2020.112036] [PMID: 32056955]

[9]   F. Zhao, J. Wu, Y. Ying, Y. She, J. Wang, and J. Ping, "Carbon nanomaterial-enabled pesticide biosensors: Design strategy, biosensing mechanism, and practical application", *Trends Analyt. Chem.,* vol. 106, pp. 62-83, 2018.
        [http://dx.doi.org/10.1016/j.trac.2018.06.017]

[10]   S. Huang, J. Yao, X. Chu, Y. Liu, Q. Xiao, and Y. Zhang, "One-step facile synthesis of nitrogen-doped carbon dots: a ratiometric fluorescent probe for evaluation of acetylcholinesterase activity and detection of organophosphorus pesticides in tap water and food", *J. Agric. Food Chem.,* vol. 67, no. 40, pp. 11244-11255, 2019.
        [http://dx.doi.org/10.1021/acs.jafc.9b03624] [PMID: 31532667]

[11]   P. Wang, H. Li, M.M. Hassan, Z. Guo, Z.Z. Zhang, and Q. Chen, "Fabricating an acetylcholinesterase modulated UCNPs-Cu2+ fluorescence biosensor for ultrasensitive detection of organophosphorus pesticides-diazinon in food", *J. Agric. Food Chem.,* vol. 67, no. 14, pp. 4071-4079, 2019.
[http://dx.doi.org/10.1021/acs.jafc.8b07201] [PMID: 30888170]

[12]   M. Arvand, and A.A. Mirroshandel, "An efficient fluorescence resonance energy transfer system from quantum dots to graphene oxide nano sheets: Application in a photoluminescence aptasensing probe for the sensitive detection of diazinon", *Food Chem.,* vol. 280, pp. 115-122, 2019.
[http://dx.doi.org/10.1016/j.foodchem.2018.12.069] [PMID: 30642476]

[13]   H. Ouyang, Q. Lu, W. Wang, Y. Song, X. Tu, C. Zhu, J.N. Smith, D. Du, Z. Fu, and Y. Lin, "Dual-readout immunochromatographic assay by utilizing MnO_2 nanoflowers as the unique colorimetric/chemiluminescent probe", *Anal. Chem.,* vol. 90, no. 8, pp. 5147-5152, 2018.
[http://dx.doi.org/10.1021/acs.analchem.7b05247] [PMID: 29590527]

[14]   R.H. Wang, C.L. Zhu, L.L. Wang, L.Z. Xu, W.L. Wang, C. Yang, and Y. Zhang, "Dual-modal aptasensor for the detection of isocarbophos in vegetables", *Talanta,* vol. 205, pp. 120094-120100, 2019.
[http://dx.doi.org/10.1016/j.talanta.2019.06.094] [PMID: 31450466]

[15]   Y. Dong, W. Zheng, D. Chen, X. Li, J. Wang, Z. Wang, and Y. Chen, "Click reaction-mediated T_2 immunosensor for ultrasensitive detection of pesticide residues *via* brush-like nanostructure-triggered coordination chemistry", *J. Agric. Food Chem.,* vol. 67, no. 35, pp. 9942-9949, 2019.
[http://dx.doi.org/10.1021/acs.jafc.9b03463] [PMID: 31403785]

[16]   L. Cervera-Chiner, M. Juan-Borrás, C. March, A. Arnau, I. Escriche, Á. Montoya, and Y. Jiménez, "High fundamental frequency quartz crystal microbalance (HFF-QCM) immunosensor for pesticide detection in honey", *Food Control,* vol. 92, pp. 1-6, 2018.
[http://dx.doi.org/10.1016/j.foodcont.2018.04.026]

[17]   Y. Dai, T. Wang, X. Hu, S. Liu, M. Zhang, and C. Wang, "Highly sensitive microcantilever-based immunosensor for the detection of carbofuran in soil and vegetable samples", *Food Chem.,* vol. 229, pp. 432-438, 2017.
[http://dx.doi.org/10.1016/j.foodchem.2017.02.093] [PMID: 28372196]

[18]   C.S. Pundir, A. Malik, and Preety, "Bio-sensing of organophosphorus pesticides: A review", *Biosens. Bioelectron.,* vol. 140, pp. 111348-111360, 2019.
[http://dx.doi.org/10.1016/j.bios.2019.111348] [PMID: 31153016]

[19]   M.K.L. da Silva, H.C. Vanzela, L.M. Defavari, and I. Cesarino, "Determination of carbamate pesticide in food using a biosensor based on reduced graphene oxide and acetylcholinesterase enzyme", *Sens. Actuators B Chem.,* vol. 277, pp. 555-561, 2018.
[http://dx.doi.org/10.1016/j.snb.2018.09.051]

[20]   L. Zhang, Z. Liu, Q. Xie, Y. Li, Y. Ying, and Y. Fu, "Bio-inspired assembly of reduced graphene oxide by fibrin fiber to prepare multi-functional conductive bio-nanocomposites as versatile electrochemical platforms", *Carbon,* vol. 153, pp. 504-512, 2019.
[http://dx.doi.org/10.1016/j.carbon.2019.06.101]

[21]   L. Zhang, Z. Liu, S. Zha, G. Liu, W. Zhu, Q. Xie, Y. Li, Y. Ying, and Y. Fu, "Bio-/nanoimmobilization platform based on bioinspired fibrin-bone@polydopamine-shell adhesive composites for biosensing", *ACS Appl. Mater. Interfaces,* vol. 11, no. 50, pp. 47311-47319, 2019.
[http://dx.doi.org/10.1021/acsami.9b15376] [PMID: 31742992]

[22]   R.K. Mishra, L.J. Hubble, A. Martín, R. Kumar, A. Barfidokht, J. Kim, M.M. Musameh, I.L. Kyratzis, and J. Wang, "Wearable flexible and stretchable glove biosensor for on-site detection of organophosphorus chemical threats", *ACS Sens.,* vol. 2, no. 4, pp. 553-561, 2017.
[http://dx.doi.org/10.1021/acssensors.7b00051] [PMID: 28723187]

[23]   X. Xu, Y. Guo, L. Wang, K. He, Y. Guo, X. Wang, and S. Gunasekaran, "Hapten-grafted programmed probe as a corecognition element for a competitive immunosensor to detect acetamiprid residue in

agricultural products", *J. Agric. Food Chem.*, vol. 66, no. 29, pp. 7815-7821, 2018.
[http://dx.doi.org/10.1021/acs.jafc.8b02487] [PMID: 29944365]

[24] M. Liu, A. Khan, Z. Wang, Y. Liu, G. Yang, Y. Deng, and N. He, "Aptasensors for pesticide detection", *Biosens. Bioelectron.*, vol. 130, pp. 174-184, 2019.
[http://dx.doi.org/10.1016/j.bios.2019.01.006] [PMID: 30738246]

[25] A. Talan, A. Mishra, S.A. Eremin, J. Narang, A. Kumar, and S. Gandhi, "Ultrasensitive electrochemical immuno-sensing platform based on gold nanoparticles triggering chlorpyrifos detection in fruits and vegetables", *Biosens. Bioelectron.*, vol. 105, pp. 14-21, 2018.
[http://dx.doi.org/10.1016/j.bios.2018.01.013] [PMID: 29346076]

[26] L. Fan, C. Zhang, W. Yan, Y. Guo, S. Shuang, C. Dong, and Y. Bi, "Design of a facile and label-free electrochemical aptasensor for detection of atrazine", *Talanta*, vol. 201, pp. 156-164, 2019.
[http://dx.doi.org/10.1016/j.talanta.2019.03.114] [PMID: 31122406]

[27] S. Feng, Y. Li, R. Zhang, and Y. Li, "A novel electrochemical sensor based on molecularly imprinted polymer modified hollow N, S-Mo2C/C spheres for highly sensitive and selective carbendazim determination", *Biosens. Bioelectron.*, vol. 142, pp. 111491-111497, 2019.
[http://dx.doi.org/10.1016/j.bios.2019.111491] [PMID: 31326864]

[28] A. Joshi, and K.H. Kim, "Recent advances in nanomaterial-based electrochemical detection of antibiotics: Challenges and future perspectives", *Biosens. Bioelectron.*, vol. 153, pp. 112046-112068, 2020.
[http://dx.doi.org/10.1016/j.bios.2020.112046] [PMID: 32056661]

[29] Y. Yun, M. Pan, L. Wang, S. Li, Y. Wang, Y. Gu, J. Yang, and S. Wang, "Fabrication and evaluation of a label-free piezoelectric immunosensor for sensitive and selective detection of amantadine in foods of animal origin", *Anal. Bioanal. Chem.*, vol. 411, no. 22, pp. 5745-5753, 2019.
[http://dx.doi.org/10.1007/s00216-019-01954-4] [PMID: 31243479]

[30] S. Li, Y. Zhang, W. Wen, W. Sheng, J. Wang, S. Wang, and J. Wang, "A high-sensitivity thermal analysis immunochromatographic sensor based on au nanoparticle-enhanced two-dimensional black phosphorus photothermal-sensing materials", *Biosens. Bioelectron.*, vol. 133, pp. 223-229, 2019.
[http://dx.doi.org/10.1016/j.bios.2019.03.039] [PMID: 30951982]

[31] S. Yan, X. Lai, Y. Wang, N. Ye, and Y. Xiang, "Label free aptasensor for ultrasensitive detection of tobramycin residue in pasteurized cow's milk based on resonance scattering spectra and nanogold catalytic amplification", *Food Chem.*, vol. 295, pp. 36-41, 2019.
[http://dx.doi.org/10.1016/j.foodchem.2019.05.110] [PMID: 31174769]

[32] B. Fang, S. Hu, C. Wang, M. Yuan, Z. Huang, K. Xing, D. Liu, J. Peng, and W. Lai, "Lateral flow immunoassays combining enrichment and colorimetry-fluorescence quantitative detection of sulfamethazine in milk based on trifunctional magnetic nanobeads", *Food Control*, vol. 98, pp. 268-273, 2019.
[http://dx.doi.org/10.1016/j.foodcont.2018.11.039]

[33] L. Cao, Q. Zhang, H. Dai, Y. Fu, and Y. Li, "Separation/concentration-signal-amplification in-one method based on electrochemical conversion of magnetic nanoparticles for electrochemical biosensing", *Electroanalysis*, vol. 30, no. 3, pp. 517-524, 2018.
[http://dx.doi.org/10.1002/elan.201700653]

[34] L. Yan, L. Dou, T. Bu, Q. Huang, R. Wang, Q. Yang, L. Huang, J. Wang, and D. Zhang, "Highly sensitive furazolidone monitoring in milk by a signal amplified lateral flow assay based on magnetite nanoparticles labeled dual-probe", *Food Chem.*, vol. 261, pp. 131-138, 2018.
[http://dx.doi.org/10.1016/j.foodchem.2018.04.016] [PMID: 29739573]

[35] W. Huang, H. Zhang, G. Lai, S. Liu, B. Li, and A. Yu, "Sensitive and rapid aptasensing of chloramphenicol by colorimetric signal transduction with a DNAzyme-functionalized gold nanoprobe", *Food Chem.*, vol. 270, pp. 287-292, 2019.
[http://dx.doi.org/10.1016/j.foodchem.2018.07.127] [PMID: 30174048]

[36]   Y.Y. Wu, B.W. Liu, P. Huang, and F.Y. Wu, "A novel colorimetric aptasensor for detection of chloramphenicol based on lanthanum ion-assisted gold nanoparticle aggregation and smartphone imaging", *Anal. Bioanal. Chem.*, vol. 411, no. 28, pp. 7511-7518, 2019.
[http://dx.doi.org/10.1007/s00216-019-02149-7] [PMID: 31641824]

[37]   Y. Chen, C. Huang, B. Hellmann, Z. Jin, X. Xu, and G. Xiao, "A new HPTLC platformed luminescent biosensor system for facile screening of captan residue in fruits", *Food Chem.*, vol. 309, pp. 125691-125695, 2020.
[http://dx.doi.org/10.1016/j.foodchem.2019.125691] [PMID: 31679853]

[38]   R. Jalili, A. Khataee, M.R. Rashidi, and A. Razmjou, "Detection of penicillin G residues in milk based on dual-emission carbon dots and molecularly imprinted polymers", *Food Chem.*, vol. 314, pp. 126172-126178, 2020.
[http://dx.doi.org/10.1016/j.foodchem.2020.126172] [PMID: 31951890]

[39]   Q. Ouyang, Y. Liu, Q. Chen, Z. Guo, J. Zhao, H. Li, and W. Hu, "Rapid and specific sensing of tetracycline in food using a novel upconversion aptasensor", *Food Control,* vol. 81, pp. 156-163, 2017.
[http://dx.doi.org/10.1016/j.foodcont.2017.06.004]

[40]   S. Liu, J. Bai, Y. Huo, B. Ning, Y. Peng, S. Li, D. Han, W. Kang, and Z. Gao, "A zirconium-porphyrin MOF-based ratiometric fluorescent biosensor for rapid and ultrasensitive detection of chloramphenicol", *Biosens. Bioelectron.*, vol. 149, pp. 111801-111808, 2020.
[http://dx.doi.org/10.1016/j.bios.2019.111801] [PMID: 31726276]

[41]   C. Wang, X. Li, T. Peng, Z. Wang, K. Wen, and H. Jiang, "Latex bead and colloidal gold applied in a multiplex immunochromatographic assay for high-throughput detection of three classes of antibiotic residues in milk", *Food Control,* vol. 77, pp. 1-7, 2017.
[http://dx.doi.org/10.1016/j.foodcont.2017.01.016]

[42]   Q. Shi, J. Huang, Y. Sun, R. Deng, M. Teng, Q. Li, Y. Yang, X. Hu, Z. Zhang, and G. Zhang, "A SERS-based multiple immuno-nanoprobe for ultrasensitive detection of neomycin and quinolone antibiotics via a lateral flow assay", *Mikrochim. Acta,* vol. 185, no. 2, pp. 84-91, 2018.
[http://dx.doi.org/10.1007/s00604-017-2556-x] [PMID: 29594367]

[43]   Y. Zhang, W. Tan, Y. Zhang, H. Mao, S. Shi, L. Duan, H. Wang, and J. Yu, "Ultrasensitive and selective detection of Staphylococcus aureus using a novel IgY-based colorimetric platform", *Biosens. Bioelectron.*, vol. 142, pp. 111570-111575, 2019.
[http://dx.doi.org/10.1016/j.bios.2019.111570] [PMID: 31401227]

[44]   R. Gao, Z. Zhong, X. Gao, and L. Jia, "Graphene oxide quantum dots assisted construction of fluorescent aptasensor for rapid detection of pseudomonas aeruginosa in food samples", *J. Agric. Food Chem.*, vol. 66, no. 41, pp. 10898-10905, 2018.
[http://dx.doi.org/10.1021/acs.jafc.8b02164] [PMID: 30247907]

[45]   H. Wang, L. Wang, Q. Hu, R. Wang, Y. Li, and M. Kidd, "Rapid and sensitive detection of campylobacter jejuni in poultry products using a nanoparticle-based piezoelectric immunosensor integrated with magnetic immunoseparation", *J. Food Prot.*, vol. 81, no. 8, pp. 1321-1330, 2018.
[http://dx.doi.org/10.4315/0362-028X.JFP-17-381] [PMID: 30019963]

[46]   S. Zhan, H. Fang, J. Fu, W. Lai, Y. Leng, X. Huang, and Y. Xiong, "Gold nanoflower-enhanced dynamic light scattering immunosensor for the ultrasensitive no-wash detection of *Escherichia coli* O157:H7 in milk", *J. Agric. Food Chem.*, vol. 67, no. 32, pp. 9104-9111, 2019.
[http://dx.doi.org/10.1021/acs.jafc.9b03400] [PMID: 31334655]

[47]   L. Jin, T. Li, B. Wu, T. Yang, D. Zou, X. Liang, L. Hu, G. Huang, and J. Zhang, "Rapid detection of *Salmonella* in milk by nuclear magnetic resonance based on membrane filtration superparamagnetic nanobiosensor", *Food Control,* vol. 110, pp. 107011-107017, 2020.
[http://dx.doi.org/10.1016/j.foodcont.2019.107011]

[48]   Y. Wang, Z. Ye, and Y. Ying, "Development of a disposable impedance biosensor and its application for determination of *Escherichia coli* O157:H7"., *Trans. ASABE,* vol. 57, no. 2, pp. 585-591, 2013.

[49]   X. Ma, X. Xu, Y. Xia, and Z. Wang, "SERS aptasensor for *Salmonella typhimurium* detection based on spiny gold nanoparticles", *Food Control,* vol. 84, pp. 232-237, 2018.
[http://dx.doi.org/10.1016/j.foodcont.2017.07.016]

[50]   Z. Zhan, H. Li, J. Liu, G. Xie, F. Xiao, X. Wu, Z.P. Aguilar, and H. Xu, "A competitive enzyme linked aptasensor with rolling circle amplification (ELARCA) assay for colorimetric detection of *Listeria monocytogenes*", *Food Control,* vol. 107, pp. 106806-106812, 2020.
[http://dx.doi.org/10.1016/j.foodcont.2019.106806]

[51]   J. Feng, Q. Shen, J. Wu, Z. Dai, and Y. Wang, "Naked-eyes detection of *Shigella flexneri* in food samples based on a novel gold nanoparticle-based colorimetric aptasensor", *Food Control,* vol. 98, pp. 333-341, 2019.
[http://dx.doi.org/10.1016/j.foodcont.2018.11.048]

[52]   T. Bu, Q. Huang, L. Yan, W. Zhang, L. Dou, L. Huang, Q. Yang, B. Zhao, B. Yang, T. Li, J. Wang, and D. Zhang, "Applicability of biological dye tracer in strip biosensor for ultrasensitive detection of pathogenic bacteria", *Food Chem.,* vol. 274, pp. 816-821, 2019.
[http://dx.doi.org/10.1016/j.foodchem.2018.09.066] [PMID: 30373015]

[53]   Q. Wang, M. Long, C. Lv, S. Xin, X. Han, and W. Jiang, "Lanthanide-labeled fluorescent-nanoparticle immunochromatographic strips enable rapid and quantitative detection of *Escherichia coli* O157:H7 in food samples", *Food Control,* vol. 109, pp. 106894-106902, 2020.
[http://dx.doi.org/10.1016/j.foodcont.2019.106894]

[54]   S. Alamer, S. Eissa, R. Chinnappan, P. Herron, and M. Zourob, "Rapid colorimetric lactoferrin-based sandwich immunoassay on cotton swabs for the detection of foodborne pathogenic bacteria", *Talanta,* vol. 185, pp. 275-280, 2018.
[http://dx.doi.org/10.1016/j.talanta.2018.03.072] [PMID: 29759200]

[55]   H. Yousefi, M.M. Ali, H.M. Su, C.D.M. Filipe, and T.F. Didar, "Sentinel wraps: real-time monitoring of food contamination by printing dnazyme probes on food packaging", *ACS Nano,* vol. 12, no. 4, pp. 3287-3294, 2018.
[http://dx.doi.org/10.1021/acsnano.7b08010] [PMID: 29621883]

[56]   Y. Hou, G. Cai, L. Zheng, and J. Lin, "A microfluidic signal-off biosensor for rapid and sensitive detection of *Salmonella* using magnetic separation and enzymatic catalysis", *Food Control,* vol. 103, pp. 186-193, 2019.
[http://dx.doi.org/10.1016/j.foodcont.2019.04.008]

[57]   X. Zhang, S. Tsuji, H. Kitaoka, H. Kobayashi, M. Tamai, K.I. Honjoh, and T. Miyamoto, "Simultaneous detection of *Escherichia coli* O157:H7, *Salmonella enteritidis*, and *Listeria monocytogenes* at a very low level using simultaneous enrichment broth and multichannel SPR biosensor", *J. Food Sci.,* vol. 82, no. 10, pp. 2357-2363, 2017.
[http://dx.doi.org/10.1111/1750-3841.13843] [PMID: 28833106]

[58]   H. Zhang, L. Xue, F. Huang, S. Wang, L. Wang, N. Liu, and J. Lin, "A capillary biosensor for rapid detection of Salmonella using Fe-nanocluster amplification and smart phone imaging", *Biosens. Bioelectron.,* vol. 127, pp. 142-149, 2019.
[http://dx.doi.org/10.1016/j.bios.2018.11.042] [PMID: 30597432]

[59]   L. Xu, Z. Lu, L. Cao, H. Pang, Q. Zhang, Y. Fu, Y. Xiong, Y. Li, X. Wang, J. Wang, Y. Ying, and Y. Li, "In-field detection of multiple pathogenic bacteria in food products using a portable fluorescent biosensing system", *Food Control,* vol. 75, pp. 21-28, 2017.
[http://dx.doi.org/10.1016/j.foodcont.2016.12.018]

[60]   S. Bu, K. Wang, C. Ju, C. Wang, Z. Li, Z. Hao, M. Shen, and J. Wan, "Point-of-care assay to detect foodborne pathogenic bacteria using a low-cost disposable medical infusion extension line as readout and MnO_2 nanoflowers", *Food Control,* vol. 98, pp. 399-404, 2019.
[http://dx.doi.org/10.1016/j.foodcont.2018.11.053]

[61]   F. Huang, H. Zhang, L. Wang, W. Lai, and J. Lin, "A sensitive biosensor using double-layer capillary

based immunomagnetic separation and invertase-nanocluster based signal amplification for rapid detection of foodborne pathogen", *Biosens. Bioelectron.,* vol. 100, pp. 583-590, 2018.
[http://dx.doi.org/10.1016/j.bios.2017.10.005] [PMID: 29032045]

[62]   X. Wei, W. Zhou, S.T. Sanjay, J. Zhang, Q. Jin, F. Xu, D.C. Dominguez, and X. Li, "Multiplexed instrument-free bar-chart spinchip integrated with nanoparticle-mediated magnetic aptasensors for visual quantitative detection of multiple pathogens", *Anal. Chem.,* vol. 90, no. 16, pp. 9888-9896, 2018.
[http://dx.doi.org/10.1021/acs.analchem.8b02055] [PMID: 30028601]

[63]   K.Y. Goud, S.K. Kailasa, V. Kumar, Y.F. Tsang, S.E. Lee, K.V. Gobi, and K.H. Kim, "Progress on nanostructured electrochemical sensors and their recognition elements for detection of mycotoxins: A review", *Biosens. Bioelectron.,* vol. 121, pp. 205-222, 2018.
[http://dx.doi.org/10.1016/j.bios.2018.08.029] [PMID: 30219721]

[64]   Y. Hui, B. Wang, R. Ren, A. Zhao, F. Zhang, S. Song, and Y. He, An electrochemical aptasensor based on DNA-AuNPs-HRP nanoprobes and exonuclease-assisted signal amplification for detection of aflatoxin B_1., *Food Control,* vol. 109, pp. 106902-106909, 2020.
[http://dx.doi.org/10.1016/j.foodcont.2019.106902]

[65]   H. Zejli, K.Y. Goud, and J.L. Marty, "Label free aptasensor for ochratoxin A detection using polythiophene-3-carboxylic acid", *Talanta,* vol. 185, pp. 513-519, 2018.
[http://dx.doi.org/10.1016/j.talanta.2018.03.089] [PMID: 29759234]

[66]   H. Bhardwaj, G. Sumana, and C.A. Marquette, "A label-free ultrasensitive microfluidic surface Plasmon resonance biosensor for Aflatoxin B1 detection using nanoparticles integrated gold chip", *Food Chem.,* vol. 307, pp. 125530-125536, 2020.
[http://dx.doi.org/10.1016/j.foodchem.2019.125530] [PMID: 31639579]

[67]   L. Hao, W. Wang, X. Shen, S. Wang, Q. Li, F. An, and S. Wu, "A fluorescent DNA hydrogel aptasensor based on the self-assembly of rolling circle amplification products for sensitive detection of ochratoxin a", *J. Agric. Food Chem.,* vol. 68, no. 1, pp. 369-375, 2020.
[http://dx.doi.org/10.1021/acs.jafc.9b06021] [PMID: 31829586]

[68]   Y. Yao, H. Wang, X. Wang, X. Wang, and F. Li, "Development of a chemiluminescent aptasensor for ultrasensitive and selective detection of aflatoxin B1 in peanut and milk", *Talanta,* vol. 201, pp. 52-57, 2019.
[http://dx.doi.org/10.1016/j.talanta.2019.03.109] [PMID: 31122460]

[69]   J. Wu, L. Zeng, N. Li, C. Liu, and J. Chen, "A wash-free and label-free colorimetric biosensor for naked-eye detection of aflatoxin B1 using G-quadruplex as the signal reporter", *Food Chem.,* vol. 298, pp. 125034-125039, 2019.
[http://dx.doi.org/10.1016/j.foodchem.2019.125034] [PMID: 31261013]

[70]   X. Li, J. Wang, C. Yi, L. Jiang, J. Wu, X. Chen, X. Shen, Y. Sun, and H. Lei, "A smartphone-based quantitative detection device integrated with latex microsphere immunochromatography for on-site detection of zearalenone in cereals and feed", *Sens. Actuators B Chem.,* vol. 290, pp. 170-179, 2019.
[http://dx.doi.org/10.1016/j.snb.2019.03.108]

[71]   J.M.D. Machado, R.R.G. Soares, V. Chu, and J.P. Conde, "Multiplexed capillary microfluidic immunoassay with smartphone data acquisition for parallel mycotoxin detection", *Biosens. Bioelectron.,* vol. 99, pp. 40-46, 2018.
[http://dx.doi.org/10.1016/j.bios.2017.07.032] [PMID: 28735045]

[72]   L. Wang, X. Peng, H. Fu, C. Huang, Y. Li, and Z. Liu, "Recent advances in the development of electrochemical aptasensors for detection of heavy metals in food", *Biosens. Bioelectron.,* vol. 147, pp. 111777-111786, 2020.
[http://dx.doi.org/10.1016/j.bios.2019.111777] [PMID: 31634804]

[73]   X. Wang, M. Cheng, Q. Yang, H. Wei, A. Xia, L. Wang, Y. Ben, Q. Zhou, Z. Yang, and X. Huang, "A living plant cell-based biosensor for real-time monitoring invisible damage of plant cells under heavy

metal stress", *Sci. Total Environ.,* vol. 697, pp. 134097-134106, 2019.
[http://dx.doi.org/10.1016/j.scitotenv.2019.134097] [PMID: 31484090]

[74]   P. Zhu, Y. Shang, W. Tian, K. Huang, Y. Luo, and W. Xu, "Ultra-sensitive and absolute quantitative detection of Cu^{2+} based on DNAzyme and digital PCR in water and drink samples", *Food Chem.,* vol. 221, pp. 1770-1777, 2017.
[http://dx.doi.org/10.1016/j.foodchem.2016.10.106] [PMID: 27979159]

[75]   T. Liu, Z. Chu, and W. Jin, "Electrochemical mercury biosensors based on advanced nanomaterials", *J. Mater. Chem. B Mater. Biol. Med.,* vol. 7, no. 23, pp. 3620-3632, 2019.
[http://dx.doi.org/10.1039/C9TB00418A]

[76]   Y. Gong, Y. Zheng, B. Jin, M. You, J. Wang, X. Li, M. Lin, F. Xu, and F. Li, "A portable and universal upconversion nanoparticle-based lateral flow assay platform for point-of-care testing", *Talanta,* vol. 201, pp. 126-133, 2019.
[http://dx.doi.org/10.1016/j.talanta.2019.03.105] [PMID: 31122402]

[77]   D. Feng, P. Li, X. Tan, Y. Wu, F. Wei, F. Du, C. Ai, Y. Luo, Q. Chen, and H. Han, "Electrochemiluminescence aptasensor for multiple determination of Hg^{2+} and Pb^{2+} ions by using the MIL-53(Al)@CdTe-PEI modified electrode", *Anal. Chim. Acta,* vol. 1100, pp. 232-239, 2020.
[http://dx.doi.org/10.1016/j.aca.2019.11.069] [PMID: 31987146]

[78]   K. Mao, H. Zhang, Z. Wang, H. Cao, K. Zhang, X. Li, and Z. Yang, "Nanomaterial-based aptamer sensors for arsenic detection", *Biosens. Bioelectron.,* vol. 148, pp. 111785-111799, 2020.
[http://dx.doi.org/10.1016/j.bios.2019.111785] [PMID: 31689596]

[79]   Y. Liu, Q. Ouyang, H. Li, M. Chen, Z. Zhang, and Q. Chen, "Turn-on fluoresence sensor for Hg^{2+} in food based on FRET between aptamers-functionalized upconversion nanoparticles and gold nanoparticles", *J. Agric. Food Chem.,* vol. 66, no. 24, pp. 6188-6195, 2018.
[http://dx.doi.org/10.1021/acs.jafc.8b00546] [PMID: 29847117]

[80]   M. Jia, Y. Lu, R. Wang, N. Ren, J. Zhang, X. Changhua, and J. Wu, "Extended GR-5 DNAzyme-based Autonomous isothermal Cascade machine: An efficient and sensitive one-tube colorimetric platform for Pb^{2+} detection", *Sens. Actuators B Chem.,* vol. 304, pp. 127366-127372, 2020.
[http://dx.doi.org/10.1016/j.snb.2019.127366]

[81]   R. Deng, H. Yang, Y. Dong, Z. Zhao, X. Xia, Y. Li, and J. Li, "Temperature-robust dnazyme biosensors confirming ultralow background detection", *ACS Sens.,* vol. 3, no. 12, pp. 2660-2666, 2018.
[http://dx.doi.org/10.1021/acssensors.8b01122] [PMID: 30457325]

[82]   M. Zhang, P. Wu, J. Wu, J. Ping, and J. Wu, "Advanced DNA-based methods for the detection of peanut allergens in processed food", *Trends Analyt. Chem.,* vol. 114, pp. 278-292, 2019.
[http://dx.doi.org/10.1016/j.trac.2019.01.021]

[83]   Y. Wang, Z. Rao, J. Zhou, L. Zheng, and L. Fu, "A chiral assembly of gold nanoparticle trimer-based biosensors for ultrasensitive detection of the major allergen tropomyosin in shellfish", *Biosens. Bioelectron.,* vol. 132, pp. 84-89, 2019.
[http://dx.doi.org/10.1016/j.bios.2019.02.038] [PMID: 30856431]

[84]   X. Sun, Y. Ye, S. He, Z. Wu, J. Yue, H. Sun, and X. Cao, "A novel oriented antibody immobilization based voltammetric immunosensor for allergenic activity detection of lectin in kidney bean by using AuNPs-PEI-MWCNTs modified electrode", *Biosens. Bioelectron.,* vol. 143, pp. 111607-111614, 2019.
[http://dx.doi.org/10.1016/j.bios.2019.111607] [PMID: 31445384]

[85]   S.P. White, C.D. Frisbie, and K.D. Dorfman, "Detection and sourcing of gluten in grain with multiple floating-gate transistor biosensors", *ACS Sens.,* vol. 3, no. 2, pp. 395-402, 2018.
[http://dx.doi.org/10.1021/acssensors.7b00810] [PMID: 29411606]

[86]   F. Malvano, D. Albanese, R. Pilloton, and M. Di Matteo, "A new label-free impedimetric aptasensor for gluten detection", *Food Control,* vol. 79, pp. 200-206, 2017.
[http://dx.doi.org/10.1016/j.foodcont.2017.03.033]

[87] M.A. Pereira-Barros, M.F. Barroso, L. Martín-Pedraza, E. Vargas, S. Benedé, M. Villalba, J.M. Rocha, S. Campuzano, and J.M. Pingarrón, "Direct PCR-free electrochemical biosensing of plant-food derived nucleic acids in genomic DNA extracts. Application to the determination of the key allergen Sola l 7 in tomato seeds", *Biosens. Bioelectron.,* vol. 137, pp. 171-177, 2019.
[http://dx.doi.org/10.1016/j.bios.2019.05.011] [PMID: 31096083]

[88] M. Angelopoulou, P.S. Petrou, E. Makarona, W. Haasnoot, I. Moser, G. Jobst, D. Goustouridis, M. Lees, K. Kalatzi, I. Raptis, K. Misiakos, and S.E. Kakabakos, "Ultrafast multiplexed-allergen detection through advanced fluidic design and monolithic interferometric silicon chips", *Anal. Chem.,* vol. 90, no. 15, pp. 9559-9567, 2018.
[http://dx.doi.org/10.1021/acs.analchem.8b02321] [PMID: 29999303]

[89] Z. Wu, D. He, E. Xu, A. Jiao, M.F.J. Chughtai, and Z. Jin, "Rapid detection of β-conglutin with a novel lateral flow aptasensor assisted by immunomagnetic enrichment and enzyme signal amplification", *Food Chem.,* vol. 269, pp. 375-379, 2018.
[http://dx.doi.org/10.1016/j.foodchem.2018.07.011] [PMID: 30100448]

[90] D. Jiang, P. Ge, L. Wang, H. Jiang, M. Yang, L. Yuan, Q. Ge, W. Fang, and X. Ju, "A novel electrochemical mast cell-based paper biosensor for the rapid detection of milk allergen casein", *Biosens. Bioelectron.,* vol. 130, pp. 299-306, 2019.
[http://dx.doi.org/10.1016/j.bios.2019.01.050] [PMID: 30776617]

[91] H.Y. Lin, C.H. Huang, J. Park, D. Pathania, C.M. Castro, A. Fasano, R. Weissleder, and H. Lee, "Integrated magneto-chemical sensor for on-site food allergen detection", *ACS Nano,* vol. 11, no. 10, pp. 10062-10069, 2017.
[http://dx.doi.org/10.1021/acsnano.7b04318] [PMID: 28792732]

[92] M. Stredansky, L. Redivo, P. Magdolen, A. Stredansky, and L. Navarini, "Rapid sucrose monitoring in green coffee samples using multienzymatic biosensor", *Food Chem.,* vol. 254, pp. 8-12, 2018.
[http://dx.doi.org/10.1016/j.foodchem.2018.01.171] [PMID: 29548475]

[93] O. Hosu, M. Lettieri, N. Papara, A. Ravalli, R. Sandulescu, C. Cristea, and G. Marrazza, "Colorimetric multienzymatic smart sensors for hydrogen peroxide, glucose and catechol screening analysis", *Talanta,* vol. 204, pp. 525-532, 2019.
[http://dx.doi.org/10.1016/j.talanta.2019.06.041] [PMID: 31357329]

[94] Z. Zhang, J. Liu, J. Fan, Z. Wang, and L. Li, "Detection of catechol using an electrochemical biosensor based on engineered *Escherichia coli* cells that surface-display laccase", *Anal. Chim. Acta,* vol. 1009, pp. 65-72, 2018.
[http://dx.doi.org/10.1016/j.aca.2018.01.008] [PMID: 29422133]

[95] K. Varmira, G. Mohammadi, M. Mahmoudi, R. Khodarahmi, K. Rashidi, M. Hedayati, H.C. Goicoechea, and A.R. Jalalvand, "Fabrication of a novel enzymatic electrochemical biosensor for determination of tyrosine in some food samples", *Talanta,* vol. 183, pp. 1-10, 2018.
[http://dx.doi.org/10.1016/j.talanta.2018.02.053] [PMID: 29567149]

[96] R. Long, Y. Guo, L. Xie, S. Shi, J. Xu, C. Tong, Q. Lin, and T. Li, "White pepper-derived ratiometric carbon dots for highly selective detection and imaging of coenzyme A", *Food Chem.,* vol. 315, pp. 126171-126176, 2020.
[http://dx.doi.org/10.1016/j.foodchem.2020.126171] [PMID: 31991253]

[97] C.A.R. Salamanca-Neto, G.G. Marcheafave, J. Scremin, E.C.M. Barbosa, P.H.C. Camargo, R.F.H. Dekker, I.S. Scarminio, A.M. Barbosa-Dekker, and E.R. Sartori, "Chemometric-assisted construction of a biosensing device to measure chlorogenic acid content in brewed coffee beverages to discriminate quality", *Food Chem.,* vol. 315, pp. 126306-126314, 2020.
[http://dx.doi.org/10.1016/j.foodchem.2020.126306] [PMID: 32035315]

[98] C. Chen, S. Feng, M. Zhou, C. Ji, L. Que, and W. Wang, "Development of a structure-switching aptamer-based nanosensor for salicylic acid detection", *Biosens. Bioelectron.,* vol. 140, pp. 111342-111349, 2019.

[http://dx.doi.org/10.1016/j.bios.2019.111342] [PMID: 31153018]

[99]   S.S. Baghbaderani, and A. Noorbakhsh, "Novel chitosan-Nafion composite for fabrication of highly sensitive impedimetric and colorimetric As(III) aptasensor", *Biosens. Bioelectron.,* vol. 131, pp. 1-8, 2019.
       [http://dx.doi.org/10.1016/j.bios.2019.01.059] [PMID: 30797108]

[100] K. Mahato, and P. Chandra, "Paper-based miniaturized immunosensor for naked eye ALP detection based on digital image colorimetry integrated with smartphone", *Biosens. Bioelectron.,* vol. 128, pp. 9-16, 2019.
       [http://dx.doi.org/10.1016/j.bios.2018.12.006] [PMID: 30616217]

CHAPTER 4

Internal Quality Grading Technologies and Applications for Agricultural Products

Aichen Wang[1,*]**, Wen Zhang**[2] **and Jiangbo Li**[3,4]

[1] *School of Agricultural Engineering, Jiangsu University, 301 Xuefu Road, Zhenjiang 212013, Jiangsu, PR China*

[2] *School of Life Science and Engineering, Southwest University of Science and Technology, Mianyang 621010, Sichuan, PR China*

[3] *Beijing Research Center of Intelligent Equipment for Agriculture, Beijing 100096, PR China*

[4] *Key Laboratory of Modern Agricultural Equipment and Technology (Jiangsu University), Ministry of Education, Zhenjiang 212013, Jiangsu, PR China*

Abstract: The internal quality of agricultural products is an important attribute that is considered by consumers when buying them. Grading agricultural products according to their internal quality, is an effective way to make the best use of the products, and thus improve the overall value. In recent years, several nondestructive, intelligent sensing techniques have been studied extensively for detecting the internal quality of agricultural products, including Vis/NIR spectroscopy, multi-/hyper-spectral imaging, nuclear magnetic resonance and imaging, X-ray and computed tomography, electrical nose and acoustic technique. In this chapter, the working principle of each technique is provided, and corresponding applications in the agricultural domain are reviewed to provide overall understanding of these techniques. The challenges and perspectives of these techniques are also analyzed.

1. INTRODUCTION

With the improvement of living standards of human beings in recent decades, the quality of agricultural products has become a key factor that consumers would consider when buying them [1]. Quality is not a single well-defined attribute and is often evaluated or represented by several attributes, including color, shape, texture, defects, sugar content, firmness, soluble solids content (SSC), acidity and nutritional contents [2]. These attributes can be grouped into two categories, external and internal quality.

Recent developed nondestructive, intelligent sensing techniques have been studied extensively for the quality detection of agricultural products, both externally and internally. However, due to the difference in detecting mechanisms of different

* **Corresponding author Aichen Wang:** School of Agricultural Engineering, Jiangsu University, 301 Xuefu Road, Zhenjiang 212013, Jiangsu, PR China; Tel: +86 15952883501; E-mail: acwang@ujs.edu.cn

sensing techniques, some of them are unable to acquire the quality information deep inside the tissue, therefore they can only be used for external quality detection, such as machine vision, while others can acquire the internal information of agricultural plant tissue, such as NIR spectroscopy, making them suitable for detecting the internal quality of agricultural products. In this chapter, the techniques that are suitable for internal quality detection of agricultural products including NIR, multi/hyper-spectral imaging, magnetic resonance imaging (MRI), X-ray, computed tomography (CT), electrical nose and acoustic technique, are reviewed, as well as applications and challenges.

2. INTERNAL QUALITY GRADING TECHNOLOGIES AND APPLICATIONS

2.1. Vis/NIR Spectroscopy

2.1.1. Principle

Vis/NIR spectroscopy is a type of vibrational spectroscopy that covers the range of the electromagnetic spectrum between 380 and 2500 nm. This region can be further divided into three parts: visible (380-780 nm), shortwave near-infrared (780-1100 nm), and longwave near-infrared (1100-2500 nm). When incident light hits a sample, the visible and near-infrared photons penetrate and interact with the biological materials. The photons in the visible region are mainly absorbed by pigments (carotenoid, anthocyanin, and chlorophyll), while the near-infrared photons have strong interactions with hydrogen bonds such as N-H, C-H and O-H [3, 4]. During the interaction with biological materials, the photons may be absorbed or scattered, resulting in reflected or transmitted light exiting from the sample that carries its internal chemical and structural information (Fig. **1**). The reflected or transmitted light is dispersed by monochromators and then received by detectors, forming the Vis/NIR spectra, which can be used for sample internal quality analysis [1].

When detecting the internal quality of agricultural products, a Vis/NIS system is necessary. A typical Vis/NIR system consists of a light source, a spectrometer, a sample compartment, and relevant optics accessories. Based on the arrangement of the detecting system, three detection modes are frequently adopted: reflectance, transmittance, and interactance [1, 5]. In the reflectance mode (Fig. **2a**), the incident light hits a sample and is reflected both diffusely and specularly. The diffuse reflectance interacts with the sample and carries effective information that can be used for internal quality analysis. However, the specular reflectance does not carry useful information about the sample, hence should be avoided [1]. In transmittance mode, it can be further classified as the full transmittance mode (Fig. **2b**) and partial transmittance mode (Fig. **2c**), according to the relative position of the detector and light source. In the full transmittance mode, the light source and detector are configured in the same line at opposite sides of a sample, while in the partial transmittance, the light source, the detector, and the sample are not in the same line. The interactance mode (Fig. **2d**) was first proposed by Conway *et al.* [6] in cases where transmittance and reflectance modes cannot be used directly. The interactance mode is similar to the diffuse reflectance mode. However, there is a light barrier between the detector and light source to block specular reflectance, and the light source is installed parallel to the detector [1].

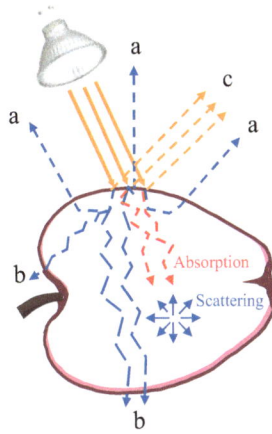

Fig. (1). Distribution of incident light in fruits: '**a**' denotes diffuse reflectance; '**b**' denotes transmittance; '**c**' denotes specular reflectance [1].

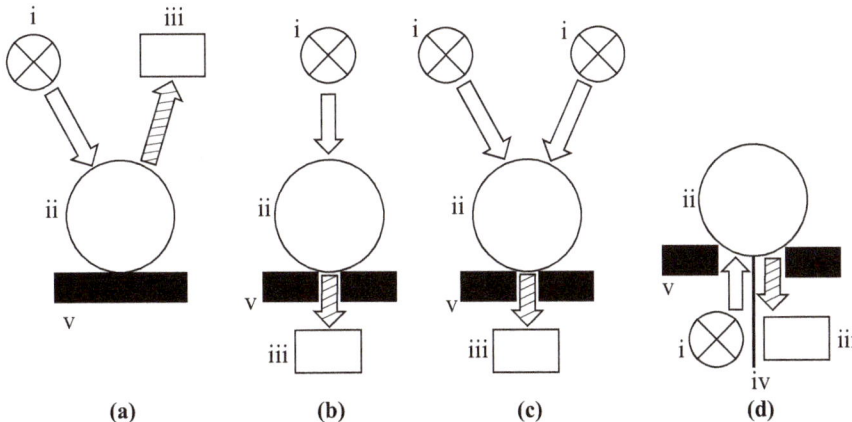

Fig. (2). Detection modes: **(a)** diffuse reflectance, **(b)** full transmittance, **(c)** partial transmittance, and **(d)** interactance, with (i) the light source, (ii) fruit, (iii) monochromator/detector, (iv) light barrier, and (v) support.

Optimal detection mode should be properly determined firstly according to actual detecting requirements, in order to obtain satisfying results. The three detection modes have been compared by researchers to analyze their advantages and disadvantages. Schaare and Fraser [7] compared the performances of the three modes to nondestructively detect the SSC, density, and internal flesh color of yellow-fleshed kiwifruit. They concluded that the interactance mode yielded the most accurate results, with SEP (Square Error of Prediction) being 0.80 °Brix, 3.6 kg•m^{-3} and 1.63° for SSC, density, and internal flesh hue angle, respectively. McGlone *et al.* (2003) reported slightly better performance of transmittance mode for intact mandarins' SSC (Soluble Solids Content) prediction compared to reflectance and interactance modes. However, Cayuela and Weiland [8] indicated that the transmittance mode was more influenced by the fruit size, since light would propagate through the whole fruit to be received by a detector hence making it difficult to acquire accurate transmission measurements on grading lines. Therefore, they used reflectance mode in their experiment to predict the internal quality of intact oranges. Fu *et al.* [9], obtained better results for detecting brown heart in pears by transmittance mode compared to reflectance mode. They concluded that the inferior results of the reflectance mode were due to the shallow penetration depth of light. Xing and Guyer [10] suggested that both transmittance and reflectance modes could be used for identifying internally insect-infested tart cherries, with transmittance, about 5% higher than reflectance in total classification accuracy. However, Wang *et al.* [11 - 13], found that the three detection modes performed almost equally for detecting internal insect infestation and predicting SSC in jujubes. For on-line sorting systems, Moghimi *et al.* [14], concluded that reflectance and transmittance modes were more applicable than interactance since stray light should be avoided, which was a difficult task for interactance mode. Wang *et al.* [15] compared the three detection modes for SSC determination of navel oranges. The results showed that the best performance was achieved by transmittance mode for both intact and peeled oranges.

These studies suggest that comparisons of detection modes do not lead to a unanimous conclusion regarding the superior performance of any, among the three detection modes [1, 16]. Different detecting objectives require specific system configurations and detection modes [17]. This should be attributed to the fact that the physical properties such as volume, peel thickness, tissue and stone size (if there is one) of different agricultural products differ greatly from each other [1]. For agricultural products with thick peels such as watermelons, transmittance mode is preferable due to its good ability to capture internal information about the sample [18, 19]. For fruits with large stones such as mangos and peaches, reflectance or interactance mode is generally a better choice since light cannot propagate through the stones [20, 21]. For citrus with relatively thick peels, both transmittance and reflectance modes have been widely used to determine the internal quality [22 - 24]. This may be due to the information included in citrus peel could, to some extent, represent the information included in its flesh [1]. This hypothesis can be supported by the research of Wang *et al.* [15], who found the spectra of orange peel having a strong correlation with the SSC of orange flesh. For portable NIR systems, the interactance and reflectance modes are widely applied because they are easy to be configured as a compact probe [25, 26]. For on-line NIR systems that require non-contact detection at high transmit speed, transmittance and reflectance modes are more applicable [1].

2.1.2. Applications

The first application of Vis/NIR spectroscopy in the agricultural domain was conducted by Norris to measure moisture in the grain [27]. Since then, the technique has been investigated for detecting the internal quality of a wide variety of agricultural products. Vis/NIR spectroscopy has been regarded as the most appropriate approach for detecting the internal quality of agricultural products and it has been put into practical applications in many fields. In terms of detecting systems and scenarios, these applications can be grouped into three categories: laboratory applications with bench-top systems, online applications with online detection systems, and in-field applications with portable systems [1]. By conducting laboratory experiments, researchers and biological engineers could master principles of the detecting method, determine instrumental setups, and optimize configurations as well as parameters of the detecting system, thus laying a solid foundation for practical applications. Vis/NIR spectroscopy is now being investigated to detect the internal quality of fruits, vegetables, meat, grain products, dairy products, fish products, and beverages. Representative applications are listed in Table 1. Compared with other agricultural products, fruits are more frequently studied using Vis/NIR spectroscopy. It can be observed that the technique has been used for detecting different internal attributes of many fruits. The fruits involved range from small ones like bayberry to large ones like watermelon, from fruits with very thin peel like apple to those with extremely thick peel like watermelon and grapefruit. And the attributes include SSC, firmness, starch, titratable acidity (TA), dry matter (DM), total sugar, pH, disease, and texture. Based on these researches and findings, commercial on-line fruit sorting systems have been developed and put into practical applications (Table 2). Regarding portable applications of Vis/NIR technique on fruits, the main purpose is to detect the maturity of the fruits, based on which the optimal harvest time can be determined in the field. The technique has also been adopted for detecting the internal quality of meat including adulteration, protein, moisture, fat, TVB-N, *etc.*, quality of milk products like melamine, quality of grain including moisture, protein, lipid, *etc.*, and the quality of beverages including glucose, fructose, sucrose, alcohol concentration, *etc.*

Table 1. Representative applications of Vis/NIR technique for detecting the internal quality of agricultural products.

Application Type	Product	Detection Mode	Wavelength (nm)	Attribute	Reference
Laboratory	Apple	Interactance	500-1100	SSC, firmness, starch, titratable acidity (TA)	[28]
	Citrus	Reflectance, Transmittance, Interactance	500-1100	SSC	[29]
	Banana, Mango	Interactance	500-1050	SSC, Dry matter (DM)	[20]
	Watermelon	Transmittance	700-1100	Total sugar, Maturity	[30]
	Kiwifruit	Interactance	500-1100	DM, SSC	[31]
	Pear	Reflectance	350-1800	SSC, pH, firmness	[32]
	Cherry	Reflectance, transmittance	350-1050	Insect infestation	[10]
	Pineapple	Reflectance	650-1000	SSC	[33]
	Grapefruit	Reflectance	200-2500	Citrus cancer	[34]
	Tomato	Interactance	1100-2500	Alcohol insoluble solids, SSC, textural parameters	[35]
	Bayberry	Transmittance	800-2632	TA, malic acid, citrus acid	[36]
	Chesnut	Reflectance	833-2500	SSC	[37]
	Olive	Reflectance	780-2500	Firmness, oil content	[38]
	Melon	Reflectance	550-950	SSC	[39]
	Beef	Reflectance	200-2600	Adulteration	[40]
	Beef	Reflectance	900-1700	Protein, intramuscular fat (IMF), moisture, myoglobin, Warner-Bratzler shear force (WBSF), water holding capacity (WHC)	[41]

(Table 1) cont.....

Application Type	Product	Detection Mode	Wavelength (nm)	Attribute	Reference
Laboratory	Pork	Reflectance	1000-2500	Total volatile basic nitrogen (TVB-N) content, WBSF	[42]
	Chicken breast	Reflectance	400-1000	Moisture, water activity, pH, TVB-N, ATP breakdown compounds, mesophilic bacteria	[43]
	Liquid milk, infant formula, milk powder	Transmittance	1110-2500	Melamine	[44]
	Poultry egg	Transmittance	300-1100	Freshness	[45]
	Sorghum grain	Reflectance	400-2500	Total phenols, condensed tannins, 3-deoxyanthocyanidins	[46]
	Flour	Reflectance	1100-2498	Gliadin, Glutenin content	[47]
	Soybean	Reflectance	1250-2500	Moisture, protein, lipid, ash content	[48]
	European sea bass	Reflectance	1100-2500	Protein, water, ether extract, energy	[49]
	Surimi	Reflectance	400-1100	Water, protein	[50]
	Bayberry juice	Transmittance	800-2400	Glucose, fructose, sucrose	[51]
	Beer	Transmittance	1100-1220	Alcohol concentration, real extract	[52]
	Apple wine	Transmittance	830-2500	Alcohol concentration, titratable acidity	[53]
On-line	Milk, powder milk, butter, cheese	Reflectance	1200-2400	Fat, protein, solids, salt, moisture	[54]
	Milk	Transmittance	600-1050	Fat, protein, lactose, somatic cell count, milk urea nitrogen	[55]
	Ground meat	Reflectance	1441, 1510, 1655, 1728, 1810	Fat, water, protein	[56]
	Pork	Reflectance	450-910	pH	[55]
	Grain	Reflectance	600-1100	Protein	[57]
	Maize	Not specified	960-1690	Dry matter, crude protein, starch	[58]
	Apple	Transmittance	650-950	Dry matter	[59]
	Watermelon	Transmittance	200-1100	SSC	[19]

Table 2. Some commercial on-line fruit sorting systems based on Vis/NIR [1].

Company	Country	Parameters Analyzed	Speed	Sample	Website
Aweta	Netherlands	Sugar content, maturity, firmness, DM, internal flaws	10 products / s	Apples, pears, kiwifruits, peaches, citrus fruits, mangos, avocados, *etc.*	http://www.aweta.nl
Compac	New Zealand	Taste (Brix)	12 products / s	Apples, avocados, citrus, pears, kiwifruits, peaches, *etc.*	http://www.compacsort.com
CVS	Australia	Sweetness, DM	10 products / s	Apples, peaches, nectarines, plums, apricots, mangoes, kiwifruits, mandarins, *etc.*	http://www.cvs.com.au
Greefa	Netherlands	Brix value, internal brownness, core rot	6 products / s	Apples, avocados, citrus fruits, kiwis, mangoes, pears, stone fruits	http://www.greefa.nl
Maf-Roda	France	Brix, DM percentage, oil percentage	Not specified	Apples, apricots, avocados, cherries, grapes, lemons, *etc.*	http://www.maf-raoda.com
Sacmi	Italy	Sugar content, ripeness, DM, starch content, acidity, brown core, water core	10 products / s	Medium-size fruits and large-size fruits	http://www.sacmi.com
Taste Tech	New Zealand	Brix value, internal defects, internal color	Not specified	Apples, citrus fruits, kiwis, mangoes, pears, avocados, tomatoes, melons, pomegranate, *etc.*	http://www.taste-technologies.com/
Unitec	Italy	Sugar content, firmness, ripeness, acidity, DM	12 products / s	Citrus, apples, peaches, *etc.*	http://www.unitec-group.com

2.1.3. Challenges and Perspectives

Vis/NIR spectroscopy has been successfully applied for practical applications in the agricultural domain. There are still several challenges that hinder the widespread use of this technique. Firstly, the process of establishing a quality detection model based on destructive measurements and Vis/NIR spectroscopy is time-consuming and tedious. When the property of the products changes, such as SSC of different cultivars of apples, the model needs to be updated or even retrained, which would require a lot of duplicating work. And the established model is difficult to be transferred to another system. For example, a SSC prediction model established using one spectrometer cannot be used directly on another spectrometer. Therefore, an effective model update and transfer strategies should be developed to make the model training process much easier. In addition, external parameters, such as the temperature of samples and detectors, stray light, and vibration, influence the robustness of the detection system and models. These parameters should be considered to improve the robustness of the systems and models.

2.2. Multi-/Hyper-spectral Imaging

2.2.1. Principle

Hyperspectral imaging is an emerging technique that integrates conventional imaging and spectroscopy to acquire both spatial and spectral information from an object [60]. It was originally developed by Goetz *et al.* [61], for earth remote sensing. Hyperspectral image, also called *hypercube*, is a three-dimensional block of data containing two spatial and one spectral dimension (Fig. **3**). In the *hypercube*, for each wavelength λ, there is a two-dimensional image, and for each pixel in one image, a spectrum can be extracted from all the pixels at the same position on all the two-dimensional images. Therefore, both spatial and spectral information of an object can be extracted from their *hypercubes*. To acquire the *hypercube* of an object, typically four approaches can be applied according to the data acquisition principle, namely, point scanning, line scanning, area scanning, and the single-shot method (Fig. **4**) [62]. In the point scanning method, a single point on the surface of an object is scanned to obtain the Vis/NIR spectrum of this point, forming one pixel of a *hypercube*. Then either the detector or the sample is moved along two spatial dimensions (x and y) to scan the rest points of the sample. The disadvantage of this approach is that it's very time-consuming to acquire the data point by point, and difficult to guarantee the repositioning accuracy. By contrast, the line scanning method, or pushbroom method, scans one line on the sample instead of a point, obtaining a line of pixels containing one dimensional spatial information y dimension in Fig. (**4b**) and corresponding spectral information. By moving either the sample or the sample in the x direction in Fig. (**4b**), the complete *hypercube* is obtained. The line scanning characteristic makes this method suitable to online detection for the quality of agricultural products with a conveyor belt system. But the transmission speed should be controlled to a low value since the line scanning takes time. The area scan method (Fig. **4c**) records a 2-D monochromatic image with full spatial information at a single wavelength, and by repeating this process at different wavelengths, a stack of single band images can be obtained forming the complete *hypercube* of the sample. To change the wavelength between each recording, a tunable filter is often used; Acouso-optic tunable filters (AOTF) and liquid crystal tunable filters (LCTF) are the two most frequently used ones [60]. These tunable filters provide instant transition from one wavelength to another and the spectral resolution can be high to sub-nanometers, enabling the acquisition of hyperspectral images. Another type of wavelength transition device is a much simpler one, with several filters installed on a filter wheel. After collecting the images at one wavelength, the filter wheel rotates to another filter. By repeating this process several images that correspond to each filter can be acquired and a multi-spectral image is formed. Compared with a hyperspectral image, a multi-spectral image usually comprises of several or tens of monochromatic images at different wavelengths, thus containing less spectral information [63]. The last data acquisition method is the single shot mode. This method records both spatial and spectral information simultaneously with only one exposure. To achieve this, a large area sensor is necessary in which pixels are divided into many identical square or rectangular areas; one area of pixels store the same spatial information of the sample but different spectral information at different wavelengths. Therefore, limited by area of the sensor at the current stage, this type of instrument cannot provide high spatial and spectral resolution. With respect to sensing mode, as similar to Vis/NIR spectroscopy technique, three modes (reflectance, transmittance, and interactance) are also three frequently modes, as shown in Figs. (**4e and f**). The advantages and disadvantages of the three detection modes are discussed in the section 'Vis/NIR spectroscopy'. For data processing, since multi-/hyper-spectral images contain both spectral and spatial information, the data can be analyzed in terms of both spectral and spatial aspects. Namely, both spectroscopy analysis and image processing methods, can be applied.

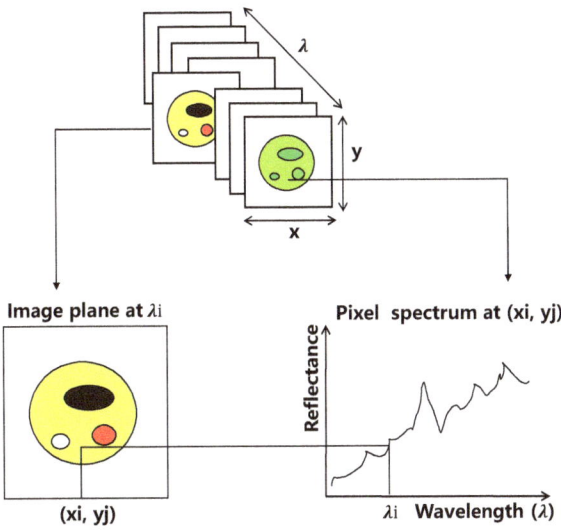

Fig. (3). Schematic representation of hyperspectral imaging *hypercube* showing the relationship between spectral and spatial dimensions [60].

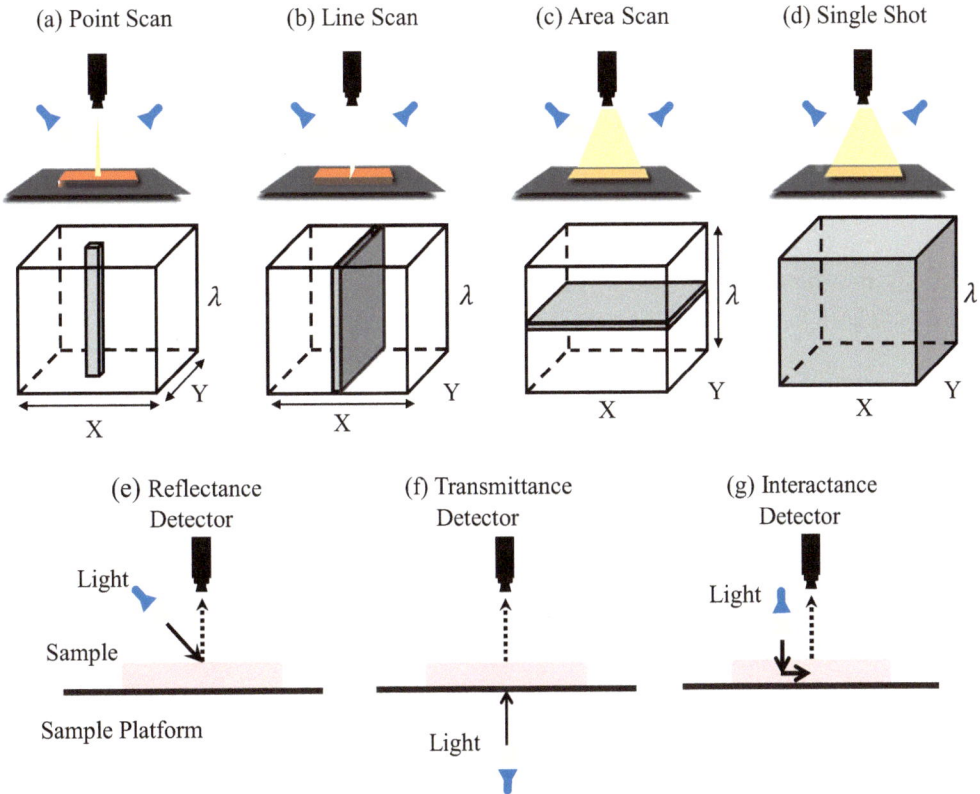

Fig. (4). Acquisition approaches of hyperspectral images (scanning directions are shown by arrows, and gray areas shows data acquired each time) and image sensing modes [62].

2.2.2. Application

Due to the merit of multi-/hyper-spectral imaging technique to acquire both the spatial and spectral information of samples, the emerging sensing technique has been explored intensively for analyzing both external and internal properties for a broad range of food and agricultural products. For internal attributes, the technique has been used for inspecting physical, chemical, and biological properties of fruits and vegetables, meat, grain, dairy products, fish and beverages, *etc.*, as shown in Table **3**. Compared with Vis/NIR spectroscopy, hyperspectral imaging is more frequently used for detecting damages and injuries caused by mechanical impact or disease infection. This is because this technique is able to acquire the spatial information of samples, while Vis/NIR spectroscopy can only collect information within a small range of a sample. The damages and injuries on a sample occur randomly, making it difficult for Vis/NIR spectroscopy to acquire useful information. With respect to image acquisition method, the line-scan reflectance mode is the most frequently utilized, followed by the area-scan mode using electronically tunable filters [64]. In contrast, the point-scan mode is seldomly used recently since it's very time-consuming and the re-position accuracy cannot be guaranteed. For single shot mode, limited by the low spatial and spectral resolution, the acquired data is multispectral images, and its application is mainly on detecting external quality of agricultural products. There are researches trying to use this quick technique to evaluate protein content of processed pork meats [65], obtaining a determination coefficient of cross-validation set (R^2_{CV}) of 0.8318.

Table 3. Representative applications of hyperspectral imaging technique for detecting the internal quality of agricultural products.

Product	Image Acquisition Method	Wavelength (nm)	Attribute	Reference
Apple	Line-scan reflectance	900-1700	Bruise	[66]
Apple	Line-scan reflectance	900-2500	Bruise	[67]
Kiwifruits	Line-scan reflectance	408-1117	Bruise	[68]
Blueberry	Area-scan reflectance	950-1650	Bruise	[69]
Onion	Area-scan reflectance	950-1650	Sour skin	[70]
Citrus	Area-scan reflectance	400-1100	Decay	[71]
Banana	Line-scan reflectance	400-1000	SSC, firmness	[72]
Strawberry	Line-scan reflectance	400-1000	Moisture content, SSC, pH	[73]
Peach	Line-scan reflectance	400-1000	Fungal diseases	[74]
Peach	Line-scan reflectance	400-1000	Cold injury	[75]

(Table 3) cont.....

Product	Image Acquisition Method	Wavelength (nm)	Attribute	Reference
Beef	Line-scan reflectance	400-1000	Tenderness	[76]
Fish fillet	Line-scan reflectance	460-1040	Fat, water content	[77]
Pork	Line-scan reflectance	430-1000	Quality, marbling level,	[78]
Pork	Single-shot reflectance	465-630, 16 bands	Protein	[65]
Wheat kernel	Area-scan reflectance	1000-1600	Fungus infection	[79]
Wheat kernel	Line-scan reflectance	400-1700	Fusarium damage	[80]
Tea	Line-scan reflectance	408-1117	Quality classification	[81]

Another range of applications of the hyperspectral imaging technique is to combine with spatially-resolved technique to measure the absorbing and scattering properties of agricultural and biological tissues. Qin and Lu [82] developed a spatially-resolved hyperspectral imaging system (Fig. **5**) to acquire the spatially-resolved reflectance profiles of samples. The system is comprised of a high-performance CCD camera, an imaging spectrograph, a zoom lens, a computer, a camera control board, and a quartz-tungsten-halogen light source with an optical fiber. This hyperspectral spatially-resolved scattering imaging technique has been used to detect firmness of apple [83], fat in milk [84], firmness and SSC of peach [85] and bruising of pickling cucumber [86], *etc.*

2.2.3. Challenges and Perspectives

Though this emerging technique has been studied intensively for detecting internal quality of different agricultural products, there are still several challenges that hinders it from practical applications, including the influence of physical and biological variability, whole surface detection, discrimination between defects and stems/calyxes, unobvious defect detection, robustness of features and algorithms, as well as the development of rapid optical detection systems [2]. Among these obstacles, the image acquisition speed is the most critical one, since a long data collecting time is not suitable for online or portable applications. The single shot instrumentation is a promising approach for putting the technique to practical applications, as long as the spatial and spectral resolution is improved in the future. Commercial multi-/hyper-spectral imaging systems like Hyperspec Inspector by Headwall Photonics (Fitchburg, MA, USA) and SisuCHE-MA by Specim (Oulu, Finland), are appearing on the market, which will broaden the practical applications of the technique.

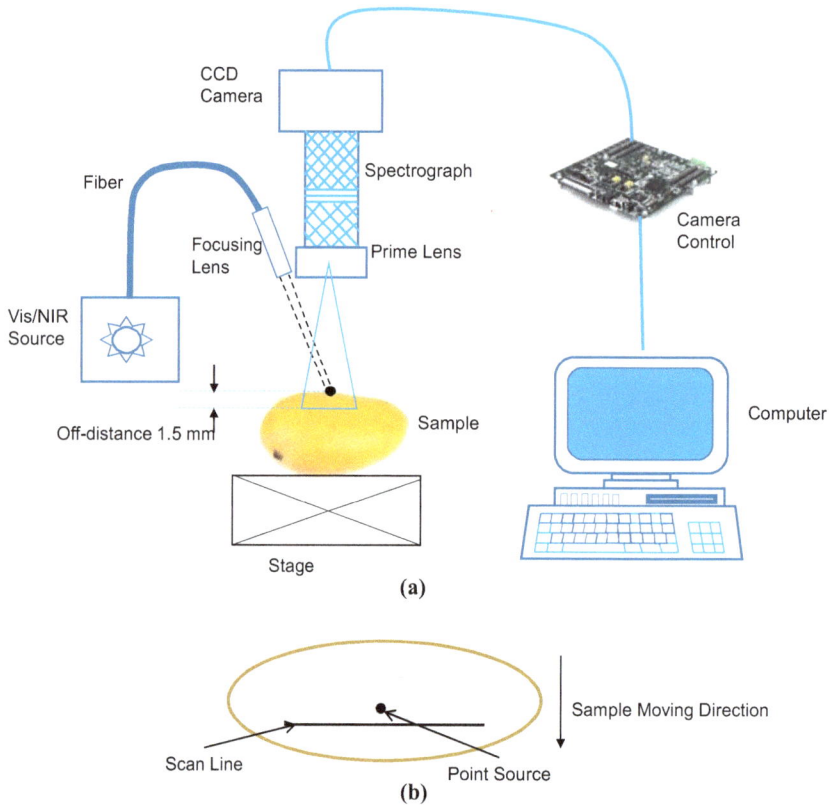

Fig. (5). Hyperspectral imaging system for acquiring spatially resolved scattering images from a fruit sample (a: system components; b: top view of line scanning mode for acquiring spatially-resolved reflectance image) [87].

2.3. Nuclear Magnetic Resonance and Imaging

2.3.1. Principle

Nuclear magnetic resonance (NMR) is a uniquely informative technique based on the interaction of the magnetic moment of atomic nuclei with external magnetic fields [88]. When nuclei in a strong constant magnetic field are placed in an external magnetic field, which is usually referred to as a radio-frequency (RF) pulse, the nuclei respond by producing an electromagnetic signal with a frequency at ω_0 which is related to the magnitude of the applied external magnetic field [89]. With the presence of the external magnetic field, nuclei will preferentially align their spin in the direction of the magnetic field, which is known as Larmor effect [90]. The frequency ω_0 at which nuclei process around the axis is termed as the Larmor frequency, defined by

$$\omega_0 = \gamma B_0$$

where, γ is a constant (the magnetogyric ratio) whose value depends on which nucleus is being investigated, and B_0 is the strength of the external magnetic field [90]. The radiation signals emitted from the nuclei in the form of a free induction decay is received by detectors and sent to computer by which they are converted into spectroscopy. NMR can be basically grouped into LF-NMR (low-field NRM) and HR-NMR (high-resolution NMR). HR-NMR needs superconducting magnets working at 200-750 MHz in which the instrumentations are expensive and complicated, making this technique less suitable for agricultural applications. In contrast, LF-NMR operates at a frequency of less than 60 MHz; the instrumentations is more cost-efficient and easier to access [91]. Thus the LF-NMR has been investigated extensively for detecting the quality of agricultural products.

Magnetic resonance imaging (MRI) was originally called nuclear magnetic resonance imaging (NMRI) because it is an extension of NMR spectroscopy. To avoid negative associations, 'nuclear' was dropped [92]. To generate images of a sample, in addition to the homogeneous external magnetic field with strength of B0, magnetic field gradients are also used to produce frequency changes and encode spatial coordinates in a sample [93]. Pulses of radio waves excite the nuclear spin energy transition, and magnetic field gradients localize the polarization in space. Magnetic gradients coils collect data spatially and create two-/three-dimensional images displaying different physicochemical characteristics and contrasts [93]. A MRI systems typically consists of (1) a large magnet to generate a homogeneous magnetic field, (2) smaller electromagnetic coils to generate magnetic field gradients, and (3) a radio transmitters and receiver with its associated transmitting and receiving antennae or coils [90], as shown in Fig. (6).

2.3.2. Applications

From the MRI images, different parameters can be extracted, such as proton density, and proton relaxation expressed by spin-lattice relation time (T1) and spin-spin relation time (T2). These parameters have proved to be able to quantify many chemical and structural characteristics of agricultural products [94]. NMR and MRI have been utilized to measure both compositional and physical properties of agricultural products, as well as some on-line applications for monitoring food processing.

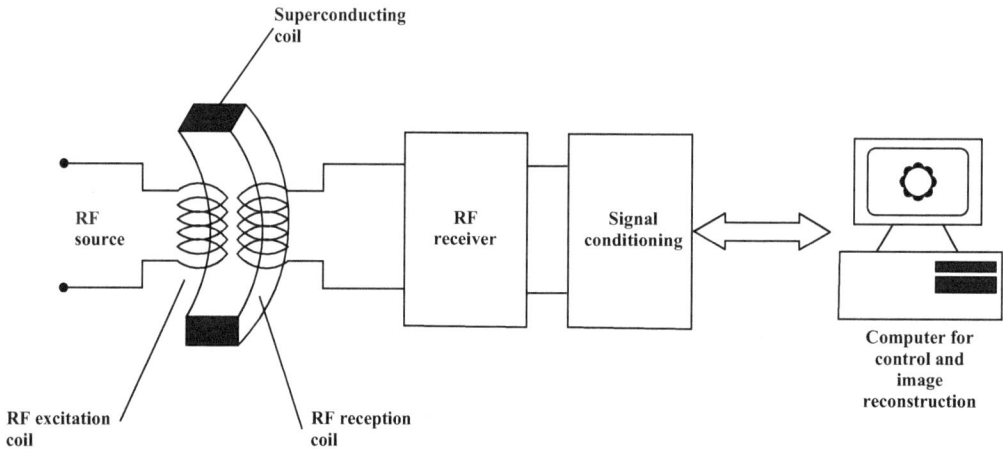

Fig. (6). Block diagram of a typical MRI system [90].

For most agricultural products especially fruits and vegetables, water is a major component that closely relates to the quality of the products. Hydrogen is a basic composition of the water atom, and the formation of NMR spectroscopy and MRI images is mainly based on the excitation of ^{1}H photons. Therefore, NMR and MRI techniques are very suitable for quantifying moisture and water content in agricultural products, as well as monitoring water mobility and distribution in these products [95]. As shown in Table **4**, one main application of the NMR/MRI technique is to monitor the water dynamics and distribution within agricultural products, especially for product processing like drying. The technique has also been used to measure water content for the drying process of different agricultural products online [91]. For example, Lv *et al.* [96] used a novel smart NMR/MRI detection system to monitor the residual moisture content, distribution, and state of water (free, immobilized, and bound) in fresh corn kernels during microwave vacuum drying. It was concluded that the NMR/MRI is a suitable tool for monitoring drying process in real time. The NMR/MRI technique has also been used for determining total fat contents, fatty acid composition of the fat, as well as protein content in agricultural products. For example, Zang *et al.* [97] used LF-NMR combined with PLSR algorithm to predict fat content in small yellow croaker, obtaining a RCV2 of 0.9054 for fat content prediction. Gudjonsdottir *et al.* [98] analyzed the relationship between NMR parameters and protein content in rehydratred cod. Results showed that the parameters T_1 and T_{21} were the most correlating factors for the changes and denaturation of protein.

Table 4. Representative applications of NMR/MRI technique for detecting the internal quality of agricultural products.

Application	Product	Attribute	Reference
Water-related detection	Soybean	Water distribution	[106]
	Strawberry	Water loss rate and decay	[107]
	Cherry tomato	Water molecular motion and distribution	[108]
	Kiwifruit	Water distribution	[109]
	Corn kernels	Moisture content, state of water	[96]
	Mushroom, potato, carrot and lotus	Moisture content	[110]
	Small yellow croaker	Water dynamics	[97]
Fat and protein detection	Small yellow croaker	Fat content	[97]
	Milk powder	Fat content	[111]
	Rehydated cod	Protein content	[98]
Physical property detection	Huanghua pear	Firmness	[99]
	Onion	Cell membrane integrity	[100]
	Kiwifruit	Texture change	[101]
Defect detection	Pear	Internal browning	[102]
	Apple	Watercore	[103]
	Avocado	Bruising	[104]
	Pear	Bruised volume	[105]

With respect to physical properties, firmness is one of the significant parameters representing the ripeness of horticultural products. NMR/MRI has been implemented to assess the firmness changes of fruits during maturation and storage as a function of relaxation times. Zhou and Li [99] analyzed the texture of magnetic resonance images to predict the firmness of Huanghua pears during storage using an artificial neural network, obtaining a correlation coefficient of 0.969 between co-occurrence matrix-derived texture parameters and firmness. NMR was also employed to study the effects of high pressure and thermal processing on membrane permeability and cell compartmentalization in onion tissue [100]. Results showed that the T_2 relaxation parameter successfully discriminated different degrees of membrane damage caused by high press and thermal processing. Taglienti *et al.* [101] evaluated the effect of storage conditions on minimal structural changes of kiwifruit tissue using MRI and concluded that the MRI was an effective tool for identifying minimal texture changes linked with water mobility.

Since the NMR/MIR technique is capable of obtaining internal images of a sample, it is an effective approach to detect internal defects of agricultural products. Hernandez-Sanchez *et al.* [102] found that at higher magnetic field strength and for long pulse spacing, disordered pear tissue with internal browning showed higher transverse relaxation rates compared to sound tissue, enabling NMR/MRI technique to detect internal browning of pears. Results showed that a minimum value of 12% of tissue affected by breakdown can be clearly identified. MRI was also applied for assessing watercore development in apples [103], bruising of avocado [104] and bruised volume of pear [105], *etc.*

2.3.3. Challenges and Perspectives

From the current research advance of using the NMR/MRI technique for detecting the internal quality of agricultural products, we can see that almost all applications are research oriented at the presented time, rather than practically oriented. This can be attributed to several disadvantages of the technique. The instruments for NMR/MRI are relatively expensive, and the data collection and processing speed is low, making this technique less preferable for practical applications. The operation and maintenance of the instruments need professional staff to conduct, and the user interface is not user-friendly. In addition, for current online application, the sensitivity of the system is far away from enough, considering acceptable prices [91]. Therefore, much effort should be focused on overcoming these shortcomings to make this technique suitable for practical applications in the agricultural domain.

2.4. X-ray and Computed Tomography

2.4.1. Principle

Like Vis/NIR, X-ray is a form of high-energy electromagnetic radiation with the wavelength ranging from 0.01 to 10 nm, corresponding to frequencies in the range of 30 to 30000 Petahertz and energies in the range 120 eV to 120 keV. X-rays have the property of both particles and waves when interacting with matter. Some basic properties of X-rays are that, they: travel in straight lines, are not deflected by electric field or magnetic fields, have high penetrating power, and can blacken a photographic film [112]. X-rays can be generated by an X-ray tube, a vacuum tube in which a high voltage is used to accelerate the electrons released by a hot cathode. When the high-velocity electrons collide with a metal target, the anode, X-rays are created [113]. When X-rays hit a target material and pass through it, the incident photons transfer energy to electrons and nuclei of the target, hence the total energy of the X-ray beam decreases at an exponential rate, called attenuation. After interacting with a matter, the X-ray beams are either transmitted, scattered or absorbed. As a result, the transmitted light is exponentially attenuated, which can be measured by the attenuation coefficient, μm. And the intensity of the transmitted light can be calculated as

$$I = I_0 e^{-\mu_m z \rho}$$

where, I is the intensity of photons exiting through a matter, μm is mass attenuation coefficient in mm^2/g, ρ is material density in g/mm^3, and z the thickness in mm of the sample. The exiting photon energy depends on the material properties of the sample, including thickness [112]. After being transmitted through a material, the exiting photons can be received by a detector and a 2D slice image can be obtained; this process is often called radiography. When multiple 2D slice images are taken from different directions of an object, these cross-sectional images can be combined into a three-dimensional image for the inside of an object. This process is called X-ray computed tomography (CT) [114]. For interpreting CT scans, a CT number which is a linear transformation of the original linear attenuation coefficient measurement, is widely used.

$$CT_{Number} = \frac{\mu - \mu_{Water}}{\mu_{Water}} 1000$$

where, μ is the linear attenuation coefficient of a sample, and μ_{Water} is the linear attenuation coefficient of water (approximately 0.195). In the definition of CT number, the radiodensity of distilled water at standard pressure and temperature is defined as zero Hounsfield units (HU) [115].

Like the procedures for processing color and grayscale images in machine vision, data analysis procedures for X-ray CT images include image enhancement, segmentation for extracting region of interests (ROIs) and decision making process. Different image enhancing methods including median filter, Gaussian filter and noise reduction technology have been utilized for reducing image noise and correcting defects [115]. Thresholding based on pixel intensity and intensity histogram have been proved to be effective approaches extracting ROIs. And machine learning algorithms have also been used to help making decisions.

2.4.2. Applications

Due to the short wavelengths, X-rays are capable to penetrate deep into a sample, and carry useful information associated with the internal quality of the sample by the transmitted signals. For applications in agricultural domain, CT is more widely investigated than radiography since the three dimensional CT images have high spatial resolution and can show more details inside the sample. X-ray CT has now been applied for detecting different attributes of agricultural products, like moisture, acidity, bruise, infection, fact and microstructure. (Table **5**). Huang *et al.* [116] applied CT to detect the internal quality of apple including moisture, acidity, and sugar content. Relationship between CT average numbers and internal quality parameters (moisture, acidity, and sugar) of apple were established, and the determination coefficients were 0.9075, 0.8233 and 0.8464, respectively. CT was also used to detect tissue bruising and physiological disorder of fruits. Diels *et al.* [117] quantified bruise volume in apples by analyzing X-ray CT images. The results showed that the shapes of bruised tissues in apples were irregular, imposing difficulties to estimate through simple geometric assumption, and the quantification of bruise volume by CT was sensitive to the threshold value selected for image segmentation. Jasarolmasjed *et al.* [118] found that it was possible to determine the bitter pit distribution from top to the bottom of an individual fruit and collectively quantify the disorder progression. In addition, classification of healthy and bitter pitted apples using CT image analysis and logistic regression resulted in false negative of 7-21%. CT was also applied for detecting the quality of grain and meat, such as kernel damage and hardness of grain [119, 120], and fat content and salt concentration of meat [121 - 123], *etc.* Another important application of CT is to observe microstructure of agricultural products. For example, Vicent *et al.* [124] applied X-ray micro-CT to visualize, characterize and quantify the 3D microstructure and the ice crystal size distribution of apple cortex tissue during freezing under different conditions.

Table 5. Representative applications of X-ray CT technique for detecting the internal quality of agricultural products.

Product	Attribute	Reference
Apple	Moisture, acidity, sugar	[116]
Apple	Bruise volume	[117]
Apple	Bitter pit infection	[118]
Pear	Bruise level	[125]
Wheat kernel	Kernel damage	[119]
Maize	Hardness	[120]
Beef	Intramuscular fat	[121]

(Table 5) cont.....

Product	Attribute	Reference
Lamb	Intramuscular fat	[122]
Park	Salt concentration	[123]
Frozen apple	Microstructure, ice crystal distribution	[124]

2.4.3. Challenges and Perspectives

X-ray CT technique has been studied a lot for application in the agricultural domain. However, all the research works are now in the laboratory. Current practical applications of the technique are mainly in the medical domain for diagnostic and therapeutic purposes in various medical disciplines. And many researchers using this technique for agricultural applications are still using medical CT instruments, which are expensive and not suitable for online detection. The data collection time also makes the current instruments not practically used for online detection. In addition, there are no standard image processing protocols for analyzing CT images. Another concern about using the technique is negative hazards. A properly designed shield can prevent human exposure. Also, steady improvements in instrumentation and data processing software should be achieved in the future. Therefore, there is still a long way before this technique can be used practically for agricultural applications.

2.5. Electrical Nose Technique

An electronic nose (E-nose), also known as an artificial olfactory system, is an electronic system that mimics a biological nose. The E-nose system consists of a sample processing module, multiple chemical sensor arrays, and a pattern recognition module. E-nose is usually used for the detection of volatile organic compounds (VOCs), and it eventually became one of the most commonly used detection methods in the food industry [126]. Different olfactory factors were detected by the human nose before 1990s. Although such sensory evaluation is convenient, the results are not accurate with poor repeatability and objectivity. Moreover, the human nose is more susceptible to uncontrollable factors such as the physical and mental health of humans [127]. By contrast, E-nose is an objective analysis method and has advantages of high efficiency, easy implementation, and low cost.

2.5.1. Principle

The technical core of the E-nose system consists of a gas sensor array, a signal preprocessing unit and a pattern recognition system, as shown in Fig. (7). The sensors generate different response signals after contacting or reacting with different gases. The response spectrum is obtained by processing these signals. After signal preprocessing and pattern recognition analysis, the gas can be detected qualitatively or quantitatively.

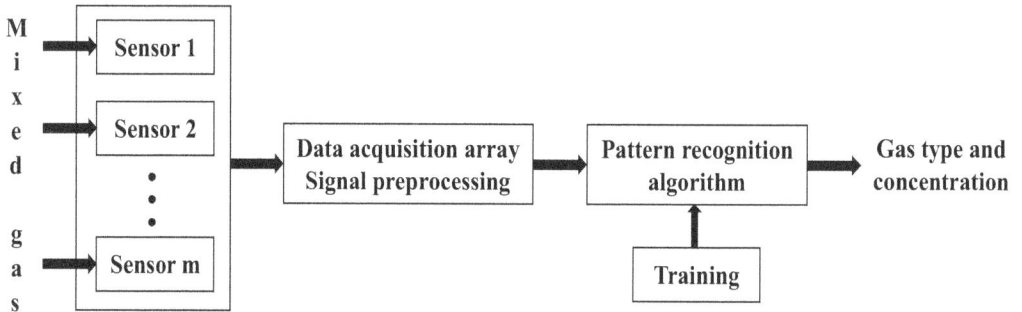

Fig. (7). The system workflow of electronic nose (E-nose).

The performance of E-nose depends on the choice of the gas sensor array. The sensitive material is the key part of the sensor. Different types of gas sensors have different sensitive materials. At present, the widely used gas sensors can be divided into several categories, like metal oxide sensor, organic polymer film gas sensor, field-effect tube gas sensor, mass sensitive gas sensor and optical gas sensor. Table **6** compares the five types of gas sensors.

Table 6. Comparison of five types of gas sensors.

Mechanism of Action	Measuring Method	Gas Sensor Type	Preparation Methods	Advantages	Disadvantages
Metal oxide sensor	Electrical conductivity	Metal oxide	Micro machining, spraying	low price, quick response, strong stability	Work at high temperature
Organic polymer film gas sensor	Conductivity, capacitance	Organic polymer	Micromachining, electroplating, screen printing, rotary coating	Work at room temperature	Sensitive to humidity
Field-effect tube gas sensor	Voltage, current	Field-effect tube gas sensor	Micromachining technique	High integration	The gas response needs to meet a threshold

(Table 6) cont.....

Mechanism of Action	Measuring Method	Gas Sensor Type	Preparation Methods	Advantages	Disadvantages
Mass sensitive gas sensor	Frequency	Quartz crystal microbalance, acoustic surface	Micromachining, electroplating, screen printing, rotary coating	High sensitivity, work at room temperature	High technological quality requirements
Optical gas sensor	Wavelength, light absorption	Fiber optic gas sensor	Micromachining, electroplating, screen printing, rotary coating	Strong noise shielding ability and adaptability	Relatively high price, narrow scope of application

2.5.2. Signal Processing Methods

Signal preprocessing and pattern recognition are crucial for E-nose technique, which is used to analyze and judge signals, similar to the human brain. The sensitivity of a gas sensor is closely related to both the quality of sensor and the method of signal processing. Filtering, exchanging and feature extraction are the main tasks of signal pretreatment. Feature extraction is the key step. Difference, normalization and relative methods are usually used for feature extraction and calibration of sensor arrays. After signal pretreatment, pattern recognition is conducted to reprocess the acquired information. The composition and concentration of mixed gas can be obtained from pattern recognition. The pattern recognition process can be divided into two stages. The first stage is the learning stage, in which the E-nose is trained to identify the characteristics of the measured gas, and the second, is the application stage, in which the measured gas is identified by pattern recognition. Pattern recognition methods have different effects with different signal preprocessing methods. The widely used pattern recognition methods included linear classification, principal component analysis (PCA), discriminant analysis (DA), template matching, and artificial neural network (ANN).

2.5.3. Applications

Food safety has received more attention in recent years. The E-nose technique is becoming one of the popular nondestructive methods for evaluating the quality of agricultural products. Compared with traditional methods, the E-nose technique has advantages in operation, detection time, portability and automation. At present, the E-nose technique has been applied to seafood, meat, fruit, vegetables, wine, milk, and tea.

2.5.3.1. Seafood and Meat

The E-nose is widely applied in the evaluation of the quality of seafood and meat, especially for evaluating freshness. Hosseini *et al.* [128], developed an E-nose method combined with the support vector machines (SVMs) to evaluate fish freshness. The method can successfully determine the freshness of fish with an accuracy of about 91%. Zhang *et al.* [129], used an E-nose with six TGS gas sensors for spoiling and formaldehyde-containing detection of seafood. Results showed that the E-nose was an efficient method for seafood quality assessment.

E-nose was also commonly used for the quality evaluation of meat. El Barbri *et al.* [130], used the E -nose and bacteriological measurements to analyze the freshness of beef and sheep meat. They proved that the E-nose method could be used for quality evaluation of red meats. Musatov *et al.* [131], detected the meat freshness by using a KAMINA E-nose combined with the linear discriminant analysis (LDA).

2.5.3.2. Fruit

The application of E-nose in fruit quality detection mainly focuses on the fruit freshness, maturity and decay degree, as well as the identification of fruit varieties, origin, damage and disease. Ren *et al.* [132], used an E-nose system to detect the impact injury of an apple falling from different heights and proved that the E-nose technique combined with ANN and multivariate analysis algorithm was an effective approach for detecting apple impact injury. Jia *et al.* [133] used the PEN3 E-nose to detect and recognize fresh and moldy apples, and demonstrated that the PEN3 E -nose could not only effectively recognize fresh and moldy apples, but also distinguish apples inoculated with different molds. Berna *et al.* [134] used a mass spectrometry-based electronic nose (MSE-nose) to investigate the shelf life and cultivar effect on tomato aroma profile.

2.5.3.3. Wine

The different contents of alcohols, lipids, acids, phenols and other organic compounds in wine lead to different aroma types. The complex composition of trace organic compounds poses several challenges to wine brand identification, flavor identification and wine age detection. Jing *et al.* [135] designed an E-nose to classify Chinese liquors of the same aroma style. A multi-linear classifier was used and the classification rate based on geographical origins and raw materials was 100% and 98.5%, respectively. Qin *et al.* [136] used the colorimetric artificial nose to characterize and identify Chinese liquors from six different geographic origins, and demonstrated that colorimetric artificial nose was able to classify Chinese liquors from different geographic origins. Yu *et al.* [137] evaluated the wine age by the gas chromatography-based electronic nose combined with chemometrics. The results showed that E-nose could classify samples with accuracy of about 96.88%.

2.5.4. Challenges and Perspectives

E-nose has the advantages of simple structure, fast analysis speed, high accuracy and objective analysis results, together with low cost. E-nose plays an increasingly important role in the field of agricultural product inspection. However, there are still some technical problems needed to be solved in the current E-nose researches. First, the quantitative analysis of E-nose is relatively poor compared with the traditional detection techniques such as electrochemical method, optical method and chromatography. Secondly, E-nose can only identify one or several specific gases, and cannot distinguish most of the existing smells as human olfactory organs do. And lastly, E-nose is not stable enough and requires a high working environment. Temperature, humidity and other factors can affect the performance of E-nose. Therefore, the development of a high-precision gas sensor, reasonable construction of multi-sensor array and selection of appropriate pattern recognition method are the ways to solve these problems.

2.6. Acoustic Technique

The acoustic vibration method is an effective nondestructive testing method, which is widely used for quality detection of agricultural products. In this method, the vibration characteristics are measured to evaluate the quality of agricultural products. The acoustic vibration characteristics of agricultural products correlate with their mechanical and structural properties, including the modulus of elasticity, Poisson's ratio, density, mass and shape [138]. In microcosm, the mechanical and structural properties of agricultural products are related to culture properties, including the size of cell, the thickness and strength of cell walls and cell turgor. Agricultural products keep ripening and softening during the growing and storage periods. Therefore, the acoustic vibration method can be used to evaluate the quality related to the mechanical and structural properties of agricultural products.

2.6.1. Principle

The acoustic vibration method is generally used to measure the acoustic vibration characteristics of agricultural products. Due to the difference between the mechanical structure and internal structure of agricultural products, the acoustic vibration characteristics are different, including resonance frequency, propagation velocity, acoustic impedance and attenuation coefficient.

As shown in Fig. (**8**), a sample is assumed to be a simple elastic body, which is comprised of two masses connected with a spring [139]. The natural frequency (*f*) of the system can be calculated by the equation

$$f = \frac{1}{2p}\sqrt{\frac{4k}{m}}$$

where, *m* is the sample mass, *k* is the spring constant of the system. The equation can be converted into:

$$k = \pi^2 f^2 m$$

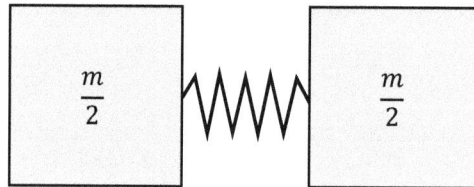

Fig. (8). Mass–spring model for an elastic body of mass *m* [139].

Spring constant *k* is related to the elastic properties of the system (the tested sample). Therefore, f^2m can be used as an index for elasticity or texture evaluation. Cooke *et al.* [140, 141] proposed a modified stiffness coefficient, $f^2m^{2/3}$, for evaluating fruit texture. They found that the improved stiffness coefficient had a better correlation with fruit texture than the parameter f^2m. Both theoretical analysis and experimental results proved that the elasticity properties of agricultural products could be reflected by the acoustic vibration characteristics.

2.6.2. General Process of the Acoustic Vibration Method

As shown in Fig. (**9**), the process of the acoustic vibration method is generally based on three steps, excitation, signal acquisition, and signal processing. A sample is excited by an excitation signal generated by the excitation part. The response signal of the sample to the excitation is recorded by the signal acquisition module, after which the signals are analyzed and used for further analysis in the signal processing module.

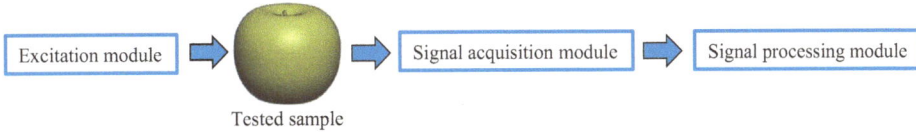

Fig. (9). The general process of acoustic vibration method for quality evaluation of agricultural products [138].

For signal excitation, the impact method and forced method are two main excitation methods. A comparison of the two excitation methods is shown in Table **7**. Generally, for the impact method, an instantaneous forward excitation is applied to the tested sample, while for the forced method, a varying force is used. Therefore, the major advantage of the impact method is its short excitation duration. The forced vibration method has the advantage of good repeatability, but it usually takes a long time for data collection and cannot meet the requirements of online detection.

Table 7. Excitation methods and commonly used excitation devices in the acoustic vibration method [138].

Excitation Methods	Advantages	Disadvantages	Excitation Devices
Impact method	● Short excitation duration ● Device is simple ● Wide frequency range	● Some methods are unrepeatable ● Possible to damage samples	Hammer, stick
			Pendulum
Swept sine wave method	● Repeatable measurement ● Energy can be concentrated in a narrow frequency band	● Device is costly ● Long excitation duration	Vibrator
			Speaker
			Piezoelectric vibration generator

With respect to sensors, there are two types of sensors that can be used to collect the response signal, contact and non-contact vibration sensors (Table **8**). Contact sensors mainly include acceleration sensor pickup and piezoelectric sensor, which directly contact the surface of samples. The contact sensor will increase the weight of the sample, affect its original vibration, and even contaminate the samples. By contrast, non-contact sensors, such as microphone and laser Doppler vibrator (LDV) have become popular instruments for vibration measurement.

Table 8. Vibration measurement methods and commonly used detection sensors in the acoustic vibration method [138].

Vibration Measurement Methods	Sensors	Advantages	Disadvantages
Contact	Accelerometer	• Stable and commonly used	• Prevent free vibration of samples • Possible to damage the surface of samples
	Piezoelectric sensor		
Non-contact	Microphone	• Low cost • Non-contact	• Affected by environmental noise
	Laser Doppler vibrometer	• Not affected by environmental noise • Non-contact	• Device is costly • Need a reflective surface

2.6.3. Applications

2.6.3.1. Texture evaluation

Texture is a key factor for evaluating the quality of ago-products, such as firmness, rigidity, adhesion, elasticity, chewability and powdery. Accurate measurement of texture is of great significance to determine the best harvest and edible time of agricultural products. Therefore, texture evaluation is the main research field of acoustic vibration method for evaluating the quality of agricultural products. Good correlations have been found for many fruits including apples [142], pears [143], kiwifruits [144] and pitayas [145]. Researchers also find that the correlation coefficients between stiffness coefficient and texture indices are affected by the variety, storage condition, size and shape of samples, as well as the excitation method [146].

2.6.3.2. Ripeness Evaluation

Agricultural products keep softening during the ripening and storage stage. Accurate evaluation of ripeness is important for consumers to determine the optimal time for eating. Therefore, it is necessary to monitor the maturity and to classify agricultural products to provide a reference basis for consumers and sellers. To determine preservation methods and consumption time [147]. Taniwaki *et al.* [148, 149] evaluated the ripeness of melons and persimmons by using a laser Doppler vibrometer. The elasticity index (EI) with the change of time was used to judge the ripening speed of fruit andΔEI/day was used to quantify the ripening speed. Results showed that the elasticity index could be used to estimate the optimum ripeness and shelf life of agricultural products. Fumuro *et al.* [145], investigated the variation of EI of pitayas harvested at different months, and determined optimal harvest ripeness for pollinated pitayas in July and September. Hongwiangjan *et al.* [150], applied an acoustic vibration method to evaluate the ripeness of pomelos harvested at different times. The discriminant accuracy was 96.7% when the pomelos were divided into immature, early-mature and late-mature groups. Vursavus *et al.* [151] used three nondestructive acoustic sensors to measure the acoustic vibration characteristics of peaches. The peaches were divided into soft, intermediate and hard groups. The results showed that the best classification accuracy was 87% when the three sensors were fused. Baltazar *et al.* [152], investigated the time-course changes in the stiffness coefficient of tomatoes. The bayesian classification was used to classify tomatoes with different ripeness, and the accuracy was 89%.

2.6.3.3. *Defect detection*

2.6.3.3.1. *Eggshell Cracks Detection*

Eggshell cracking is a major problem for the egg industry. Eggs are vulnerable to bacterial infections such as salmonella during storage, and broken eggs are more susceptible to harmful microbial contamination. Therefore, accurate identification and removal of broken eggs is of great significance.

The acoustic vibration method is an effective non-destructive method for eggshell cracks detection. In present researches, eggs are excited by the impact method, and the response signals of the eggs are collected by a microphone or piezoelectric sensor. Jin *et al.* [153, 154] developed a novel device for eggshell crack detection. The egg rolled down from a corrugated plate, and the response signal was collected by an electret microphone attached to the back of the plate. Pan *et al.* [155] proposed an acoustic vibration method combined with computer vision for the detection of eggshell cracks. All these studies proved that the eggshell cracks detection by the acoustic vibration method is feasible.

2.6.3.3.2. *Hollow Heart Detection*

Hollow heart is a serious quality problem that often happens in fruits. For example, the hollow heart of watermelon will seriously affect the taste and shelf life of watermelon. The acoustic vibration characteristics of the hollow watermelon are different from those of intact watermelon Various devices were designed and developed for the nondestructive measurement of internal hollow or creases based on the acoustic vibration method [156 - 158]. Besides, the hollow heart of potatoes could also be detected by the acoustic vibration method [159].

2.6.4. Challenges and Perspectives

The acoustic vibration method is a commonly used non-destructive method for quality evaluation of agricultural products. In order to achieve the commercial application of the acoustic vibration method, portable instruments and online detecting systems should be developed in the future. However, some challenges still exist. For the excitation and signal acquisition modules, contact measurement may prevent the free vibration and even damage the surface of samples, and it is not suitable for online detection. Non-contact measurement is becoming popular, but the detection accuracy and speed need to be improved before it can be put into practical applications. Another problem lies in the excessive use of texture indices including hardness, stiffness, fragility, chewy, and juiciness. This is not a standard index for representing the quality of tested samples, hence makes it difficult to explain and compare the detection results.

3. SUMMARY

This chapter provides an overview of the emerging techniques for detecting the internal quality of agricultural products, including Vis/NIR spectroscopy, multi-/hyper-spectral imaging, nuclear magnetic resonance and imaging, X-ray and computed tomography, electrical nose and acoustic technique. The working principle of each technique is provided, and corresponding applications in the agricultural domain are reviewed to provide an overall understanding of these techniques. Among these techniques, Vis/NIR spectroscopy has been successfully implemented for practical applications in agriculture, while most of the other techniques are still limited within laboratory researches. The price and operation of instrumentations, as well as data processing algorithms, are still the main challenges for these techniques to be widely adopted in the agricultural domain. In addition, the working principle of some techniques, for example, X-ray CT, does not allow the technique to be used for quick online detection scenarios. Therefore, understanding the advantages and disadvantages of these techniques will help to select the most appropriate technique for specific applications in agriculture.

CONSENT FOR PUBLICATION

Not applicable

CONFLICT OF INTEREST

The authors confirm that this chapter contents have no conflict of interest.

ACKNOWLEDGEMENTS

The authors gratefully acknowledge the financial support provided by Key Laboratory of Modern Agricultural Equipment and Technology (Jiangsu University) of Ministry of Education for the Open fund (No. JNZ201918 and No. JNZ201919).

REFERENCES

[1]   L. Xie, A. Wang, H. Xu, X. Fu, and Y. Ying, "Applications of near-infrared systems for quality evaluation of fruits: A review", *Trans. ASABE,* vol. 59, no. 2, pp. 399-419, 2016.
[http://dx.doi.org/10.13031/trans.59.10655]

[2]   B. Zhang, B. Gu, G. Tian, J. Zhou, J. Huang, and Y. Xiong, "Challenges and solutions of optical-based nondestructive quality inspection for robotic fruit and vegetable grading systems: A technical review", *Trends Food Sci. Technol.,* vol. 81, no. July, pp. 213-231, 2018.
[http://dx.doi.org/10.1016/j.tifs.2018.09.018]

[3]   H. Huang, H. Yu, H. Xu, and Y. Ying, "Near infrared spectroscopy for on/in-line monitoring of quality in foods and beverages: A review", *J. Food Eng.,* vol. 87, no. 3, pp. 303-313, 2008.
[http://dx.doi.org/10.1016/j.jfoodeng.2007.12.022]

[4]   R. Lu, *Light scattering technology for food property, quality and safety assessment.* Crc Press: Boca Raton, 2016.

[5]   B.M. Nicolaï, "Nondestructive measurement of fruit and vegetable quality by means of NIR spectroscopy: A review", *Postharvest Biol. Technol.,* vol. 46, no. 2, pp. 99-118, 2007.
[http://dx.doi.org/10.1016/j.postharvbio.2007.06.024]

[6]   J.M. Conway, K.H. Norris, and C.E. Bodwell, "A new approach for the estimation of body composition: infrared interactance", *Am. J. Clin. Nutr.,* vol. 40, no. 6, pp. 1123-1130, 1984.
[http://dx.doi.org/10.1093/ajcn/40.6.1123] [PMID: 6507337]

[7]   P.N. Schaare, and D.G. Fraser, "Comparison of reflectance, interactance and transmission modes of visible-near infrared spectroscopy for measuring internal properties of kiwifruit (*Actinidia chinensis*)", *Postharvest Biol. Technol.,* vol. 20, no. 2, pp. 175-184, 2000.
[http://dx.doi.org/10.1016/S0925-5214(00)00130-7]

[8]   J.A. Cayuela, and C. Weiland, "Intact orange quality prediction with two portable NIR spectrometers", *Postharvest Biol. Technol.,* vol. 58, no. 2, pp. 113-120, 2010.
[http://dx.doi.org/10.1016/j.postharvbio.2010.06.001]

[9]   X. Fu, Y. Ying, H. Lu, and H. Xu, "Comparison of diffuse reflectance and transmission mode of visible-near infrared spectroscopy for detecting brown heart of pear", *J. Food Eng.,* vol. 83, no. 3, pp. 317-323, 2007.
[http://dx.doi.org/10.1016/j.jfoodeng.2007.02.041]

[10] J. Xing, and D. Guyer, "Comparison of transmittance and reflectance to detect insect infestation in Montmorency tart cherry", *Comput. Electron. Agric.,* vol. 64, no. 2, pp. 194-201, 2008.
[http://dx.doi.org/10.1016/j.compag.2008.04.012]

[11] J. Wang, K. Nakano, and S. Ohashi, "Nondestructive detection of internal insect infestation in jujubes using visible and near-infrared spectroscopy", *Postharvest Biol. Technol.,* vol. 59, no. 3, pp. 272-279, 2011.
[http://dx.doi.org/10.1016/j.postharvbio.2010.09.017]

[12] J. Wang, K. Nakano, and S. Ohashi, "Nondestructive evaluation of jujube quality by visible and near-infrared spectroscopy", *Lebensm. Wiss. Technol.,* vol. 44, no. 4, pp. 1119-1125, 2011.
[http://dx.doi.org/10.1016/j.lwt.2010.11.012]

[13] J. Wang, K. Nakano, S. Ohashi, K. Takizawa, and J.G. He, "Comparison of different modes of visible and near-infrared spectroscopy for detecting internal insect infestation in jujubes", *J. Food Eng.*, vol. 101, no. 1, pp. 78-84, 2010.
[http://dx.doi.org/10.1016/j.jfoodeng.2010.06.011]

[14] A. Moghimi, M.H. Aghkhani, A. Sazgarnia, and M. Sarmad, "Vis/NIR spectroscopy and chemometrics for the prediction of soluble solids content and acidity (pH) of kiwifruit", *Biosyst. Eng.*, vol. 106, no. 3, pp. 295-302, 2010.
[http://dx.doi.org/10.1016/j.biosystemseng.2010.04.002]

[15] A. Wang, D. Hu, and L. Xie, "Comparison of detection modes in terms of the necessity of visible region (VIS) and influence of the peel on soluble solids content (SSC) determination of navel orange using VIS–SWNIR spectroscopy", *J. Food Eng.*, vol. 126, pp. 126-132, 2014.
[http://dx.doi.org/10.1016/j.jfoodeng.2013.11.011]

[16] L.E. Agelet, and C.R. Hurburgh Jr, "A tutorial on near infrared spectroscopy and its calibration", *Crit. Rev. Anal. Chem.*, vol. 40, no. 4, pp. 246-260, 2010.
[http://dx.doi.org/10.1080/10408347.2010.515468]

[17] B.A. Izneid, M.I. Fadhel, T. Al-Kharazi, M. Ali, and S. Miloud, "Design and develop a nondestructive infrared spectroscopy instrument for assessment of mango (*Mangifera indica*) quality", *J. Food Sci. Technol.*, vol. 51, no. 11, pp. 3244-3252, 2014.
[http://dx.doi.org/10.1007/s13197-012-0880-z] [PMID: 26396317]

[18] D. Jie, L. Xie, X. Fu, X. Rao, and Y. Ying, "Variable selection for partial least squares analysis of soluble solids content in watermelon using near-infrared diffuse transmission technique", *J. Food Eng.*, vol. 118, no. 4, pp. 387-392, 2013.
[http://dx.doi.org/10.1016/j.jfoodeng.2013.04.027]

[19] D. Jie, L. Xie, X. Rao, and Y. Ying, "Using visible and near infrared diffuse transmittance technique to predict soluble solids content of watermelon in an on-line detection system", *Postharvest Biol. Technol.*, vol. 90, pp. 1-6, 2014.
[http://dx.doi.org/10.1016/j.postharvbio.2013.11.009]

[20] P.P. Subedi, and K.B. Walsh, "Assessment of sugar and starch in intact banana and mango fruit by SWNIR spectroscopy", *Postharvest Biol. Technol.*, vol. 62, no. 3, pp. 238-245, 2011.
[http://dx.doi.org/10.1016/j.postharvbio.2011.06.014]

[21] P.P. Subedi, K.B. Walsh, and G Owens, "Prediction of mango eating quality at harvest using short-wave near infrared spectrometry", *Postharvest Biol. Technol.*, vol. 43, no. 3, pp. 326-334, 2007.
[http://dx.doi.org/10.1016/j.postharvbio.2006.09.012]

[22] J.A. Cayuela, "Vis/NIR soluble solids prediction in intact oranges (*Citrus sinensis* L.) cv. Valencia Late by reflectance", *Postharvest Biol. Technol.*, vol. 47, no. 1, pp. 75-80, 2008.
[http://dx.doi.org/10.1016/j.postharvbio.2007.06.005]

[23] K.S. Chia, H.A. Rahim, and R.A. Rahim, "Evaluation of common pre-processing approaches for visible (VIS) and shortwave near infrared (SWNIR) spectroscopy in soluble solids content (SSC) assessment", *Biosyst. Eng.*, vol. 115, no. 1, pp. 82-88, 2013.
[http://dx.doi.org/10.1016/j.biosystemseng.2013.02.008]

[24] B. Jamshidi, S. Minaei, E. Mohajerani, and H. Ghassemian, "Reflectance Vis/NIR spectroscopy for nondestructive taste characterization of Valencia oranges", *Comput. Electron. Agric.*, vol. 85, pp. 64-69, 2012.
[http://dx.doi.org/10.1016/j.compag.2012.03.008]

[25] F. Antonucci, F. Pallottino, G. Paglia, A. Palma, S. D'Aquino, and P. Menesatti, "Non-destructive estimation of mandarin maturity status through portable VIS-NIR spectrophotometer", *Food Bioprocess Technol.*, vol. 4, no. 5, pp. 809-813, 2011.
[http://dx.doi.org/10.1007/s11947-010-0414-5]

[26] C. Camps, and D. Christen, "Non-destructive assessment of apricot fruit quality by portable visible-near infrared spectroscopy", *Lebensm. Wiss. Technol.*, vol. 42, no. 6, pp. 1125-1131, 2009.
[http://dx.doi.org/10.1016/j.lwt.2009.01.015]

[27] K.H. Norris, "Design and development of a new moisture meter", *Agric. Eng.*, vol. 45, no. 7, pp. 370-372, 1964.

[28] V.A. McGlone, R.B. Jordan, and P.J. Martinsen, "Vis/NIR estimation at harvest of pre-and post-storage quality indices for 'Royal Gala'apple", *Postharvest Biol. Technol.*, vol. 25, no. 2, pp. 135-144, 2002.
[http://dx.doi.org/10.1016/S0925-5214(01)00180-6]

[29] A. Wang, and L. Xie, "Technology using near infrared spectroscopic and multivariate analysis to determine the soluble solids content of citrus fruit", *J. Food Eng.*, vol. 143, pp. 17-24, 2014.
[http://dx.doi.org/10.1016/j.jfoodeng.2014.06.023]

[30] A.T. Abebe, "Total sugar and maturity evaluation of intact watermelon using near infrared spectroscopy", *J. Near Infrared Spectrosc.*, vol. 14, no. 1, pp. 67-70, 2006.
[http://dx.doi.org/10.1255/jnirs.588]

[31] V.A. McGlone, C.J. Clark, and R.B. Jordan, "Comparing density and VNIR methods for predicting quality parameters of yellow-fleshed kiwifruit (*Actinidia chinensis*)", *Postharvest Biol. Technol.*, vol. 46, no. 1, pp. 1-9, 2007.
[http://dx.doi.org/10.1016/j.postharvbio.2007.04.003]

[32] J. Li, W. Huang, C. Zhao, and B. Zhang, "A comparative study for the quantitative determination of soluble solids content, pH and firmness of pears by Vis/NIR spectroscopy", *J. Food Eng.*, vol. 116, no. 2, pp. 324-332, 2013.
[http://dx.doi.org/10.1016/j.jfoodeng.2012.11.007]

[33] K.S. Chia, H.A. Rahim, and R.A. Rahim, "Prediction of soluble solids content of pineapple via non-invasive low cost visible and shortwave near infrared spectroscopy and artificial neural network", *Biosyst. Eng.*, vol. 113, no. 2, pp. 158-165, 2012.
[http://dx.doi.org/10.1016/j.biosystemseng.2012.07.003]

[34] D. Balasundaram, T.F. Burks, D.M. Bulanon, T. Schubert, and W.S. Lee, "Spectral reflectance characteristics of citrus canker and other peel conditions of grapefruit", *Postharvest Biol. Technol.*, vol. 51, no. 2, pp. 220-226, 2009.
[http://dx.doi.org/10.1016/j.postharvbio.2008.07.014]

[35] P. Sirisomboon, M. Tanaka, T. Kojima, and P. Williams, "Nondestructive estimation of maturity and textural properties on tomato 'Momotaro'by near infrared spectroscopy", *J. Food Eng.*, vol. 112, no. 3, pp. 218-226, 2012.
[http://dx.doi.org/10.1016/j.jfoodeng.2012.04.007]

[36] L. Xie, X. Ye, D. Liu, and Y. Ying, "Prediction of titratable acidity, malic acid, and citric acid in bayberry fruit by near-infrared spectroscopy", *Food Res. Int.*, vol. 44, no. 7, pp. 2198-2204, 2011.
[http://dx.doi.org/10.1016/j.foodres.2010.11.024]

[37] J. Liu, "Non-destructive measurement of sugar content in chestnuts using near-infrared spectroscopy", In: *International Conference on Computer and Computing Technologies in Agriculture*, 2010, pp. 246-254.

[38] I. Kavdir, M.B. Buyukcan, R. Lu, H. Kocabiyik, and M. Seker, "Prediction of olive quality using FT-NIR spectroscopy in reflectance and transmittance modes", *Biosyst. Eng.*, vol. 103, no. 3, pp. 304-312, 2009.
[http://dx.doi.org/10.1016/j.biosystemseng.2009.04.014]

[39] D. Zhang, L. Xu, Q. Wang, X. Tian, and J. Li, "The optimal local model selection for robust and fast evaluation of soluble solid content in melon with thick peel and large size by Vis-NIR spectroscopy", *Food Anal. Methods,* vol. 12, no. 1, pp. 136-147, 2019.

[http://dx.doi.org/10.1007/s12161-018-1346-3]

[40] P.A. Picouet, P. Gou, R. Hyypiö, and M. Castellari, "Implementation of NIR technology for at-line rapid detection of sunflower oil adulterated with mineral oil", *J. Food Eng.,* vol. 230, pp. 18-27, 2018.
[http://dx.doi.org/10.1016/j.jfoodeng.2018.01.011]

[41] G. Ripoll, P. Albertí, B. Panea, J.L. Olleta, and C. Sañudo, "Near-infrared reflectance spectroscopy for predicting chemical, instrumental and sensory quality of beef", *Meat Sci.,* vol. 80, no. 3, pp. 697-702, 2008.
[http://dx.doi.org/10.1016/j.meatsci.2008.03.009] [PMID: 22063585]

[42] J. Cai, Q. Chen, X. Wan, and J. Zhao, "Determination of total volatile basic nitrogen (TVB-N) content and Warner-Bratzler shear force (WBSF) in pork using Fourier transform near infrared (FT-NIR) spectroscopy", *Food Chem.,* vol. 126, no. 3, pp. 1354-1360, 2011.
[http://dx.doi.org/10.1016/j.foodchem.2010.11.098]

[43] R. Grau, A.J. Sánchez, J. Girón, E. Iborra, A. Fuentes, and J.M. Barat, "Nondestructive assessment of freshness in packaged sliced chicken breasts using SW-NIR spectroscopy", *Food Res. Int.,* vol. 44, no. 1, pp. 331-337, 2011.
[http://dx.doi.org/10.1016/j.foodres.2010.10.011]

[44] R.M. Balabin, and S.V. Smirnov, "Melamine detection by mid- and near-infrared (MIR/NIR) spectroscopy: a quick and sensitive method for dairy products analysis including liquid milk, infant formula, and milk powder", *Talanta,* vol. 85, no. 1, pp. 562-568, 2011.
[http://dx.doi.org/10.1016/j.talanta.2011.04.026] [PMID: 21645742]

[45] M. Aboonajmi, and T.A. Najafabadi, "Prediction of poultry egg freshness using vis-nir spectroscopy with maximum likelihood method", *Int. J. Food Prop.,* vol. 17, no. 10, pp. 2166-2176, 2014.
[http://dx.doi.org/10.1080/10942912.2013.784330]

[46] L. Dykes, L. Hoffmann, O. Portillo-Rodriguez, W.L. Rooney, and L.W. Rooney, "Prediction of total phenols, condensed tannins, and 3-deoxyanthocyanidins in sorghum grain using near-infrared (NIR) spectroscopy", *J. Cereal Sci.,* vol. 60, no. 1, pp. 138-142, 2014.
[http://dx.doi.org/10.1016/j.jcs.2014.02.002]

[47] I.J. Wesley, O. Larroque, B.G. Osborne, N. Azudin, H. Allen, and J.H. Skerritt, "Measurement of gliadin and glutenin content of flour by NIR spectroscopy", *J. Cereal Sci.,* vol. 34, no. 2, pp. 125-133, 2001.
[http://dx.doi.org/10.1006/jcrs.2001.0378]

[48] D.S. Ferreira, O.F. Galão, J.A.L. Pallone, and R.J. Poppi, "Comparison and application of near-infrared (NIR) and mid-infrared (MIR) spectroscopy for determination of quality parameters in soybean samples", *Food Control,* vol. 35, no. 1, pp. 227-232, 2014.
[http://dx.doi.org/10.1016/j.foodcont.2013.07.010]

[49] G. Xiccato, A. Trocino, F. Tulli, and E. Tibaldi, "Prediction of chemical composition and origin identification of european sea bass (*Dicentrarchus labrax* L.) by near infrared reflectance spectroscopy (NIRS)", *Food Chem.,* vol. 86, no. 2, pp. 275-281, 2004.
[http://dx.doi.org/10.1016/j.foodchem.2003.09.026]

[50] M. Uddin, E. Okazaki, H. Fukushima, S. Turza, Y. Yumiko, and Y. Fukuda, "Nondestructive determination of water and protein in surimi by near-infrared spectroscopy", *Food Chem.,* vol. 96, no. 3, pp. 491-495, 2006.
[http://dx.doi.org/10.1016/j.foodchem.2005.04.017]

[51] L. Xie, X. Ye, D. Liu, and Y. Ying, "Quantification of glucose, fructose and sucrose in bayberry juice by NIR and PLS", *Food Chem.,* vol. 114, no. 3, pp. 1135-1140, 2009.
[http://dx.doi.org/10.1016/j.foodchem.2008.10.076]

[52] S. Castritius, A. Kron, T. Schäfer, M. Rädle, and D. Harms, "Determination of alcohol and extract concentration in beer samples using a combined method of near-infrared (NIR) spectroscopy and refractometry", *J. Agric. Food Chem.,* vol. 58, no. 24, pp. 12634-12641, 2010.

[http://dx.doi.org/10.1021/jf1030604] [PMID: 21090679]

[53] B. Peng, N. Ge, L. Cui, and H. Zhao, "Monitoring of alcohol strength and titratable acidity of apple wine during fermentation using near-infrared spectroscopy", *Lebensm. Wiss. Technol.,* vol. 66, pp. 86-92, 2016.
[http://dx.doi.org/10.1016/j.lwt.2015.10.018]

[54] G. Asimopoylos, S. Savvidis, and N. Asimopoylos, "On – line monitoring of dairy products with the use of NIR technology", *Acta Montan. Slovaca,* vol. 9, no. 1, pp. 36-40, 2004.

[55] M. Kawasaki, S. Kawamura, M. Tsukahara, S. Morita, M. Komiya, and M. Natsuga, "Near-infrared spectroscopic sensing system for on-line milk quality assessment in a milking robot", *Comput. Electron. Agric.,* vol. 63, no. 1, pp. 22-27, 2008.
[http://dx.doi.org/10.1016/j.compag.2008.01.006]

[56] G. Tøgersen, T. Isaksson, B.N. Nilsen, E.A. Bakker, and K.I. Hildrum, "On-line NIR analysis of fat, water and protein in industrial scale ground meat batches", *Meat Sci.,* vol. 51, no. 1, pp. 97-102, 1999.
[http://dx.doi.org/10.1016/S0309-1740(98)00106-5] [PMID: 22061541]

[57] D.S. Long, R.E. Engel, and M.C. Siemens, "Measuring grain protein concentration with in-line near infrared reflectance spectroscopy", *Agron. J.,* vol. 100, no. 2, pp. 247-252, 2008.
[http://dx.doi.org/10.2134/agronj2007.0052]

[58] J.M. Montes, "Near-infrared spectroscopy on combine harvesters to measure maize grain dry matter content and quality parameters", *Plant Breed.,* vol. 125, no. 6, pp. 591-595, 2006.
[http://dx.doi.org/10.1111/j.1439-0523.2006.01298.x]

[59] V.A. McGlone, and P.J. Martinsen, "Transmission measurements on intact apples moving at high speed", *J. Near Infrared Spectrosc.,* vol. 12, no. 1, pp. 37-43, 2004.
[http://dx.doi.org/10.1255/jnirs.406]

[60] A.A. Gowen, C.P. O'Donnell, P.J. Cullen, G. Downey, and J.M. Frias, "Hyperspectral imaging - an emerging process analytical tool for food quality and safety control", *Trends Food Sci. Technol.,* vol. 18, no. 12, pp. 590-598, 2007.
[http://dx.doi.org/10.1016/j.tifs.2007.06.001]

[61] A.F.H. Goetz, G. Vane, J.E. Solomon, and B.N. Rock, "Imaging spectrometry for earth remote sensing", *Science (80-.),* vol. 228, no. 4704, pp. 1147-1153, 1985.
[http://dx.doi.org/10.1126/science.228.4704.1147]

[62] D. Wu, and D.W. Sun, "Advanced applications of hyperspectral imaging technology for food quality and safety analysis and assessment: A review - Part I: Fundamentals", *Innov. Food Sci. Emerg. Technol.,* vol. 19, pp. 1-14, 2013.
[http://dx.doi.org/10.1016/j.ifset.2013.04.014]

[63] B. Zhang, L. Liu, B. Gu, J. Zhou, J. Huang, and G. Tian, "From hyperspectral imaging to multispectral imaging: Portability and stability of HIS-MIS algorithms for common defect detection", *Postharvest Biol. Technol.,* vol. 137, pp. 95-105, 2018.
[http://dx.doi.org/10.1016/j.postharvbio.2017.11.004]

[64] J. Qin, K. Chao, M.S. Kim, R. Lu, and T.F. Burks, "Hyperspectral and multispectral imaging for evaluating food safety and quality", *J. Food Eng.,* vol. 118, no. 2, pp. 157-171, 2013.
[http://dx.doi.org/10.1016/j.jfoodeng.2013.04.001]

[65] J. Ma, D. W. Sun, H. Pu, Q. Wei, and X. Wang, "Protein content evaluation of processed pork meats based on a novel single shot (snapshot) hyperspectral imaging sensor", *J. Food Eng,* vol. 240, pp. 207-213, 2019.
[http://dx.doi.org/10.1016/j.jfoodeng.2018.07.032]

[66] R. Lu, "Detection of bruises on apples using near–infrared hyperspectral imaging", *Trans. ASAE,* vol. 46, no. 2, p. 523, 2003.

[67] J.C. Keresztes, M. Goodarzi, and W. Saeys, "Real-time pixel based early apple bruise detection using

short wave infrared hyperspectral imaging in combination with calibration and glare correction techniques", *Food Control,* vol. 66, pp. 215-226, 2016.
[http://dx.doi.org/10.1016/j.foodcont.2016.02.007]

[68]   Q. Lü, M.J. Tang, J.R. Cai, J.W. Zhao, and S. Vittayapadung, "Vis/NIR hyperspectral imaging for detection of hidden bruises on kiwifruits", *Czech J. Food Sci.,* vol. 29, no. 6, pp. 595-602, 2011.
[http://dx.doi.org/10.17221/69/2010-CJFS]

[69]   Y. Jiang, C. Li, and F. Takeda, "Nondestructive detection and quantification of blueberry bruising using near-infrared (NIR) hyperspectral reflectance imaging", *Sci. Rep.,* vol. 6, no. October, p. 35679, 2016.
[http://dx.doi.org/10.1038/srep35679] [PMID: 27767050]

[70]   W. Wang, C. Li, E.W. Tollner, R.D. Gitaitis, and G.C. Rains, "Shortwave infrared hyperspectral imaging for detecting sour skin (Burkholderia cepacia)-infected onions", *J. Food Eng.,* vol. 109, no. 1, pp. 38-48, 2012.
[http://dx.doi.org/10.1016/j.jfoodeng.2011.10.001]

[71]   D. Lorente, J. Blasco, A.J. Serrano, E. Soria-Olivas, N. Aleixos, and J. Gómez-Sanchis, "Comparison of ROC feature selection method for the detection of decay in citrus fruit using hyperspectral images", *Food Bioprocess Technol.,* vol. 6, no. 12, pp. 3613-3619, 2013.
[http://dx.doi.org/10.1007/s11947-012-0951-1]

[72]   P. Rajkumar, and N. Wang, "G. Elmasry, G. S. V. Raghavan, and Y. Gariepy, "Studies on banana fruit quality and maturity stages using hyperspectral imaging", *J. Food Eng.,* vol. 108, no. 1, pp. 194-200, 2012.
[http://dx.doi.org/10.1016/j.jfoodeng.2011.05.002]

[73]   G. ElMasry, N. Wang, A. ElSayed, and M. Ngadi, "Hyperspectral imaging for nondestructive determination of some quality attributes for strawberry", *J. Food Eng.,* vol. 81, no. 1, pp. 98-107, 2007.
[http://dx.doi.org/10.1016/j.jfoodeng.2006.10.016]

[74]   Y. Sun, K. Wei, Q. Liu, L. Pan, and K. Tu, "Classification and discrimination of different fungal diseases of three infection levels on peaches using hyperspectral reflectance imaging analysis", *Sensors (Basel),* vol. 18, no. 4, 2018.E1295
[http://dx.doi.org/10.3390/s18041295] [PMID: 29690625]

[75]   L. Pan, Q. Zhang, W. Zhang, Y. Sun, P. Hu, and K. Tu, "Detection of cold injury in peaches by hyperspectral reflectance imaging and artificial neural network", *Food Chem.,* vol. 192, pp. 134-141, 2016.
[http://dx.doi.org/10.1016/j.foodchem.2015.06.106] [PMID: 26304330]

[76]   G.K. Naganathan, L.M. Grimes, J. Subbiah, C.R. Calkins, A. Samal, and G.E. Meyer, "Visible/near-infrared hyperspectral imaging for beef tenderness prediction", *Comput. Electron. Agric.,* vol. 64, no. 2, pp. 225-233, 2008.
[http://dx.doi.org/10.1016/j.compag.2008.05.020]

[77]   G. ElMasry, and J.P. Wold, "High-speed assessment of fat and water content distribution in fish fillets using online imaging spectroscopy", *J. Agric. Food Chem.,* vol. 56, no. 17, pp. 7672-7677, 2008.
[http://dx.doi.org/10.1021/jf801074s] [PMID: 18656933]

[78]   J. Qiao, M.O. Ngadi, N. Wang, C. Gariépy, and S.O. Prasher, "Pork quality and marbling level assessment using a hyperspectral imaging system", *J. Food Eng.,* vol. 83, no. 1, pp. 10-16, 2007.
[http://dx.doi.org/10.1016/j.jfoodeng.2007.02.038]

[79]   H. Zhang, J. Paliwal, D.S. Jayas, and N.D.G. White, "Classification of fungal infected wheat kernels using near-infrared reflectance hyperspectral imaging and support vector machine", *Trans. ASABE,* vol. 50, no. 5, pp. 1779-1785, 2007.
[http://dx.doi.org/10.13031/2013.23935]

[80]   S.R. Delwiche, M.S. Kim, and Y. Dong, "Fusarium damage assessment in wheat kernels by Vis/NIR

hyperspectral imaging", *Sens. Instrum. Food Qual. Saf.,* vol. 5, no. 2, pp. 63-71, 2011.
[http://dx.doi.org/10.1007/s11694-011-9112-x]

[81] J. Zhao, Q. Chen, J. Cai, and Q. Ouyang, "Automated tea quality classification by hyperspectral imaging", *Appl. Opt.,* vol. 48, no. 19, pp. 3557-3564, 2009.
[http://dx.doi.org/10.1364/AO.48.003557] [PMID: 19571909]

[82] J. Qin, and R. Lu, "Hyperspectral diffuse reflectance imaging for rapid, noncontact measurement of the optical properties of turbid materials", *Appl. Opt.,* vol. 45, no. 32, pp. 8366-8373, 2006.
[http://dx.doi.org/10.1364/AO.45.008366] [PMID: 17068584]

[83] H. Cen, R. Lu, F. Mendoza, and R.M. Beaudry, "Relationship of the optical absorption and scattering properties with mechanical and structural properties of apple tissue", *Postharvest Biol. Technol.,* vol. 85, pp. 30-38, 2013.
[http://dx.doi.org/10.1016/j.postharvbio.2013.04.014]

[84] J. Qin, and R. Lu, "Measurement of the absorption and scattering properties of turbid liquid foods using hyperspectral imaging", *Appl. Spectrosc.,* vol. 61, no. 4, pp. 388-396, 2007.
[http://dx.doi.org/10.1366/000370207780466190] [PMID: 17456257]

[85] H. Cen, R. Lu, F.A. Mendoza, and D.P. Ariana, "Assessing multiple quality attributes of peaches using optical absorption and scattering properties", *Trans. ASABE,* vol. 55, no. 2, pp. 647-657, 2012.
[http://dx.doi.org/10.13031/2013.41366]

[86] R. Lu, D.P. Ariana, and H. Cen, "Optical absorption and scattering properties of normal and defective pickling cucumbers for 700–1000 nm", *Sens. Instrum. Food Qual. Saf.,* vol. 5, no. 2, pp. 51-56, 2011.
[http://dx.doi.org/10.1007/s11694-011-9108-6]

[87] A. Wang, and X. Wei, "A sequential method for estimating the optical properties of two-layer agro-products from spatially-resolved diffuse reflectance: simulation", *Artif. Intell. Agric,* 2019.
[http://dx.doi.org/10.1016/j.aiia.2019.12.003]

[88] C. Simoneau, M.J. McCarthy, and J.B. German, "Magnetic resonance imaging and spectroscopy for food systems", *Food Res. Int.,* vol. 26, no. 5, pp. 387-398, 1993.
[http://dx.doi.org/10.1016/0963-9969(93)90082-T]

[89] S.M. Kim, P. Chen, M.J. McCarthy, and B. Zion, "Fruit internal quality evaluation using on-line nuclear magnetic resonance sensors", *J. Agric. Eng. Res.,* vol. 74, no. 3, pp. 293-301, 1999.
[http://dx.doi.org/10.1006/jaer.1999.0465]

[90] K.K. Patel, M.A. Khan, and A. Kar, "Recent developments in applications of MRI techniques for foods and agricultural produce—an overview", *J. Food Sci. Technol.,* vol. 52, no. 1, pp. 1-26, 2015.
[http://dx.doi.org/10.1007/s13197-012-0917-3] [PMID: 26787928]

[91] M.C. Ezeanaka, J. Nsor-Atindana, and M. Zhang, "Online low-field nuclear magnetic resonance (LF-NMR) and magnetic resonance imaging (MRI) for food quality optimization in food processing", *Food Bioprocess Technol.,* vol. 12, no. 9, pp. 1435-1451, 2019.
[http://dx.doi.org/10.1007/s11947-019-02296-w]

[92] D.W. McRobbie, E.A. Moore, M.J. Graves, and M.R. Prince, *MRI from Picture to Proton.* Cambridge university press: England, 2017.
[http://dx.doi.org/10.1017/9781107706958]

[93] T. Kamal, "Potential uses of LF-NMR and MRI in the study of water dynamics and quality measurement of fruits and vegetables", *J. Food Process. Preserv.,* vol. 43, no. 11, pp. 1-21, 2019.
[http://dx.doi.org/10.1111/jfpp.14202]

[94] Q. Chen, C. Zhang, J. Zhao, and Q. Ouyang, ""Recent advances in emerging imaging techniques for non-destructive detection of food quality and safety," TrAC -", *Trends Analyt. Chem.,* vol. 52, pp. 261-274, 2013.
[http://dx.doi.org/10.1016/j.trac.2013.09.007]

[95] K. Fan, and M. Zhang, "Recent developments in the food quality detected by non-invasive nuclear

magnetic resonance technology", *Crit. Rev. Food Sci. Nutr.,* vol. 59, no. 14, pp. 2202-2213, 2019.
[http://dx.doi.org/10.1080/10408398.2018.1441124] [PMID: 29451810]

[96]   W. Lv, M. Zhang, Y. Wang, and B. Adhikari, "Online measurement of moisture content, moisture distribution, and state of water in corn kernels during microwave vacuum drying using novel smart NMR/MRI detection system", *Dry. Technol.,* vol. 36, no. 13, pp. 1592-1602, 2018.
[http://dx.doi.org/10.1080/07373937.2017.1418751]

[97]   X. Zang, "Non-destructive measurement of water and fat contents, water dynamics during drying and adulteration detection of intact small yellow croaker by low field NMR", *J. Food Meas. Charact.,* vol. 11, no. 4, pp. 1550-1558, 2017.
[http://dx.doi.org/10.1007/s11694-017-9534-1]

[98]   M. Gudjónsdóttir, S. Arason, and T. Rustad, "The effects of pre-salting methods on water distribution and protein denaturation of dry salted and rehydrated cod–A low-field NMR study", *J. Food Eng.,* vol. 104, no. 1, pp. 23-29, 2011.
[http://dx.doi.org/10.1016/j.jfoodeng.2010.11.022]

[99]   R. Zhou, and Y. Li, "Texture analysis of MR image for predicting the firmness of Huanghua pears (*Pyrus pyrifolia* Nakai, cv. Huanghua) during storage using an artificial neural network", *Magn. Reson. Imaging,* vol. 25, no. 5, pp. 727-732, 2007.
[http://dx.doi.org/10.1016/j.mri.2006.09.011] [PMID: 17540285]

[100]   M.E. Gonzalez, D.M. Barrett, M.J. McCarthy, F.J. Vergeldt, E. Gerkema, A.M. Matser, and H. Van As, "1H-NMR study of the impact of high pressure and thermal processing on cell membrane integrity of onions", *J. Food Sci.,* vol. 75, no. 7, pp. E417-E425, 2010.
[http://dx.doi.org/10.1111/j.1750-3841.2010.01766.x] [PMID: 21535535]

[101]   A. Taglienti, R. Massantini, R. Botondi, F. Mencarelli, and M. Valentini, "Postharvest structural changes of Hayward kiwifruit by means of magnetic resonance imaging spectroscopy", *Food Chem.,* vol. 114, no. 4, pp. 1583-1589, 2009.
[http://dx.doi.org/10.1016/j.foodchem.2008.11.066]

[102]   N. Hernández-Sánchez, B.P. Hills, P. Barreiro, and N. Marigheto, "An NMR study on internal browning in pears", *Postharvest Biol. Technol.,* vol. 44, no. 3, pp. 260-270, 2007.
[http://dx.doi.org/10.1016/j.postharvbio.2007.01.002]

[103]   A. Melado-Herreros, M.A. Muñoz-García, A. Blanco, J. Val, M.E. Fernández-Valle, and P. Barreiro, "Assessment of watercore development in apples with MRI: Effect of fruit location in the canopy", *Postharvest Biol. Technol.,* vol. 86, pp. 125-133, 2013.
[http://dx.doi.org/10.1016/j.postharvbio.2013.06.030]

[104]   M. Mazhar, "Non-destructive 1H-MRI assessment of flesh bruising in avocado (*Persea americana* M.) cv. Hass", *Postharvest Biol. Technol.,* vol. 100, pp. 33-40, 2015.
[http://dx.doi.org/10.1016/j.postharvbio.2014.09.006]

[105]   M. S. Razavi, A. Asghari, M. Azadbakht, and H. A. Shamsabadi, "Analyzing the pear bruised volume after static loading by Magnetic Resonance Imaging (MRI)", *Sci. Hortic. (Amsterdam),* vol. 229, pp. 33-39, 2018.
[http://dx.doi.org/10.1016/j.scienta.2017.10.011]

[106]   F.L. Chen, Y.M. Wei, and B. Zhang, "Characterization of water state and distribution in textured soybean protein using DSC and NMR", *J. Food Eng.,* vol. 100, no. 3, pp. 522-526, 2010.
[http://dx.doi.org/10.1016/j.jfoodeng.2010.04.040]

[107]   Z.Q. Jin, J.S. Zhang, X.Y. Lin, R.S. Ruan, N. Wang, and W.J. Chen, "Study on water loss rate and decay of strawberry by NMR and MRI", *Shipin Kexue,* vol. 28, no. 8, pp. 108-111, 2007.

[108]   A. Ciampa, M.T. Dell'Abate, O. Masetti, M. Valentini, and P. Sequi, "Seasonal chemical–physical changes of PGI Pachino cherry tomatoes detected by magnetic resonance imaging (MRI)", *Food Chem.,* vol. 122, no. 4, pp. 1253-1260, 2010.
[http://dx.doi.org/10.1016/j.foodchem.2010.03.078]

[109] U. Tylewicz, "Effect of pulsed electric field treatment on water distribution of freeze-dried apple tissue evaluated with DSC and TD-NMR techniques", *Innov. Food Sci. Emerg. Technol.,* vol. 37, pp. 352-358, 2016.
[http://dx.doi.org/10.1016/j.ifset.2016.06.012]

[110] W. Lv, M. Zhang, B. Bhandari, L. Li, and Y. Wang, "Smart NMR method of measurement of moisture content of vegetables during microwave vacuum drying", *Food Bioprocess Technol.,* vol. 10, no. 12, pp. 2251-2260, 2017.
[http://dx.doi.org/10.1007/s11947-017-1991-3]

[111] P.A.M. Nascimento, P.L. Barsanelli, A.P. Rebellato, J.A.L. Pallone, L.A. Colnago, and F.M.V. Pereira, "Time-Domain Nuclear Magnetic Resonance (TD-NMR) and chemometrics for determination of fat content in commercial products of milk powder", *J. AOAC Int.,* vol. 100, no. 2, pp. 330-334, 2017.
[http://dx.doi.org/10.5740/jaoacint.16-0408] [PMID: 28055818]

[112] N. Kotwaliwale, K. Singh, A. Kalne, S.N. Jha, N. Seth, and A. Kar, "X-ray imaging methods for internal quality evaluation of agricultural produce", *J. Food Sci. Technol.,* vol. 51, no. 1, pp. 1-15, 2014.
[http://dx.doi.org/10.1007/s13197-011-0485-y] [PMID: 24426042]

[113] E. Whaites, *Essentials of dental radiography and radiology.* Churchill Livingstone: London, United Kingdom, 2002.

[114] Z. Du, X. Zeng, X. Li, X. Ding, J. Cao, and W. Jiang, "Recent advances in imaging techniques for bruise detection in fruits and vegetables", *Trends Food Sci. Technol.,* 2020.
[http://dx.doi.org/10.1016/j.tifs.2020.02.024]

[115] Z. Du, Y. Hu, N. Ali Buttar, and A. Mahmood, "X-ray computed tomography for quality inspection of agricultural products: A review", *Food Sci. Nutr.,* vol. 7, no. 10, pp. 3146-3160, 2019.
[http://dx.doi.org/10.1002/fsn3.1179] [PMID: 31660129]

[116] H. Taotao, S. Teng, and Z. Jingping, ""Non-destructive detection of internal quality of apple based on CT image," J. Zhejiang Univ", *Agriculture Life Sci.,* vol. 39, no. 1, p. 92, 2013.

[117] E. Diels, "Assessment of bruise volumes in apples using X-ray computed tomography", *Postharvest Biol. Technol.,* vol. 128, pp. 24-32, 2017.
[http://dx.doi.org/10.1016/j.postharvbio.2017.01.013]

[118] S. Jarolmasjed, C.Z. Espinoza, S. Sankaran, and L.R. Khot, "Postharvest bitter pit detection and progression evaluation in 'Honeycrisp'apples using computed tomography images", *Postharvest Biol. Technol.,* vol. 118, pp. 35-42, 2016.

[119] P. Boniecki, H. Piekarska-Boniecka, K. Świerczyński, K. Koszela, M. Zaborowicz, and J. Przybył, "Detection of the granary weevil based on X-ray images of damaged wheat kernels", *J. Stored Prod. Res.,* vol. 56, pp. 38-42, 2014.
[http://dx.doi.org/10.1016/j.jspr.2013.11.001]

[120] A. Guelpa, A. du Plessis, M. Kidd, and M. Manley, "Non-destructive estimation of maize (*Zea mays* L.) kernel hardness by means of an X-ray micro-computed tomography (μCT) density calibration", *Food Bioprocess Technol.,* vol. 8, no. 7, pp. 1419-1429, 2015.
[http://dx.doi.org/10.1007/s11947-015-1502-3]

[121] P. Frisullo, R. Marino, J. Laverse, M. Albenzio, and M.A. Del Nobile, "Assessment of intramuscular fat level and distribution in beef muscles using X-ray microcomputed tomography", *Meat Sci.,* vol. 85, no. 2, pp. 250-255, 2010.
[http://dx.doi.org/10.1016/j.meatsci.2010.01.008] [PMID: 20374894]

[122] N.R. Lambe, K.A. McLean, J. Gordon, D. Evans, N. Clelland, and L. Bunger, "Prediction of intramuscular fat content using CT scanning of packaged lamb cuts and relationships with meat eating quality", *Meat Sci.,* vol. 123, pp. 112-119, 2017.

[http://dx.doi.org/10.1016/j.meatsci.2016.09.008] [PMID: 27701028]

[123] C. Vestergaard, J. Risum, and J. Adler-Nissen, "Quantification of salt concentrations in cured pork by computed tomography", *Meat Sci.,* vol. 68, no. 1, pp. 107-113, 2004.
[http://dx.doi.org/10.1016/j.meatsci.2004.02.011] [PMID: 22062013]

[124] V. Vicent, P. Verboven, F-T. Ndoye, G. Alvarez, and B. Nicolaï, "A new method developed to characterize the 3D microstructure of frozen apple using X-ray micro-CT", *J. Food Eng.,* vol. 212, pp. 154-164, 2017.
[http://dx.doi.org/10.1016/j.jfoodeng.2017.05.028]

[125] M. Azadbakht, M. Vahedi Torshizi, and M.J. Mahmoodi, "The use of CT scan imaging technique to determine pear bruise level due to external loads", *Food Sci. Nutr.,* vol. 7, no. 1, pp. 273-280, 2018.
[http://dx.doi.org/10.1002/fsn3.882] [PMID: 30680181]

[126] A. Berna, "Metal oxide sensors for electronic noses and their application to food analysis", *Sensors (Basel),* vol. 10, no. 4, pp. 3882-3910, 2010.
[http://dx.doi.org/10.3390/s100403882] [PMID: 22319332]

[127] H. Shi, M. Zhang, and B. Adhikari, "Advances of electronic nose and its application in fresh foods: A review", *Crit. Rev. Food Sci. Nutr.,* vol. 58, no. 16, pp. 2700-2710, 2018.
[http://dx.doi.org/10.1080/10408398.2017.1327419] [PMID: 28665685]

[128] H.G. Hosseini, D. Luo, G. Xu, H. Liu, and D. Benjamin, "Intelligent Fish Freshness Assessment", *J. Sens.,* 2008.

[129] S. Zhang, C. Xie, Z. Bai, M. Hu, H. Li, and D. Zeng, "Spoiling and formaldehyde-containing detections in octopus with an E-nose", *Food Chem.,* vol. 113, no. 4, pp. 1346-1350, 2009.
[http://dx.doi.org/10.1016/j.foodchem.2008.08.090]

[130] N. El Barbri, E. Llobet, N. El Bari, X. Correig, and B. Bouchikhi, "Electronic nose based on metal oxide semiconductor sensors as an alternative technique for the spoilage classification of red meat", *Sensors (Basel),* vol. 8, no. 1, pp. 142-156, 2008.
[http://dx.doi.org/10.3390/s8010142] [PMID: 27879699]

[131] V.Y. Musatov, V.V. Sysoev, M. Sommer, and I. Kiselev, "Assessment of meat freshness with metal oxide sensor microarray electronic nose: A practical approach", *Sens. Actuators B Chem.,* vol. 144, no. 1, pp. 99-103, 2010.
[http://dx.doi.org/10.1016/j.snb.2009.10.040]

[132] Y. Ren, H.S. Ramaswamy, Y. Li, C. Yuan, and X. Ren, "Classification of impact injury of apples using electronic nose coupled with multivariate statistical analyses", *J. Food Process Eng.,* vol. 41, no. 5, 2018.e12698
[http://dx.doi.org/10.1111/jfpe.12698]

[133] W. Jia, G. Liang, H. Tian, J. Sun, and C. Wan, "Electronic nose-based technique for rapid detection and recognition of moldy apples", *Sensors (Basel),* vol. 19, no. 7, p. 1526, 2019.
[http://dx.doi.org/10.3390/s19071526] [PMID: 30934812]

[134] A.Z. Berna, J. Lammertyn, S. Saevels, C. Di Natale, and B.M. Nicolaï, "Electronic nose systems to study shelf life and cultivar effect on tomato aroma profile", *Sens. Actuators B Chem.,* vol. 97, no. 2–3, pp. 324-333, 2004.
[http://dx.doi.org/10.1016/j.snb.2003.09.020]

[135] Y. Jing, Q. Meng, P. Qi, M. Zeng, W. Li, and S. Ma, "Electronic nose with a new feature reduction method and a multi-linear classifier for Chinese liquor classification", *Rev. Sci. Instrum.,* vol. 85, no. 5, 2014.055004
[http://dx.doi.org/10.1063/1.4874326] [PMID: 24880405]

[136] H. Qin, "Colorimetric artificial nose for identification of Chinese liquor with different geographic origins", *Food Res. Int.,* vol. 45, no. 1, pp. 45-51, 2012.
[http://dx.doi.org/10.1016/j.foodres.2011.09.008]

[137] H. Yu, X. Dai, G. Yao, and Z. Xiao, "Application of gas chromatography-based electronic nose for classification of Chinese rice wine by wine age", *Food Anal. Methods,* vol. 7, no. 7, pp. 1489-1497, 2014.
[http://dx.doi.org/10.1007/s12161-013-9778-2]

[138] W. Zhang, Z. Lv, and S. Xiong, "Nondestructive quality evaluation of agro-products using acoustic vibration methods-A review", *Crit. Rev. Food Sci. Nutr.,* vol. 58, no. 14, pp. 2386-2397, 2018.
[http://dx.doi.org/10.1080/10408398.2017.1324830] [PMID: 28613932]

[139] M. Taniwaki, and N. Sakurai, "Evaluation of the internal quality of agricultural products using acoustic vibration techniques", *J. Jpn. Soc. Hortic. Sci.,* vol. 79, no. 2, pp. 113-128, 2010.
[http://dx.doi.org/10.2503/jjshs1.79.113]

[140] J.R. Cooke, and R.H. Rand, "A mathematical study of resonance in intact fruits and vegetables using a 3-media elastic sphere model", *J. Agric. Eng. Res.,* vol. 18, no. 2, pp. 141-157, 1973.
[http://dx.doi.org/10.1016/0021-8634(73)90023-1]

[141] J.R. Cooke, "An interpretation of the resonant behavior of intact fruits and vegetables", *Trans. ASAE,* vol. 15, no. 6, pp. 1075-1080, 1972.
[http://dx.doi.org/10.13031/2013.38074]

[142] J.A. Abbott, H.A. Affeldt, and L.A. Liljedahl, "Firmness measurement of stored delicious apples by sensory methods, magness-taylor, and sonic transmission", *J. Am. Soc. Hortic. Sci.,* vol. 117, no. 4, pp. 590-595, 1992.
[http://dx.doi.org/10.21273/JASHS.117.4.590]

[143] W. Zhang, D. Cui, and Y. Ying, "Nondestructive measurement of texture of three pear varieties and variety discrimination by the laser Doppler vibrometer method", *Food Bioprocess Technol.,* vol. 8, no. 9, pp. 1974-1981, 2015.
[http://dx.doi.org/10.1007/s11947-015-1547-3]

[144] J.A. Abbott, and D.R. Massie, "Nondestructive sonic measurement of kiwifruit firmness", *J. Am. Soc. Hortic. Sci.,* vol. 123, no. 2, pp. 317-322, 1998.
[http://dx.doi.org/10.21273/JASHS.123.2.317]

[145] M. Fumuro, N. Sakurai, and N. Utsunomiya, "Improved accuracy in determining optimal harvest time for pitaya (*Hylocereus undatus*) using the elasticity index", *J. Jpn. Soc. Hortic. Sci.,* vol. 82, no. 4, pp. 354-361, 2013.
[http://dx.doi.org/10.2503/jjshs1.82.354]

[146] J.A. Abbott, "Quality measurement of fruits and vegetables", *Postharvest Biol. Technol.,* vol. 15, no. 3, pp. 207-225, 1999.
[http://dx.doi.org/10.1016/S0925-5214(98)00086-6]

[147] M. Taniwaki, T. Hanada, and N. Sakurai, "Postharvest quality evaluation of 'Fuyu' and 'Taishuu' persimmons using a nondestructive vibrational method and an acoustic vibration technique", *Postharvest Biol. Technol.,* vol. 51, no. 1, pp. 80-85, 2009.
[http://dx.doi.org/10.1016/j.postharvbio.2008.05.014]

[148] M. Taniwaki, M. Takahashi, and N. Sakurai, "Determination of optimum ripeness for edibility of postharvest melons using nondestructive vibration", *Food Res. Int.,* vol. 42, no. 1, pp. 137-141, 2009.
[http://dx.doi.org/10.1016/j.foodres.2008.09.007]

[149] M. Taniwaki, M. Tohro, and N. Sakurai, "Measurement of ripening speed and determination of the optimum ripeness of melons by a nondestructive acoustic vibration method", *Postharvest Biol. Technol.,* vol. 56, no. 1, pp. 101-103, 2010.
[http://dx.doi.org/10.1016/j.postharvbio.2009.11.007]

[150] J. Hongwiangjan, A. Terdwongworakul, and K. Krisanapook, "Evaluation of pomelo maturity based on acoustic response and peel properties", *Int. J. Food Sci. Technol.,* vol. 50, no. 3, pp. 782-789, 2015.
[http://dx.doi.org/10.1111/ijfs.12700]

[151] K.K. Vursavus, Y.B. Yurtlu, B. Diezma-Iglesias, L. Lleo-Garcia, and M. Ruiz-Altisent, "Classification of the firmness of peaches by sensor fusion", *Int. J. Agric. Biol. Eng.,* vol. 8, no. 6, pp. 104-115, 2015.

[152] A. Baltazar, J.I. Aranda, and G. González-Aguilar, "Bayesian classification of ripening stages of tomato fruit using acoustic impact and colorimeter sensor data", *Comput. Electron. Agric.,* vol. 60, no. 2, pp. 113-121, 2008.
[http://dx.doi.org/10.1016/j.compag.2007.07.005]

[153] C. Jin, and Y. Ying, "Eggshell crack detection with rolling eggs on a corrugated plate and LibSVM", *Trans. ASABE,* vol. 57, no. 3, pp. 871-879, 2014.

[154] C. Jin, L. Xie, and Y. Ying, "Eggshell crack detection based on the time-domain acoustic signal of rolling eggs on a step-plate", *J. Food Eng.,* vol. 153, pp. 53-62, 2015.
[http://dx.doi.org/10.1016/j.jfoodeng.2014.12.011]

[155] L. Pan, K. Tu, G. Zhan, M. Liu, and X. Zou, "Eggshell crack detection based on information fusion between computer vision and acoustic response", *Nongye Gongcheng Xuebao (Beijing),* vol. 26, no. 11, pp. 332-337, 2010.

[156] R.X.Y.Y.L. Feiling, and J. Bo, "Development of a fruit quality inspecting system based on acoustic properties", *Trans. Chinese Soc. Agric. Mach,* vol. 2, 2004.

[157] B.D. Iglesias, M.R. Altisent, and P. Jancsók, "Vibrational analysis of seedless watermelons: use in the detection of internal hollows", *Span. J. Agric. Res.,* vol. 3, no. 1, pp. 52-60, 2005.
[http://dx.doi.org/10.5424/sjar/2005031-124]

[158] P.R. Armstrong, M.L. Stone, and G.H. Brusewitz, "Nondestructive acoustic and compression measurements of watermelon for internal damage detection", *Appl. Eng. Agric.,* vol. 13, no. 5, pp. 641-645, 1997.
[http://dx.doi.org/10.13031/2013.21638]

[159] I.E. Elbatawi, "An acoustic impact method to detect hollow heart of potato tubers", *Biosyst. Eng.,* vol. 100, no. 2, pp. 206-213, 2008.
[http://dx.doi.org/10.1016/j.biosystemseng.2008.02.009]

CHAPTER 5

Hyperspectral Imaging and Machine Learning for Rapid Assessment of Deoxynivalenol of Barley Kernels

Wen-Hao Su[1], Ce Yang[1,*], Yanhong Dong[2], Ryan Johnson[2], Rae Page[2], Tamas Szinyei[2], Cory D. Hirsch[2] and **Brian J. Steffenson[2]**

[1] *Department of Bioproducts and Biosystems Engineering, University of Minnesota, Saint Paul, MN 55108, USA*

[2] *Department of Plant Pathology, University of Minnesota, Saint Paul, MN 55108, USA*

Abstract: Imaging techniques can be used to evaluate the quality and safety of agricultural products. Fusarium head blight (FHB) results in reduced barley yields and also diminished value of harvested barley. Deoxynivalenol (DON) is a mycotoxin produced by the causal Fusarium species that pose health risks to humans and livestock. DON has currently measured *via* gas chromatography (GC) methods that are time-consuming and expensive. We seek to apply imaging technology to rapidly and non-destructively quantify DON in high throughput and less expensive method. The feasibility of hyperspectral imaging to determine DON contents of barley kernels was evaluated using machine learning algorithms. Partial least square discriminant analysis (PLSDA) was able to discriminate kernels into four separate classes corresponding to their DON levels. Barley kernels could be classified as having low (<5 ppm) or high DON levels, with Matthews's correlation coefficient in cross-validation (M-RCV) of as high as 0.823. PLSR showed good performance in linear algorithms for DON detection, but higher accuracy was obtained by non-linear algorithms, including weighted partial least squares regression (LWPLSR), support vector machine regression (SVMR), and artificial neural network (ANN). Among all algorithms, the non-linear LWPLSR achieved the highest accuracy, with the coefficient of determination in prediction (R^2_p) of 0.728 and root mean square error of prediction (RMSEP) of 3.802. The results demonstrate that hyperspectral imaging and machine learning algorithms have the potential to assist the FHB resistance breeding process by accelerating the quantification of DON in barley samples.

Keywords: Deoxynivalenol, Food safety, Hyperspectral imaging, Machine learning.

* **Corresponding author Ce Yang:** Department of Bioproducts and Biosystems Engineering, University of Minnesota, Saint Paul, MN 55108, USA; Tel: +1 6126266419; E-mail: ceyang@umn.edu.

Jiangbo Li & Zhao Zhang (Eds.)

1. INTRODUCTION

Imaging techniques have become valuable tools in the quality and safety assessments of all kinds of agricultural products. The quality and safety of foods are related to their sensorial (such as shape, size, smell, color), chemical compositions (such as protein, starch), and textural (mechanical) properties [1 - 5]. Imaging technologies including color imaging, ultrasound imaging, Raman imaging, thermal imaging, magnetic resonance imaging, X-ray imaging, fluorescence imaging and hyperspectral imaging have been introduced into food research as nondestructive methods [6 - 8]. Each of these imaging modes has its own unique characteristics. For example, color imaging is appropriate for sensing the shape, size or external defect of a specimen at a pixel level [9]. Fluorescence imaging is extensively used to capture the image of fluorescence emission with lower energy from an object excited by higher-energy light [10 - 12]. Among them, hyperspectral imaging records images of continuous spectral bands with high spatial and spectral resolutions [2, 13 - 17]. This technique integrating the characteristics of imaging and spectroscopy is one of the most advanced and widely adopted non-destructive imaging methods for the rapid evaluation of cereal food qualities [18 - 21]. There are three approaches to generate a hyperspectral image, which are point (whiskbroom) scanning, area scanning (tunable filter or staredown), and line (pushbroom) scanning [22, 23]. The point scanning method captures an image point by point, which is not feasible for fast image acquisition [24]. The area scan method can collect images of a fixed scene at different wavebands. The line scan method obtains images of moving samples line by line, which is more suitable for online inspection. Hyperspectral imaging has already been widely used in the quality and safety evaluations of numerous agricultural products, such as pesticide (chlorpyrifos and imidacloprid) prediction in jujube fruit [25], salt stress tolerance assessment in wheat [26], moisture distribution analysis in potato and sweet potato tubers [27], and maturity stage classification in blueberry fruit [2].

Barley (Hordeum vulgare L.) is an important crop for both human and animal consumption worldwide. In the United States (U.S.), barley contributes not only to the domestic food and feed production but also to export markets and trade balance [28]. Fusarium head blight (FHB), caused by a number of different species in the genus Fusarium, is a major disease of barley in many production regions around the world [29]. The infection of Fusarium head blight (FHB) is initiated by airborne spores, which occurs during flowering with warm temperatures and high relative humidity [30]. The initial symptoms appear as water spots on infected spikelets. Then, chlorophyll breaks down and the entire spikelet is bleached. FHB is associated with the production of highly toxic mycotoxins that can significantly impact public health [31]. The yield and quality losses due to FHB are mainly attributed to the accumulation of deoxynivalenol (DON) [32]. DON is a mycotoxin secreted by several Fusarium species to the grain [33]. A positive relationship between FHB severity and DON accumulation has been found, indicating the development of FHB resistant lines with lower DON accumulation is likely [34]. Although crop protection strategies such as crop rotation and fungicide application have been used, the breeding of resistant barley varieties is the most beneficial to reduce FHB severity [35, 36]. Effective resistance breeding requires interdisciplinary research that combines informatics, plant pathology, plant genetics, and years' worth of time. The combinations of different resistance genes can be achieved based on the knowledge of the location and role of each [37]. Currently, a single source of resistance has only partial resistance to FHB, thus breeders continue to combine genes from multiple sources to develop sufficiently resistant cultivars [37]. After new genetic variants are discovered, an important step to obtain high disease-resistant varieties in a breeding program is phenotyping.

Hyperspectral imaging is economical, cost-effective, and non-destructive for determinations of DON in barley, as it does not need additional costs for reagents and the labor compared to conventional chemical methods. This advanced imaging technique obtains full-wavelength spectral images over a wide spectral range and is a very promising way to detect a foreign metabolites produced by a fungus [38 - 40]. The assessment of infected barley has traditionally relied on visual inspection of the disease severity after artificial inoculation [41], which is time-consuming, expensive, and inaccurate. Conventional methods for detecting DON are gas chromatograph-mass spectrometry (GC–MS), and immunological methods such as enzyme-linked immunosorbent assay (ELISA) [6, 42, 43]. However, such approaches require tedious extraction and cleaning steps, and are destructive. In addition, it takes a long time to get the results, and the analysis of the sample is expensive. A more effective means to screen for disease resistance would be simple and rapid. Grain buyers/processors and breeders need such a technique to enhance the detection efficiency for DON content in barley grain. To avoid health risks, it is important to have an effective automated method to assay the metabolite. As far as we know, no research has been carried out using hyperspectral techniques to reliably measure DON content of barley samples. Thus, this study will use hyperspectral imaging to detect DON levels of barley grain. The specific objectives of this study were to: (1) build a discrimination method to classify samples of lower DON levels from higher DON levels; (2) evaluate the performance of different linear and non-linear algorithms for quantitative determination of DON contents.

2. MATERIALS AND METHODS

2.1. Samples Preparation

A total of 1052 barley samples (about 10 g per sample) were selected for imaging. After, the DON concentration (ppm) of the samples were assayed by GC-MS [44]. 931 samples were used for calibration by random selection (with 43 outliers), while the remaining 121 samples were selected for prediction with (5 outliers). After removing outliers, the numbers of samples in the correction set and prediction set were 888 and 166, respectively. Detailed statistics of DON content (without outliers) measured by GC-MS are shown in Table **1**. The upper limit of DON on grains and grain by-products allowed by U.S. Food and Drug Administration (FDA) is 5 ppm for animals with these ingredients less than 40% of their diet and 10 ppm for ruminating dairy cattle, beef, and feedlot cattle older than 4 months [45]. Based on the DON levels, the samples in the calibration set were further classified into 4 categories (I:0–5 ppm, II:5–10 ppm, III:10–15 ppm, and IV:15–35 ppm).

Table 1. Reference values of DON contents measured by GC-MS.

Sample Set	No.	Max	Min	Mean ± SD
Calibration	888	33.9	0	12.105 ± 6.599
Prediction	116	33.1	0	12.224 ± 7.282

Min: Minimum value, Max: Maximum value, SD: Standard Deviation.

2.2. Instrumentation Systems

The hyperspectral images of kernel samples were acquired by a hyperspectral imaging system (382–1030 nm). The hyperspectral system (PlantSpec 10, Middleton Spectral Vision, Middleton, WI, USA) consisted of a dual-line halogen light source (2×13×35W) (MSV Series 13 Illumination, Middleton Spectral Vision, Middleton, WI, USA), a spectrograph (V10E, Specim, Oulu, Finland) with an optical resolution of 1.2 nm, a complementary metal-oxide-semiconductor (CMOS) (MSV500, Middleton Research, Middleton, WI, USA), a conveyer belt system operated by a stepper motor. The spectra of captured hyperspectral images of barley samples were calibrated and extracted by the encoding algorithms [46, 47].

2.3. Machine Learning Algorithms

In this study, machine learning models were developed based on partial least squares regression (PLSR), PLS-based discriminant analysis (PLSDA), locally weighted partial least squares regression (LWPLSR), principal component regression (PCR), support vector machine regression (SVMR), and artificial neural network (ANN). Among them, PLSR and PCR are linear algorithms and SVMR is a method with both linear and non-linear characteristics, while ANN and LWPLSR are non-linear methods. PLSR is a classic regression method that strives to analyze different variables to determine the linear relationship between independent and dependent variables [48]. This method has the advantage of processing spectral data with strong collinearity [49, 50]. LWPLSR is one of several "just in time" modeling methods that can compensate for the lack of linear relationships [51]. The LWPLSR is an effective strategy to overcome the lack of linear relationships and to facilitate the selection of suitable calibration sets [52]. In addition, PCR can extract characteristic representations from high-dimension and collinear data [53]. SVMR is a learning technique in reproducing kernel Hilbert spaces with an ε-insensitive loss function, and this method can deal with high-dimensional input vectors [54, 55]. ANN is useful for solving the same nonlinear problem, but it requires a large number of samples to acquire a reliable model [56]. All models were cross-validated based on 'venetian blinds' with 10 splits.

2.4. Analysis of Outliers

It is critical to detect and remove outliers from a sample set to achieve accurate and reliable estimations. Outliers are samples that deviate from most other samples and may therefore interfere with the accuracy of the model. Many factors can cause outliers, such as measurement errors, mislabeling, noise, extreme samples, and biased processes. Samples can be rated as outliers based on either the X-variables or the Y-variables alone, or both. The samples may not be outliers for two sets of variables. However, after the relationship between X and Y is considered, the samples can be identified as outliers. There are many approaches (*e.g.*, leverages and scores plots of PLSR) to evaluate outliers [57 - 59]. In this study, outliers were flagged according to the limits of studentized residuals and leverage [60]. Samples with high values in studentized residuals and leverage were more likely to become outliers [61, 62]. High residuals mean that samples have large out of model residuals. The leverage is a measure of sample distance from the center of a model. Because the presence of outliers will reduce the accuracy of a model, they should be removed [63].

2.5. Model Assessment

The model performances were assessed by the root mean square error of calibration (RMSEC), prediction (RMSEP) and cross-validation (RMSECV), as well as coefficients of determination in calibration (R^2_C), cross-validation (R^2_{CV}), and prediction (R^2_p). Other parameters including the number of misclassifications, true negatives (refer to specificity), true positives (refer to sensitivity), false negatives, false positives, and the area under curve (AUC) from the receiver operating characteristic (ROC) curve were adopted to evaluate PLSDA [64].

3. RESULT AND DISCUSSION

3.1. Spectral Features

Fig. (**1a**) shows the true color of a representative barley sample. Each spectrum shown in Fig. (**1b**) is the average of hundreds of kernels from each barley line. The spectral signatures of the samples followed a similar pattern but differed in the vibration magnitude of reflectance across the whole spectral region. The molecular structure of a compound is associated with spectral reflectance since chemical bonds can absorb light of specific wavelengths. Spectra in the range of 425 to 780 nm contain the perceptual information related to pigments and chlorophyll [18, 65]. The near-infrared (NIR) wavelength values (833–1030 nm) have been ascribed to water, starch, and cellulose [66]. However, it is difficult to visualize the intrinsic variations of these samples, which means that advanced machine learning methods were required.

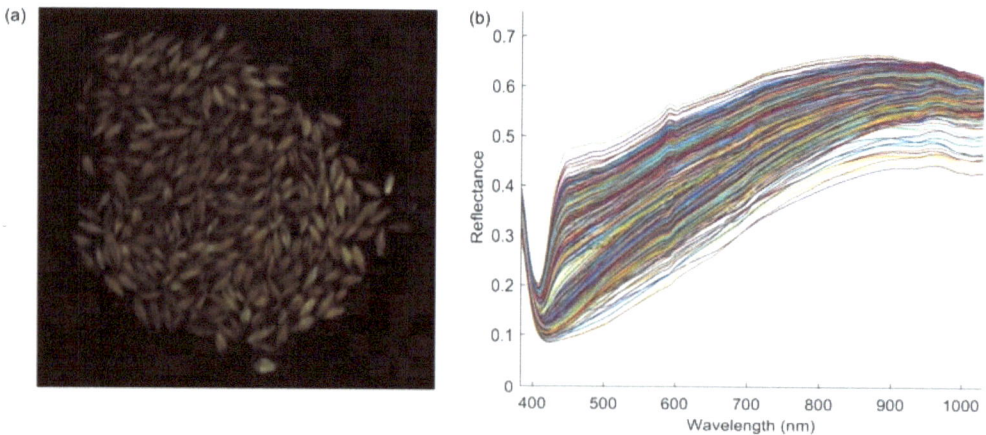

Fig. 1. (a) The true color of a representative barley sample, **(b)** the mean reflectance spectra of the ROI of samples with different levels of DON.

3.2. Outlier Evaluation

Samples with high residual variance or leverage are identified as outliers because they may have a negative influence on modeling. The plot of the standardized residuals *versus* leverage (Figs. **2a** and **b**) confirmed that the majority of the samples in the calibration and prediction sets were concentrated in a squared area within ±3 standard deviation units and the leverage threshold. This result was similar to that mentioned by Gajewicz, *et al.* [67]. The leverage threshold value is defined as 3*L/m, where m the number of samples and L is the number of latent variables [68]. The samples with leverage values above the threshold were inconsistent with most spectral samples and considered potential outliers that destabilized the model [69]. Outliers to be removed can account for up to 22% of the total samples according to guidelines [70]. In this study, 43 samples in the calibration set and 5 samples in the prediction set were determined as potential outliers. These 48 samples were removed, and then the remaining samples were re-evaluated in the model. The new PLSR model was constructed with 888 samples in the calibration set and 116 samples in the prediction set. The results showed that the overall performance of the model significantly improved after potential outliers were eliminated. Different models were then developed, and their performances were evaluated.

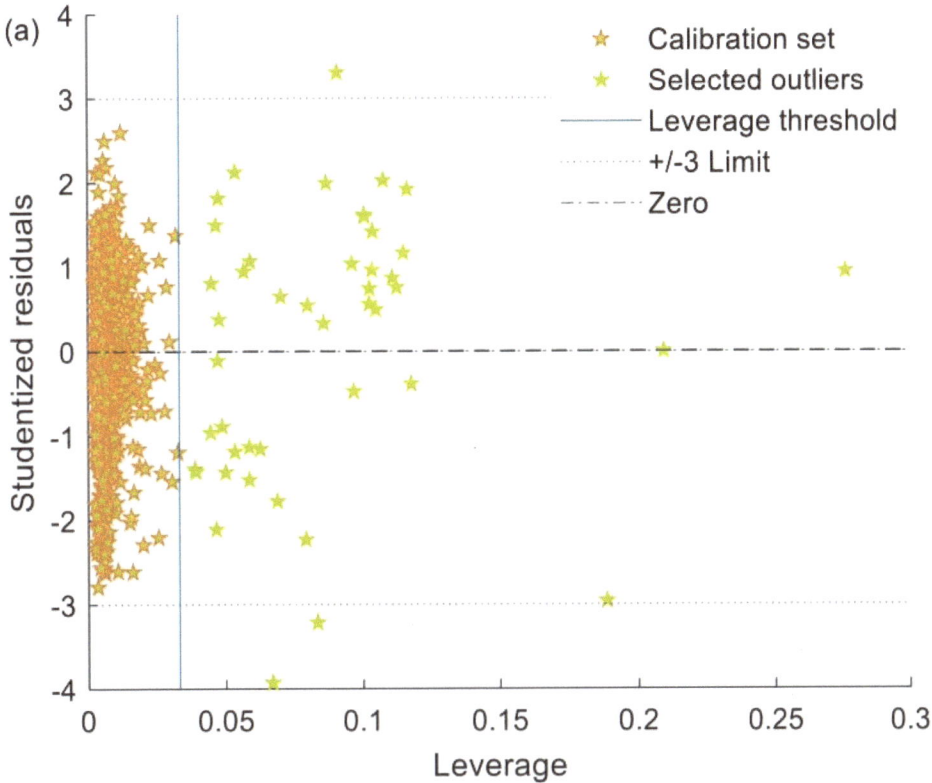

(Fig. 2) contd.....

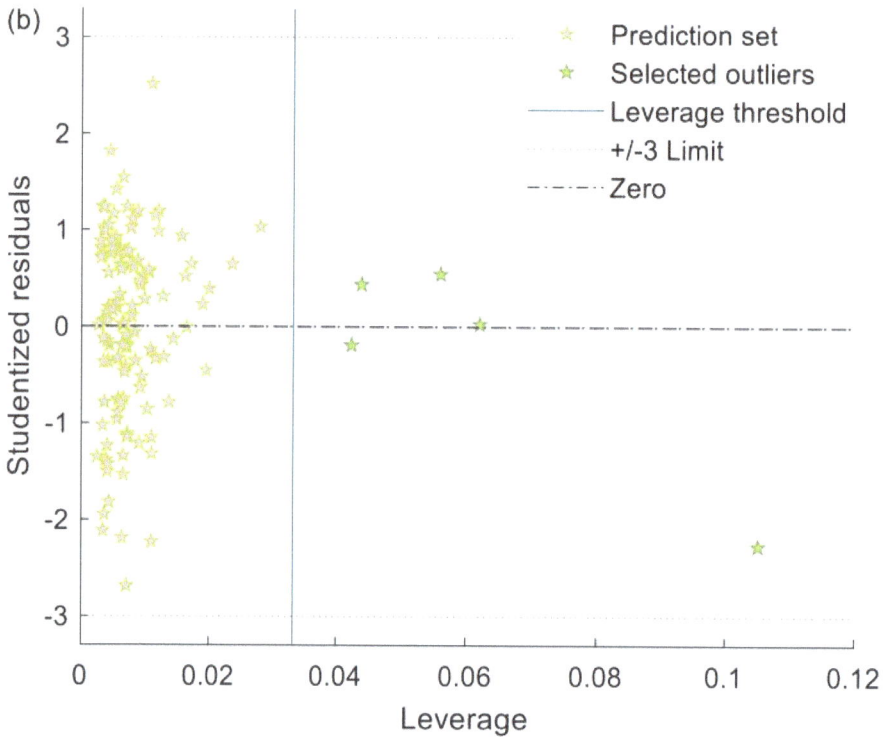

Fig. (2). The plot of residual *versus* leverage for detection of outliers in **(a)** calibration set and **(b)** prediction set.

3.3. Classification of DON Levels

In this study, PLSDA was used as a tool for the discrimination of DON levels. Fig. (**3**) shows the ROC curves, threshold plots, and final classification results. As can be seen in Figs. **3 (a1, a2, a3, a4,)**, each plot presents the sensitivity of the model *versus* the specificity as a function of the selected threshold. The AUC value (0.958) in Fig. (**3(a4)**) was the largest, followed by the values of 0.940 (Fig. **3a3**), 0.917 (Fig. **3(a1)**), and 0.845 (Fig. **3(a2)**). The inflection point of the ROC curve was closer to the upper left corner, indicating that the specificity of the model was higher with less loss of sensitivity. A detailed description of model sensitivity and specificity can be found elsewhere [71]. Figs. **3 (b1, b2, b3, b4,)** shows the model thresholds calculated based on sensitivity and specificity. Figs. **3 (c1, c2, c3, c4,)** display the classification results of different categories of samples based on model thresholds. As can be seen, most samples can be correctly distinguished from the correct class.

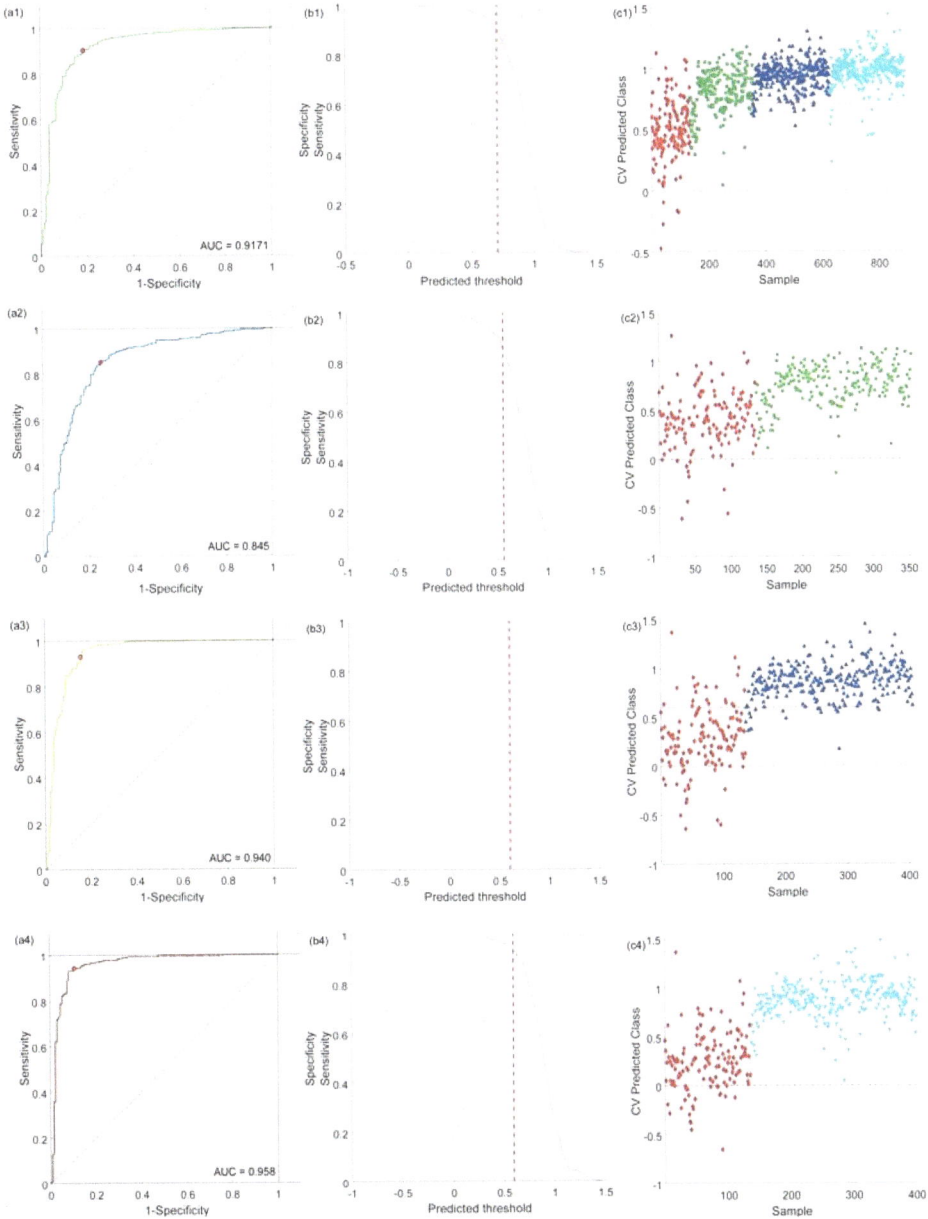

Fig. (3). (a1, a2, a3, a4) Receiver operating characteristic (ROC) curves with the area under curve (AUC) (where the red circle and dotted line denotes the model threshold and 1:1 line, respectively). **(b1, b2, b3, b4)** Threshold plots (where the thin dotted line, solid line, and thick dotted line denote specificity, sensitivity, and model threshold, respectively). **(c1, c2, c3, c4)** Classification predictions (where red diamond, green square, upper triangle, lower triangle, upper dotted line and lower dotted line denote I (0–5 ppm), II (5–10 ppm), III (10–15 ppm), IV (15–35 ppm), threshold and midline, respectively).

Table **2** describes the detailed results of classifying the four categories of DON levels using the PLSDA cross-validation model. The model was evaluated using precision (P) value, F1-score, misclassification error (Err), and Matthews correlation coefficient in cross-validation (M-RCV). More information about these parameters can be found elsewhere [25]. It was noticed that samples in class I were more easily identified from the samples in class III (10–15 ppm) or class IV (15–35 ppm), with accuracies of 0.773 or 0.823, respectively. However, the models containing the samples of class II tended to show worse performance than those without using the samples from class II. This demonstrates that the samples in class I (DON: 0–5 ppm) had the smallest difference from the sample in class II (DON: 5–10 ppm), while the difference between the samples in class I and the samples in class III (DON: 10–15 ppm) or class IV (DON: 15–30 ppm) was larger. This difference gradually increased as the difference in DON content between different groups increased. Overall, PLSDA could be used to non-invasively screen DON of lower levels (<5 ppm) from higher levels.

Table 2. Performance of the cross-validation model of PLSDA for classification of four categories of DON levels.

Class	N	Err	P	F1	Predicted as Class A	Predicted as Class B	Predicted as Unassigned	M-R_{cv}
A (I)	135	0.108	0.610	0.692	108	24	3	
B (II,III, and IV)	753	0.113	0.966	0.931	69	677	7	0.635
A (I)	135	0.184	0.782	0.749	97	34	4	
B (II)	218	0.190	0.845	0.847	27	185	6	0.598
A (I)	135	0.094	0.876	0.856	113	20	2	
B (III)	270	0.101	0.926	0.924	16	249	5	0.773
A (I)	135	0.073	0.896	0.892	120	14	1	
B (IV)	265	0.080	0.946	0.939	14	247	4	0.823

N: Number of samples belonging to each class, Err: Misclassification error, P: Precision, F1: F1 Score, M-RCV: Matthews correlation coefficient in cross-validation.

3.4. Quantitative Determination of DON Contents

The performances of five algorithms, including one linear/nonlinear model (SVMR), two linear models (PLSR and PCR), and two nonlinear models (ANN and LWPLSR) for quantitative detection of DON content are described in Table **3**. Among models with linear features, PLSR had the highest accuracy for DON detection. As can be seen, the PLSR model achieved higher performance than SVMR and PCR in both the cross-validation set ($R^2_{cv} = 0.517$, RMSECV = 4.585) and prediction set (R^2_p = 0.579, RMSEP = 4.746). Compared with PLSR, better results were achieved by nonlinear models, including ANN and LWPLSR. When only nonlinear models were considered, the LWPLSR model obtained the best prediction result in the determination of DON content of grain samples (Fig. **4**), yielding higher R^2_p (0.728) and lower RMSEP (3.802). Therefore, the predictive performance of nonlinear models was significantly higher than that of linear models. LWPLSR was a more suitable model for detecting DON content than PLSR and ANN.

Table 3. Performance of full wavelength models for determination of DON concentration.

Model	No. LV	Calibration			Cross-validation		Prediction		
		NO.	R^2_C	RMSEC	R^2_{CV}	RMSECV	NO.	R^2_p	RMSEP
SVMR	-	888	0.539	4.528	0.468	4.866	116	0.543	5.023
PLSR	10	888	0.533	4.510	0.517	4.585	116	0.579	4.746
PCR	7	888	0.412	5.062	0.400	5.110	116	0.491	5.240
ANN	-	888	0.617	4.242	0.518	4.772	116	0.645	4.384
LWPLSR	-	888	0.800	2.958	0.623	4.083	116	0.728	3.802

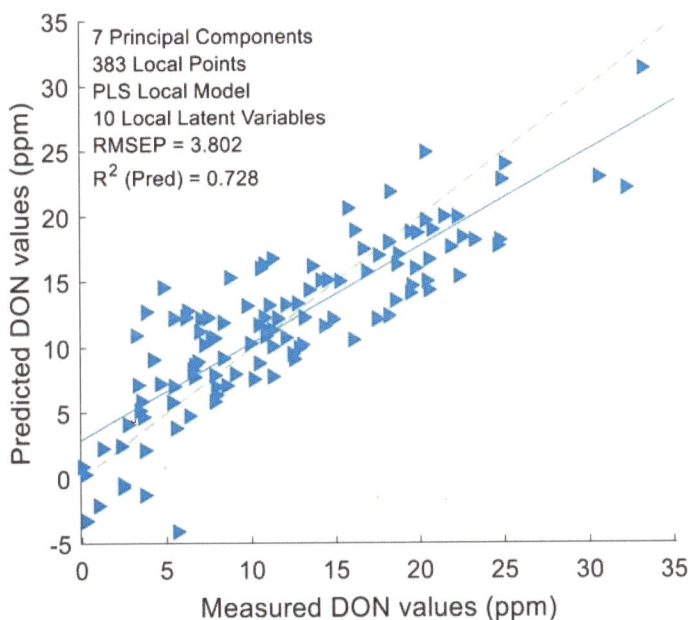

Fig. (4). Scatter plot of predicted *versus* GC-MS measured DON for the full-wavelength LWPLSR model with respect to the fitted line (continuous line) and ideal line (dotted line).

4. DISCUSSION

Previous research has also shown that hyperspectral imaging can assess symptoms caused by fungi in various food products such as citrus fruits [72, 73], peanuts [74], and maize kernels [75, 76]. Effective methods are needed to analyze hyperspectral data with high complexity. Non-linear algorithms, including ANN and LWPLSR, showed stronger prediction performance than algorithms with linear characteristics. Other studies have also confirmed the superiority of non-linear algorithms [20, 77]. In the present study, the accuracy of the PLSDA discriminant model increased as the gap of DON among different categories (I, II, III, and IV) enlarged. This proved the algorithm was feasible to classify different levels of DON. The accuracy to classify samples in group I from II was lower than that of classifying group I from the group containing II, III, and IV. This result could be due to the similarities between group I and group II more than the similarities between group I and the group, which included samples of II, III, and IV. For selection purposes, it would be helpful to have greater resolution and predictive power with lines that are lower DON (*e.g.* < 5 ppm in inoculate nurseries, or < 10 ppm). Samples greater than 10 ppm are likely to be discarded, but the variability of the sample concentration (0-10 ppm) is relatively narrow, which is not conducive to the construction of a robust model. In our study, any sample of unknown concentrations within 35 ppm (including DON contents of less than 10 ppm) can be quickly and accurately detected.

CONCLUSION

This study thoroughly evaluated the feasibility of combining hyperspectral imaging with machine learning to sort barley kernels into different classes depending on their DON levels and to reliably assess DON in barley kernels. This technique can be used for discriminating DON content of kernels in the low-level range from those in the high-level ranges. Although PLSR was more efficient than other models (SVMR and PCR) with linear regression characteristics, nonlinear models (LWPLSR and ANN) performed much better than all linear models used. LWPLSR worked best of all for the determination of DON content. However, since the results obtained were based on preliminary exploration, future studies will use samples with large variability from different origins, batches, and harvest seasons to verify the reliability of the technique.

CONSENT FOR PUBLICATION

Not applicable.

CONFLICT OF INTEREST

The authors confirm that this chapter contents have no conflict of interest.

ACKNOWLEDGEMENTS

The authors would like to acknowledge the USDA-ARS U.S. Wheat and Barley Scab Initiative (Funding No. 58-5062-8-018) supported this research.

REFERENCES

[1]   W-H. Su, S. Bakalis, and D-W. Sun, "Fourier transform mid-infrared-attenuated total reflectance (FTMIR-ATR) microspectroscopy for determining textural property of microwave baked tuber", *J. Food Eng.*, vol. 218, pp. 1-13, 2018.
[http://dx.doi.org/10.1016/j.jfoodeng.2017.08.016]

[2]   C. Yang, W.S. Lee, and P. Gader, "Hyperspectral band selection for detecting different blueberry fruit maturity stages", *Comput. Electron. Agric.*, vol. 109, pp. 23-31, 2014.
[http://dx.doi.org/10.1016/j.compag.2014.08.009]

[3]   W-H. Su, and D-W. Sun, "Chemical imaging for measuring the time series variations of tuber dry matter and starch concentration", *Comput. Electron. Agric.*, vol. 140, pp. 361-373, 2017.
[http://dx.doi.org/10.1016/j.compag.2017.06.013]

[4]   W-H. Su, S. Bakalis, and D-W. Sun, "Fingerprinting study of tuber ultimate compressive strength at different microwave drying times using mid-infrared imaging spectroscopy", *Dry. Technol.*, vol. 37, no. 9, pp. 1113-1130, 2019.
[http://dx.doi.org/10.1080/07373937.2018.1487450]

[5]   W. Su, "Detection of external defects on potatoes by hyperspectral imaging technology and image processing method", *Journal of Zhejiang University [Agriculture and Life Sciences].*, vol. 40, no. 2, pp. 188-196, 2014.

[6]   W-H. Su, I.S. Arvanitoyannis, and D-W. Sun, "Trends in food authentication", In: *Modern Techniques for Food Authentication.*, D-W. Su, Ed., Elsevier, 2018, pp. 731-758.
[http://dx.doi.org/10.1016/B978-0-12-814264-6.00018-9]

[7]   P.M. Falcone, A. Baiano, A. Conte, L. Mancini, G. Tromba, F. Zanini, and M.A. Del Nobile, "Imaging techniques for the study of food microstructure: a review", *Adv. Food Nutr. Res.*, vol. 51, pp. 205-263, 2006.
[http://dx.doi.org/10.1016/S1043-4526(06)51004-6] [PMID: 17011477]

[8]   W-H. Su, S.A. Fennimore, and D.C. Slaughter, "Fluorescence imaging for rapid monitoring of translocation behaviour of systemic markers in snap beans for automated crop/weed discrimination", *Biosyst. Eng.*, vol. 186, pp. 156-167, 2019.
[http://dx.doi.org/10.1016/j.biosystemseng.2019.07.009]

[9]   R. Qiu, C. Yang, A. Moghimi, M. Zhang, B.J. Steffenson, and C.D. Hirsch, "Detection of fusarium head blight in wheat using a deep neural network and color imaging", *Remote Sens.*, vol. 11, no. 22, p. 2658, 2019.
[http://dx.doi.org/10.3390/rs11222658]

[10]   W-H. Su, S.A. Fennimore, and D.C. Slaughter, "Development of a systemic crop signalling system for automated real-time plant care in vegetable crops", *Biosyst. Eng.*, vol. 193, pp. 62-74, 2020.
[http://dx.doi.org/10.1016/j.biosystemseng.2020.02.011]

[11]   W-H. Su, S.A. Fennimore, and D.C. Slaughter, "Computer vision technology for identification of snap bean crops using systemic rhodamine B", In: *2019 ASABE Annual International Meeting.* American Society of Agricultural and Biological Engineers., 2019, p. 1.

[http://dx.doi.org/10.13031/aim.201900075]

[12] W-H. Su, D.C. Slaughter, and S.A. Fennimore, "Non-destructive evaluation of photostability of crop signaling compounds and dose effects on celery vigor for precision plant identification using computer vision", *Comput. Electron. Agric.,* vol. 168, 2020.105155
[http://dx.doi.org/10.1016/j.compag.2019.105155]

[13] C. Xie, C. Yang, and Y. He, "Hyperspectral imaging for classification of healthy and gray mold diseased tomato leaves with different infection severities", *Comput. Electron. Agric.,* vol. 135, pp. 154-162, 2017.
[http://dx.doi.org/10.1016/j.compag.2016.12.015]

[14] W-H. Su, and D-W. Sun, "Multivariate analysis of hyper/multi-spectra for determining volatile compounds and visualizing cooking degree during low-temperature baking of tubers", *Comput. Electron. Agric.,* vol. 127, pp. 561-571, 2016.
[http://dx.doi.org/10.1016/j.compag.2016.07.007]

[15] W-H. Su, and D-W. Sun, "Evaluation of spectral imaging for inspection of adulterants in terms of common wheat flour, cassava flour and corn flour in organic Avatar wheat (Triticum spp.) flour", *J. Food Eng.,* vol. 200, pp. 59-69, 2017.
[http://dx.doi.org/10.1016/j.jfoodeng.2016.12.014]

[16] W-H. Su, S. Bakalis, and D-W. Sun, "Advanced applications of near/mid-infrared (NIR/MIR) imaging spectroscopy for rapid prediction of potato and sweet potato moisture contents", In: *2019 ASABE Annual International Meeting.* American Society of Agricultural and Biological Engineers., 2019, p. 1.
[http://dx.doi.org/10.13031/aim.201900121]

[17] W-H. Su, S. Bakalis, and D-W. Sun, "NIR/MIR spectroscopy in tandem with chemometrics for rapid identification and evaluation of potato variety and doneness degree", In: *2019 ASABE Annual International Meeting.* American Society of Agricultural and Biological Engineers., 2019, p. 1.

[18] W-H. Su, H-J. He, and D-W. Sun, "Non-Destructive and rapid evaluation of staple foods quality by using spectroscopic techniques: A review", *Crit. Rev. Food Sci. Nutr.,* vol. 57, no. 5, pp. 1039-1051, 2017.
[http://dx.doi.org/10.1080/10408398.2015.1082966] [PMID: 26480047]

[19] W.H. Su, and D.W. Sun, "Multispectral imaging for plant food quality analysis and visualization", *Compr. Rev. Food Sci. Food Saf.,* vol. 17, no. 1, pp. 220-239, 2018.
[http://dx.doi.org/10.1111/1541-4337.12317]

[20] W.H. Su, and D.W. Sun, "Fourier transform infrared and Raman and hyperspectral imaging techniques for quality determinations of powdery foods: a review", *Compr. Rev. Food Sci. Food Saf.,* vol. 17, no. 1, pp. 104-122, 2018.
[http://dx.doi.org/10.1111/1541-4337.12314]

[21] W-H. Su, and D-W. Sun, "Potential of hyperspectral imaging for visual authentication of sliced organic potatoes from potato and sweet potato tubers and rapid grading of the tubers according to moisture proportion", *Comput. Electron. Agric.,* vol. 125, pp. 113-124, 2016.
[http://dx.doi.org/10.1016/j.compag.2016.04.034]

[22] W-H. Su, and D-W. Sun, "Advanced analysis of roots and tubers by hyperspectral techniques", In: *Advances in Food and Nutrition Research.* vol. 87. Elsevier, 2019, pp. 255-303.

[23] W-H. Su, and D-W. Sun, "Mid-infrared (MIR) spectroscopy for quality analysis of liquid foods", *Food Eng. Rev.,* pp. 1-17, 2019.
[http://dx.doi.org/10.1007/s12393-019-09191-2]

[24] W-H. Su, S. Bakalis, and D-W. Sun, "Potato hierarchical clustering and doneness degree determination by near-infrared (NIR) and attenuated total reflectance mid-infrared (ATR-MIR) spectroscopy", *J. Food Meas. Charact.,* vol. 13, no. 2, pp. 1218-1231, 2019.
[http://dx.doi.org/10.1007/s11694-019-00037-3]

[25] W-H. Su, D-W. Sun, J-G. He, and L-B. Zhang, "Variation analysis in spectral indices of volatile chlorpyrifos and non-volatile imidacloprid in jujube (*Ziziphus jujuba* Mill.) using near-infrared hyperspectral imaging (NIR-HSI) and gas chromatograph-mass spectrometry (GC–MS)", *Comput. Electron. Agric.,* vol. 139, pp. 41-55, 2017.
[http://dx.doi.org/10.1016/j.compag.2017.04.017]

[26] A. Moghimi, C. Yang, M.E. Miller, S.F. Kianian, and P.M. Marchetto, "A novel approach to assess salt stress tolerance in wheat using hyperspectral imaging", *Front. Plant Sci.,* vol. 9, p. 1182, 2018.
[http://dx.doi.org/10.3389/fpls.2018.01182] [PMID: 30197650]

[27] W-H. Su, S. Bakalis, and D-W. Sun, "Chemometric determination of time series moisture in both potato and sweet potato tubers during hot air and microwave drying using near/mid-infrared (NIR/MIR) hyperspectral techniques", *Dry. Technol.,* vol. 38, no. 5-6, pp. 806-823, 2020.
[http://dx.doi.org/10.1080/07373937.2019.1593192]

[28] M. McMullen, G. Bergstrom, E. De Wolf, R. Dill-Macky, D. Hershman, G. Shaner, and D. Van Sanford, "A unified effort to fight an enemy of wheat and barley: Fusarium head blight", *Plant Dis.,* vol. 96, no. 12, pp. 1712-1728, 2012.
[http://dx.doi.org/10.1094/PDIS-03-12-0291-FE] [PMID: 30727259]

[29] S. Stenglein, "Fusarium poae: a pathogen that needs more attention", *J. Plant Pathol.,* pp. 25-36, 2009.

[30] E. Alisaac, J. Behmann, M. Kuska, H-W. Dehne, and A-K. Mahlein, "Hyperspectral quantification of wheat resistance to Fusarium head blight: comparison of two Fusarium species", *Eur. J. Plant Pathol.,* vol. 152, no. 4, pp. 869-884, 2018.
[http://dx.doi.org/10.1007/s10658-018-1505-9]

[31] J. Pedersen, "Distribution of deoxynivalenol and zearalenone in milled fractions of wheat'", *Cereal Chem.,* vol. 73, no. 3, pp. 388-391, 1996.

[32] R.C. de la Pena, K.P. Smith, F. Capettini, G.J. Muehlbauer, M. Gallo-Meagher, R. Dill-Macky, D.A. Somers, and D.C. Rasmusson, "Quantitative trait loci associated with resistance to Fusarium head blight and kernel discoloration in barley", *Theor. Appl. Genet.,* vol. 99, no. 3-4, pp. 561-569, 1999.
[http://dx.doi.org/10.1007/s001220051269] [PMID: 22665190]

[33] S.V. Pathre, and C.J. Mirocha, "Analysis of deoxynivalenol from cultures of Fusarium species", *Appl. Environ. Microbiol.,* vol. 35, no. 5, pp. 992-994, 1978.
[http://dx.doi.org/10.1128/AEM.35.5.992-994.1978] [PMID: 655717]

[34] C.A. Urrea, R.D. Horsley, B.J. Steffenson, and P.B. Schwarz, "Heritability of fusarium head blight resistance and deoxynivalenol accumulation from barley accession CIho 4196", *Crop Sci.,* vol. 42, no. 5, pp. 1404-1408, 2002.
[http://dx.doi.org/10.2135/cropsci2002.1404]

[35] H. Buerstmayr, T. Ban, and J.A. Anderson, "QTL mapping and marker-assisted selection for Fusarium head blight resistance in wheat: a review", *Plant Breed.,* vol. 128, no. 1, pp. 1-26, 2009.
[http://dx.doi.org/10.1111/j.1439-0523.2008.01550.x]

[36] R.D. Horsley, "Identification of QTLs associated with Fusarium head blight resistance in barley accession CIho 4196", *Crop Sci.,* vol. 46, no. 1, pp. 145-156, 2006.
[http://dx.doi.org/10.2135/cropsci2005.0247]

[37] L.S. Dahleen, H.A. Agrama, R.D. Horsley, B.J. Steffenson, P.B. Schwarz, A. Mesfin, and J.D. Franckowiak, "Identification of QTLs associated with Fusarium head blight resistance in Zhedar 2 barley", *Theor. Appl. Genet.,* vol. 108, no. 1, pp. 95-104, 2003.
[http://dx.doi.org/10.1007/s00122-003-1409-7] [PMID: 14556050]

[38] D. Wu, and D.W. Sun, "The use of hyperspectral techniques in evaluating quality and safety of meat and meat products", In: *Emerging Technologies in Meat Processing: Production.,* E. J. Cummins, J.G. Lyng, Eds., Processing and Technology, 2016, pp. 345-374.
[http://dx.doi.org/10.1002/9781118350676.ch13]

[39]   D-W. Sun, *Computer Vision Technology for Food Quality Evaluation.* Academic Press: Cambridge, Massachusetts, United States, 2016.

[40]   F. Zhu, H. Zhang, Y. Shao, Y. He, and M. Ngadi, "Mapping of fat and moisture distribution in Atlantic salmon using near-infrared hyperspectral imaging", *Food Bioprocess Technol.,* vol. 7, no. 4, pp. 1208-1214, 2014.
[http://dx.doi.org/10.1007/s11947-013-1228-z]

[41]   T.G. Fetch Jr, and B.J. Steffenson, "Rating scales for assessing infection responses of barley infected with *Cochliobolus sativus*", *Plant Dis.,* vol. 83, no. 3, pp. 213-217, 1999.
[http://dx.doi.org/10.1094/PDIS.1999.83.3.213] [PMID: 30845496]

[42]   J. Gilbert, M.J. Shepherd, and J.R. Startin, "A survey of the occurrence of the trichothecene mycotoxin deoxynivalenol (vomitoxin) in UK grown barley and in imported maize by combined gas chromatography-mass spectrometry", *J. Sci. Food Agric.,* vol. 34, no. 1, pp. 86-92, 1983.
[http://dx.doi.org/10.1002/jsfa.2740340113] [PMID: 6843094]

[43]   Y-C. Xu, G.S. Zhang, and F.S. Chu, "Enzyme-linked immunosorbent assay for deoxynivalenol in corn and wheat", *J. Assoc. Off. Anal. Chem.,* vol. 71, no. 5, pp. 945-949, 1988.
[http://dx.doi.org/10.1093/jaoac/71.5.945] [PMID: 3235414]

[44]   Y. Dong, B.J. Steffenson, and C.J. Mirocha, "Analysis of ergosterol in single kernel and ground grain by gas chromatography-mass spectrometry", *J. Agric. Food Chem.,* vol. 54, no. 12, pp. 4121-4125, 2006.
[http://dx.doi.org/10.1021/jf060149f] [PMID: 16756335]

[45]   R. H. Proctor, T. M. Hohn, and S. P. McCormick, "Reduced virulence of *Gibberella zeae* caused by disruption of a trichothecene toxin biosynthetic gene",

[46]   W-H. Su, and D-W. Sun, "Comparative assessment of feature-wavelength eligibility for measurement of water binding capacity and specific gravity of tuber using diverse spectral indices stemmed from hyperspectral images", *Comput. Electron. Agric.,* vol. 130, pp. 69-82, 2016.
[http://dx.doi.org/10.1016/j.compag.2016.09.015]

[47]   M. Kamruzzaman, Y. Makino, and S. Oshita, "Rapid and non-destructive detection of chicken adulteration in minced beef using visible near-infrared hyperspectral imaging and machine learning", *J. Food Eng.,* vol. 170, pp. 8-15, 2016.
[http://dx.doi.org/10.1016/j.jfoodeng.2015.08.023]

[48]   P. Geladi, and B.R. Kowalski, "Partial least-squares regression: a tutorial", *Anal. Chim. Acta,* vol. 185, pp. 1-17, 1986.
[http://dx.doi.org/10.1016/0003-2670(86)80028-9]

[49]   V. Gaydou, J. Kister, and N. Dupuy, "Evaluation of multiblock NIR/MIR PLS predictive models to detect adulteration of diesel/biodiesel blends by vegetal oil", *Chemom. Intell. Lab. Syst.,* vol. 106, no. 2, pp. 190-197, 2011.
[http://dx.doi.org/10.1016/j.chemolab.2010.05.002]

[50]   N. Dupuy, O. Galtier, D. Ollivier, P. Vanloot, and J. Artaud, "Comparison between NIR, MIR, concatenated NIR and MIR analysis and hierarchical PLS model. Application to virgin olive oil analysis", *Anal. Chim. Acta,* vol. 666, no. 1-2, pp. 23-31, 2010.
[http://dx.doi.org/10.1016/j.aca.2010.03.034] [PMID: 20433960]

[51]   H. Nakagawa, T. Tajima, M. Kano, S. Kim, S. Hasebe, T. Suzuki, and H. Nakagami, "Evaluation of infrared-reflection absorption spectroscopy measurement and locally weighted partial least-squares for rapid analysis of residual drug substances in cleaning processes", *Anal. Chem.,* vol. 84, no. 8, pp. 3820-3826, 2012.
[http://dx.doi.org/10.1021/ac202443a] [PMID: 22449097]

[52]   S. Kim, M. Kano, H. Nakagawa, and S. Hasebe, "Estimation of active pharmaceutical ingredients content using locally weighted partial least squares and statistical wavelength selection", *Int. J.*

Pharm., vol. 421, no. 2, pp. 269-274, 2011.
[http://dx.doi.org/10.1016/j.ijpharm.2011.10.007] [PMID: 22001843]

[53]   X. Yuan, B. Huang, Z. Ge, and Z. Song, "Double locally weighted principal component regression for soft sensor with sample selection under supervised latent structure", *Chemom. Intell. Lab. Syst.,* vol. 153, pp. 116-125, 2016.
[http://dx.doi.org/10.1016/j.chemolab.2016.02.014]

[54]   X. Peng, T. Shi, A. Song, Y. Chen, and W. Gao, "Estimating soil organic carbon using VIS/NIR spectroscopy with SVMR and SPA methods", *Remote Sens.,* vol. 6, no. 4, pp. 2699-2717, 2014.
[http://dx.doi.org/10.3390/rs6042699]

[55]   H. Tong, D-R. Chen, and L. Peng, "Analysis of support vector machines regression", *Found. Comput. Math.,* vol. 9, no. 2, pp. 243-257, 2009.
[http://dx.doi.org/10.1007/s10208-008-9026-0]

[56]   S. Agatonovic-Kustrin, and R. Beresford, "Basic concepts of artificial neural network (ANN) modeling and its application in pharmaceutical research", *J. Pharm. Biomed. Anal.,* vol. 22, no. 5, pp. 717-727, 2000.
[http://dx.doi.org/10.1016/S0731-7085(99)00272-1] [PMID: 10815714]

[57]   S. Verboven, and M. Hubert, "LIBRA: a MATLAB library for robust analysis", *Chemom. Intell. Lab. Syst.,* vol. 75, no. 2, pp. 127-136, 2005.
[http://dx.doi.org/10.1016/j.chemolab.2004.06.003]

[58]   J.F. Pierna, F. Wahl, O. De Noord, and D. Massart, "Methods for outlier detection in prediction", *Chemom. Intell. Lab. Syst.,* vol. 63, no. 1, pp. 27-39, 2002.
[http://dx.doi.org/10.1016/S0169-7439(02)00034-5]

[59]   D. Cousineau, and S. Chartier, "Outliers detection and treatment: a review", *Int. J. Psychol. Res. (Medellin),* vol. 3, no. 1, pp. 58-67, 2010.
[http://dx.doi.org/10.21500/20112084.844]

[60]   R.J. Pell, "Multiple outlier detection for multivariate calibration using robust statistical techniques", *Chemom. Intell. Lab. Syst.,* vol. 52, no. 1, pp. 87-104, 2000.
[http://dx.doi.org/10.1016/S0169-7439(00)00082-4]

[61]   D. Blatná, "Outliers in regression", *Trutnov,* vol. 30, pp. 2006-03.09, 2006.

[62]   E. Ranganai, "On studentized residuals in the quantile regression framework", *Springerplus,* vol. 5, no. 1, p. 1231, 2016.
[http://dx.doi.org/10.1186/s40064-016-2898-6] [PMID: 27536515]

[63]   N. Reis, A.S. Franca, and L.S. Oliveira, "Quantitative evaluation of multiple adulterants in roasted coffee by Diffuse Reflectance Infrared Fourier Transform Spectroscopy (DRIFTS) and chemometrics", *Talanta,* vol. 115, pp. 563-568, 2013.
[http://dx.doi.org/10.1016/j.talanta.2013.06.004] [PMID: 24054633]

[64]   J.A. Westerhuis, "Assessment of PLSDA cross validation", *Metabolomics,* vol. 4, no. 1, pp. 81-89, 2008.
[http://dx.doi.org/10.1007/s11306-007-0099-6]

[65]   G. Tiwari, D.C. Slaughter, and M. Cantwell, "Nondestructive maturity determination in green tomatoes using a handheld visible and near infrared instrument", *Postharvest Biol. Technol.,* vol. 86, pp. 221-229, 2013.
[http://dx.doi.org/10.1016/j.postharvbio.2013.07.009]

[66]   C. Jantra, D.C. Slaughter, P-S. Liang, and S. Pathaveerat, "Nondestructive determination of dry matter and soluble solids content in dehydrator onions and garlic using a handheld visible and near infrared instrument", *Postharvest Biol. Technol.,* vol. 133, pp. 98-103, 2017.
[http://dx.doi.org/10.1016/j.postharvbio.2017.07.007]

[67]   A. Gajewicz, M. Haranczyk, and T. Puzyn, "Predicting logarithmic values of the subcooled liquid

vapor pressure of halogenated persistent organic pollutants with QSPR: How different are chlorinated and brominated congeners?", *Atmos. Environ.,* vol. 44, no. 11, pp. 1428-1436, 2010.
[http://dx.doi.org/10.1016/j.atmosenv.2010.01.041]

[68] A.M. Pedro, and M.M. Ferreira, "Simultaneously calibrating solids, sugars and acidity of tomato products using PLS2 and NIR spectroscopy", *Anal. Chim. Acta,* vol. 595, no. 1-2, pp. 221-227, 2007.
[http://dx.doi.org/10.1016/j.aca.2007.03.036] [PMID: 17606004]

[69] P.E. Flecher, J.B. Cooper, T.M. Vess, and W.T. Welch, "Remote fiber optic Raman analysis of benzene, toulene, and ethylbenzene in mock petroleum fuels using partial least squares regression analysis", *Spectrochim. Acta A Mol. Biomol. Spectrosc.,* vol. 52, no. 10, pp. 1235-1244, 1996.
[http://dx.doi.org/10.1016/0584-8539(96)01660-1]

[70] B.G. Botelho, B.A. Mendes, and M.M. Sena, "Development and analytical validation of robust near-infrared multivariate calibration models for the quality inspection control of mozzarella cheese", *Food Anal. Methods,* vol. 6, no. 3, pp. 881-891, 2013.
[http://dx.doi.org/10.1007/s12161-012-9498-z]

[71] W-H. Su, and D-W. Sun, "Facilitated wavelength selection and model development for rapid determination of the purity of organic spelt (*Triticum spelta* L.) flour using spectral imaging", *Talanta,* vol. 155, pp. 347-357, 2016.
[http://dx.doi.org/10.1016/j.talanta.2016.04.041] [PMID: 27216692]

[72] D. Lorente, J. Blasco, A. Serrano, E. Soria-Olivas, N. Aleixos, and J. Gómez-Sanchis, "Comparison of ROC feature selection method for the detection of decay in citrus fruit using hyperspectral images", *Food Bioprocess Technol.,* vol. 6, no. 12, pp. 3613-3619, 2013.
[http://dx.doi.org/10.1007/s11947-012-0951-1]

[73] J. Li, W. Huang, X. Tian, C. Wang, S. Fan, and C. Zhao, "Fast detection and visualization of early decay in citrus using Vis-NIR hyperspectral imaging", *Comput. Electron. Agric.,* vol. 127, pp. 582-592, 2016.
[http://dx.doi.org/10.1016/j.compag.2016.07.016]

[74] J. Jiang, X. Qiao, and R. He, "Use of Near-Infrared hyperspectral images to identify moldy peanuts", *J. Food Eng.,* vol. 169, pp. 284-290, 2016.
[http://dx.doi.org/10.1016/j.jfoodeng.2015.09.013]

[75] W. Wang, G.W. Heitschmidt, X. Ni, W.R. Windham, S. Hawkins, and X. Chu, "Identification of aflatoxin B1 on maize kernel surfaces using hyperspectral imaging", *Food Control,* vol. 42, pp. 78-86, 2014.
[http://dx.doi.org/10.1016/j.foodcont.2014.01.038]

[76] H. Yao, Z. Hruska, R. Kincaid, R.L. Brown, D. Bhatnagar, and T.E. Cleveland, "Detecting maize inoculated with toxigenic and atoxigenic fungal strains with fluorescence hyperspectral imagery", *Biosyst. Eng.,* vol. 115, no. 2, pp. 125-135, 2013.
[http://dx.doi.org/10.1016/j.biosystemseng.2013.03.006]

[77] Q. Dai, J-H. Cheng, D-W. Sun, Z. Zhu, and H. Pu, "Prediction of total volatile basic nitrogen contents using wavelet features from visible/near-infrared hyperspectral images of prawn (*Metapenaeus ensis*)", *Food Chem.,* vol. 197, no. Pt A, pp. 257-265, 2016.
[http://dx.doi.org/10.1016/j.foodchem.2015.10.073] [PMID: 26616948]

CHAPTER 6

Evaluation of Fungal Contaminants in Agricultural Products by Hyperspectral Imaging

Feifei Tao[1], Haibo Yao[1,*], Zuzana Hruska[1] and Kanniah Rajasekaran[2]

[1] *Geosystems Research Institute, Mississippi State University, Building 1021, Stennis Space Center, Hancock, MS 39529, USA*

[2] *USDA-ARS, Southern Regional Research Center, New Orleans, LA 70124, USA*

Abstract: Optical-based technologies offer significant advantages compared with conventional methods for detecting mycotoxin and fungal contamination in agricultural and food commodities, such as rapidness and non-destructiveness. Hyperspectral imaging (HSI) integrates traditional imaging and spectroscopy technologies and thus makes it possible for high-throughput screening analysis in an onsite or on-line manner. Currently, HSI, in tandem with modern chemometrics, has demonstrated interesting and promising results for the detection of mycotoxin and fungal contamination in varieties of agricultural products. Therefore, the objective of this chapter is to give an overview of current research advances of HSI in both fluorescence and reflectance modes for the evaluation of mycotoxin and fungal contamination in agricultural and food commodities. Advances of HSI in evaluation of the main mycotoxins, including aflatoxins, ochratoxins, deoxynivalenol, fumonisins and their related fungal contaminants, are reviewed, and the results obtained from different studies are compared and discussed. Perspectives on its future trends and challenges concerning mycotoxin and fungal evaluation are also addressed.

Keywords: Chemometrics, Fluorescence, Fungus, Hyperspectral imaging, Mycotoxin, Rapid and nondestructive detection, Reflectance.

1. INTRODUCTION

1.1. Major Fungal Contaminants in Agricultural Products

Mycotoxins are toxic secondary metabolites mainly produced by different fungal species such as *Aspergillus* (*A.*), *Penicillium* (*P.*) and *Fusarium* (*F.*) [1]. Mycotoxin contamination frequently occurs in various food and feed commodities, leading to human and animal health risks at the global level. If ingested, mycotoxins may cause acute or chronic disease episodes, with carcino-

---

* **Corresponding author Haibo Yao:** Geosystems Research Institute, Mississippi State University, Building 1021, Stennis Space Center, Hancock, MS 39529, USA; Tel: +1 2286883742; E-mail: haibo@gri.msstate.edu

Jiangbo Li & Zhao Zhang (Eds.)

genic, mutagenic, teratogenic, estrogenic, hemorrhagic, nephrotoxic, hepatotoxic, neurotoxic and/or immunosuppressive effects [2]. To date, hundreds of mycotoxins have already been identified, but the most important ones regarding their prevalence in contaminated agricultural products are aflatoxins (AFs), ochratoxins (OTs), deoxynivalenol (DON), fumonisins (FMs), zearalenone (ZEN) and patulin [1]. Some of these have been classified by the world health organization (WHO) as human carcinogens. For instance, AFs are identified as human carcinogens (Group 1); OTs and FMs are classified as possible human carcinogens (Group 2B) [3]. The chemical structures of these main mycotoxins can be found in the study by Agriopoulou *et al.* [4].

AFs are among the most poisonous mycotoxins and are produced by certain fungi of the genus *Aspergillus*, predominantly *A. flavus* and *A. parasiticus* [5]. Among the 18 identified types of AFs, the naturally occurring and well-known types are aflatoxin B_1 (AFB$_1$), aflatoxin B_2 (AFB$_2$), aflatoxin G_1 (AFG$_1$) and aflatoxin G_2 (AFG$_2$) [6]. AFs are mycotoxins largely related to agricultural products produced in the tropics and subtropics under humid climate, such as cereals, oilseeds, spices and tree nuts. Contamination with AFs can occur both pre-harvest and post-harvest. OTs are primarily produced by *Aspergillus* and *Penicillium* species [7]. OTs may contaminate cereals (barley, corn, oats, rice, rye, wheat) and other plant products (coffee beans, nuts, dried peanuts, spices, dried fruits, raisins, wine, grape juice, and beer). Among all the OTs, Ochratoxin A (OTA) is the most prevalent and toxic [8]. DON is a toxic fungal metabolite primarily produced by *F. graminearum* and *F. culmorum* common in grains, such as wheat and wheat-based products. DON is also known as vomitoxin due to its strong emetic effects after consumption [9]. FMs are mycotoxins produced in cereals by pathogenic fungi, namely *F. verticillioides, F. proliferatum*, and related species [10]. Moreover, *A. nigri* also produces FMs in peanut, corn and grape plants [11 - 15]. Corn and corn-based products are most commonly infected with FMs, however, their presence also appears in several other grains (rice, wheat, barley, corn, rye, oat and millet) and grain products [16, 17]. More than 15 fumonisin homologues are known and characterized as fumonisin A, B, C, and P [18, 19]. Further, among fumonisin B (FB), FB$_1$, FB$_2$, and FB$_3$ are most abundant, with FB$_1$ being the most toxic form.

The toxicity of these mycotoxins has led many countries to set up strict regulations for their control in food and feed and the consequent establishment of legislation to control their possible contamination [20]. Effective analytical methods play a key role in reducing the risk of mycotoxin contamination in the food and feed chains. The traditional culture method and microscopic identification of fungal infections, is a tedious and time-consuming process requiring a significant amount of expertise. Conventional analytical methods for mycotoxins include thin-layer chromatography (TLC), high-performance liquid chromatography (HPLC), gas chromatography (GC), capillary electrophoresis (CE) and immunoassay-based technique like enzyme-linked immunosorbent assay (ELISA) [2, 21, 22]. Determination of mycotoxins using these methods is generally a long process which involves extraction procedures using solvents and an identification process based upon chromatographic or immuno-techniques. In addition, the uneven presence of mycotoxins in large-scale products often causes the traditional sample-based analyses to present a limited view of the degree of contamination, *i.e.*, they are subject to sampling error. Therefore, there is a great need for a more rapid technique for high-throughput detection of fungal infection and mycotoxins in agricultural and food commodities in a nondestructive manner, before they enter the supply chain.

Among currently emerging technologies, the optical-based methods have been demonstrated to have great potential for rapid and nondestructive determination of various quality and safety attributes of agricultural and food commodities. As one of the promising optical detection technologies in agriculture, hyperspectral imaging (HSI) technology has provided interesting and encouraging results for the detection of mycotoxin and fungal contamination in varieties of agricultural products. Therefore, the main goal of this chapter is to give an overview of the current research progress in the application of HSI technique for rapid and nondestructive evaluation of mycotoxin and fungal contaminants in different agricultural products. Specifically, applications of both fluorescence and reflectance HSI in the detection of the main mycotoxin contamination, including AFs, OTA, DON, FMs and their related fungal infection, are covered in this chapter. Perspectives of its future trends and challenges concerning mycotoxin and fungal contamination evaluation are also discussed.

1.2. HSI Technology

In the past two decades, HSI technology has seen a significant increase in agricultural and food research applications. By integrating conventional imaging and spectroscopy technologies for data acquisition, HSI technology can produce three-dimensional (3D) hyperspectral images with two spatial dimensions and one spectral dimension. Therefore, HSI makes it possible to obtain the spectral information at each pixel of a hyperspectral image, and also the image information at each covered wavelength. There are three main approaches used for 3D hyperspectral image acquisition. One approach is a point-based method which involves recording spectrum of pixel one at a time until all pixels in an image are accounted for. This approach generally uses a spectrometer and linearly moves the fiber optic head across the target area for data acquisition and the hyperspectral image data are accumulated pixel by pixel. Another approach employs a line-scanning mechanism such as pushbroom scanning [23], where one line of target reflectance is dispersed by a prism to generate full spectral information on the camera's detector array. Successive line scans eventually generate the hyperspectral cube. Fig. (**1**) illustrates a system utilizing the pushbroom line scanning approach in data acquisition [24]. The HSI system incorporates a patented line scanning technique [25] that requires no relative movement between the target and the sensor. The line scanning movement for data collection is achieved by moving the lens across the focal plane of the camera on an internal motorized stage. Through this implementation, the hyperspectral focal plane scanner removes the requirement of an external mobile platform in a pushbroom scanning system. The third approach is an area-based method that takes one wavelength at a time over the entire image area until all wavelengths are collected. This method uses tunable wavelength devices such as an acousto-optic tunable filter (AOTF) [26] or a liquid crystal tunable filter (LCTF) [27]. In this approach, the 3D datacube is acquired through sequentially varying wavelengths *via* the wavelength tuning device.

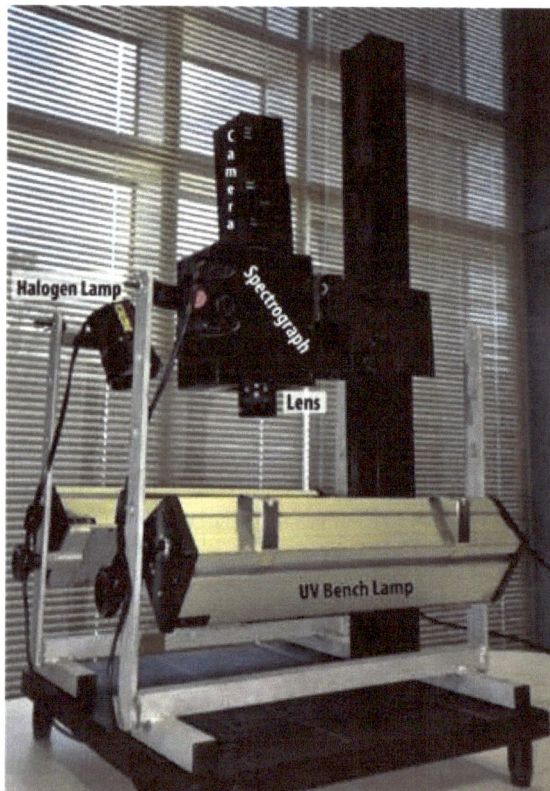

Fig. (1). Illustration of an HSI system with a build-in pushbroom line scanner for hyperspectral data acquisition [24].

HSI can be carried out in reflectance, fluorescence, transmission and scattering modes in the field of food analysis. For mycotoxin and fungal contamination detection, fluorescence and reflectance modes are commonly used. Reflectance measures the fraction of incident electromagnetic power reflected by a sample. Reflectance measurement is the traditional and most common way of using hyperspectral technology. The typical reflectance information is in the visible (Vis) and near-infrared (NIR) (400 – 2,500 nm) region of the electromagnetic spectrum. Data recorded by the hyperspectral camera are raw digital counts. When processing reflectance data, the raw digital counts are generally converted into relative reflectance as part of the image radiometric calibration. The following equation can be used for relative reflectance conversion:

$$Reflectance_\lambda = \frac{S_\lambda - D_\lambda}{R_\lambda - D_\lambda} \times 100\% \qquad (1)$$

where *Reflectance*$_\lambda$ is the reflectance at wavelength λ, S_λ is the sample intensity at wavelength λ, D_λ is the dark intensity at wavelength λ, and R_λ is the white reference intensity at wavelength λ. In this process, the calibrated reflectance value is in the range from 0% to 100%. More on the topic of hyperspectral image calibration can be found in Yao and Lewis [28].

Another approach of using hyperspectral technology for mycotoxin detection is through the measurement of fluorescence. Fluorescence is the emission of light subsequent to absorption of ultraviolet (UV) or Vis light of a fluorescent molecule or substructure, called a fluorophore. The fluorophore absorbs energy in the form of light at a specific wavelength and liberates energy in the form of emission of light at a longer wavelength [29]. The basic principles were illustrated by a Jabloński diagram [30]. Many organic and inorganic substances exhibit natural, intrinsic fluorescence when excited under a UV light source. For instance, under UV excitation, plants emit a fluorescence spectrum ranging from about 400 to 800 nm. Thus, fluorescence spectra may provide useful information for researching the properties of sample constituents and chemical compositions related to fungal contamination inspections. In the past two decades, fluorescence HSI technology has been developed to enable the acquisition of fluorescence image data with both high spectral and spatial resolutions [31]. A fluorescence HSI system is generally based on an excitation unit of UV light and a hyperspectral sensor.

2. APPLICATION OF HSI IN EVALUATION OF FUNGAL CONTAMINANTS IN AGRICULTURAL PRODUCTS

2.1. Application of Fluorescence HSI

2.1.1. Evaluation of Aflatoxin Contamination

The fluorescence of AFs has been a valuable indicator of AF contamination, and this property has been widely exploited in various analytical techniques by the utilization of fluorescence detectors [32]. Fluorescence of pure AF under long-wave UV light (UV-A, 365 nm) is reported to be bright blue (B_1, B_2; 425–480 nm) or blue-green (G_1, G_2; 480–500 nm) [33, 34]. However, in living plant tissue, AFs interact with other plant constituents and the typical fluorescence associated with AF contaminated commodities tends to be greenish-yellow [35]. Several relatively recent studies have attempted to advance the performance of the bright greenish-yellow fluorescence (BGYF) test by implementing fluorescence HSI techniques for non-destructive detection of AFs. Most of the studies concentrated on detecting AFs in corn samples [5], and a few applied the fluorescence HSI technology to address AF contamination in other commodities, including peanuts [36], pistachios, cashews [33], figs [37, 38], hazelnuts and chili peppers [39]. Yao and colleagues at Mississippi State University were the first to utilize fluorescence HSI to study the association between fluorescence emissions and AF contamination in corn [40]. Their early experiments utilized a laboratory-based fluorescence HSI instrument in the Vis range (400-600 nm) with UV excitation centered at 365 nm, to study the extent of AF contamination in single corn kernels artificially inoculated with *A. flavus*. Notably, the fluorescence spectra of contaminated kernels exhibited a peak shift to a longer (blue) wavelength compared to the control, uncontaminated kernels (Fig. **2**). The results were promising, showing a generally negative correlation between the fluorescence data and the AF concentrations determined with chemical analysis, with a determination coefficient of calibration (R_{C2}) of 0.72 using multiple linear regression (MLR) model. The classification accuracy for AF threshold set at either 20 or 100 ppb ranged from 84% to 91% based on discriminant analysis, indicating a moderate potential of fluorescence HSI for predicting AF content in individual corn kernels. To reduce data dimensionality and acquisition time, the authors implemented the genetic algorithm (GA) and selective principal component regression (SPCR) algorithm for selecting the most relevant fluorescence features. These data transformations effectively reduced the number of wavebands needed for quantifying AF contamination without compromising accuracy, attaining a comparable calibration correlation coefficient (R_C =0.80) to the standard principal component regression (PCR) analysis, based on all wavebands (R_C =0.82) [41]. Support vector machine (SVM) classification method was chosen for classification validation of the two classes (contaminated *vs.* clean) with thresholds at 20 and 100 ppb, obtaining accuracies of 87.7% and 90.5%, respectively, with 36 and 11 bands selected by the GA [42]. Many different *Aspergillus* strains co-exist in the same field, some of which produce AFs (toxigenic), and some do not (atoxigenic). To assess the differences in fluorescence emissions between corn kernels infected in the field with toxigenic or atoxigenic *A. flavus*, the fluorescence HSI system was used to image both sides of all kernels (germ and endosperm) over the visible near-infrared (VNIR) spectral range [24]. As expected, all the contaminated kernels exhibited longer peak wavelengths than did the control kernels, with a greater strain difference with respect to the controls, noted on the germ side. The best classification accuracy achieved using discriminant analysis was 94.4% with the classification algorithm threshold set at 100 ppb, however, the classification was not as good for the 20 ppb threshold (78.9%). Better overall classification accuracies were achieved when a different group of AF-contaminated kernels was similarly imaged in the 400-900 nm range, and binary encoding (BE) analysis was implemented to classify fluorescence hyperspectral images of AF-contaminated and uncontaminated corn artificially inoculated in the field [43]. With thresholds set at 20 and 100 ppb, the BE achieved classification accuracies of 87% and 88%, respectively. Additionally, the authors developed a normalized difference fluorescence index (NDFI) algorithm based on two optimal bands (437nm and 537nm), which correlated AF contamination in single corn kernels with the calculated index values, achieving a maximum correlation of -0.81 [43]. The NDFI algorithm was applied in all consecutive screening studies and incorporated in a high-speed dual-camera system developed for batch screening of AF contaminated corn using multispectral fluorescence imaging [44]. In an effort to enhance the accuracy of predicting and classifying AF-contaminated and uncontaminated corn artificially inoculated in the field, fluorescence and reflectance hyperspectral VNIR image data were analyzed separately and combined, using the least-squares support vector machine (LS-SVM) models [45]. With thresholds set at 20 and 100 ppb, the overall classification accuracies for fluorescence spectra ranged from 90.0 to 93.3%.

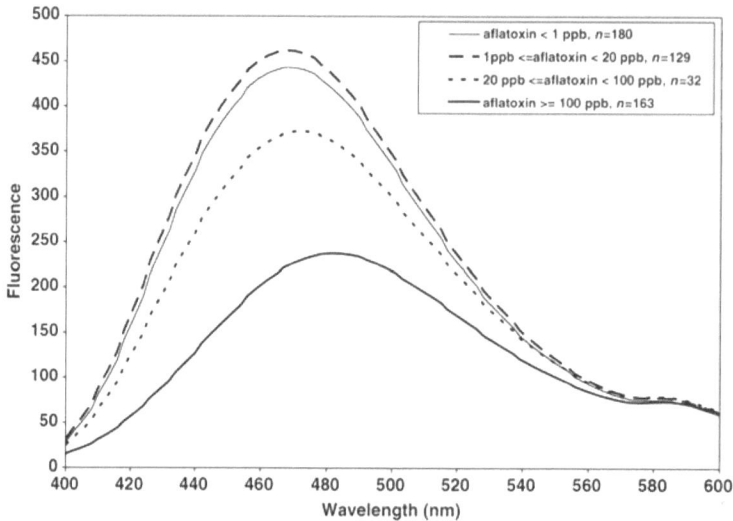

Fig. (2). Mean fluorescence spectra of corn kernels with different contamination levels [40].

To assess the effectiveness of fluorescence HSI for detecting AFs in naturally contaminated corn ears, Hruska *et al.* [46] investigated the spectral differences of corn kernels artificially and naturally infected in the field using the same fluorescence VNIR system (400-900nm). Analysis of the fluorescence spectra employed "hot" pixel classification to reveal the expected spectral shift ($p > 0.01$) between contaminated and clean kernels, with fluorescence peaks of contaminated and uncontaminated kernels at 501 and 478 nm, respectively. As both the naturally and artificially inoculated kernels exhibited peak fluorescence at 501 nm, the viability of the fluorescence HSI method for detecting AFs in naturally contaminated field corn was substantiated. In a different study, the same research group imaged (fluorescence HSI 400-700 nm) cross-sections of corn kernels lab-inoculated with toxigenic and atoxigenic strains of *A. flavus* at different time points (3, 5, 9 days), to evaluate the effect of time on the internal fluorescence spectral emissions of infected kernels compared to controls [47]. Fluorescence peak data were analyzed with analysis of variance (ANOVA) in an effort to differentiate kernels inoculated with the toxigenic (SRRC-AF13) and atoxigenic (SRRC-AF36) strains from the control kernels and from each other. The study found no difference between the AF-producing *A. flavus* and the non-producing strain in corn cross-sections on the basis of a fluorescence peak shift. However, the difference in fluorescence intensity revealed between the two strains on different post-inoculation days indicated that, in addition to the peak shift, fluorescence intensity may play a significant role in locating AF contamination in corn kernels. A related study aimed to differentiate fluorescence spectral signatures of representative resistant and susceptible corn hybrids infected by a toxigenic (SRRC-AF13) and an atoxigenic (SRRC-AF36) strain of *A. flavus*, at several time points (5, 7, 10 and 14 days), in order to evaluate fluorescence HSI (400-1000 nm) as a viable technique for early, non-invasive AF screening in resistant and susceptible corn lines [48]. Fluorescence emissions were compared for all samples over 14 days. The emission peaks of the resistant hybrid and the susceptible hybrid were significantly different from each other ($p<0.01$); also, a significant difference in fluorescence intensity was found between the treated and control kernels of both fungal strains. These results indicate a possible role of fluorescence HSI for screening corn lines, which exhibit different degrees of resistance to fungal colonization and subsequent AF contamination. Additionally, the study provides a viable and practical means for early detection of AFs with fluorescence HSI, which, in turn, supports food safety.

Other than detecting AFs in corn, fluorescence HSI has been applied with limited success to detecting AFs in nuts, dried fruits and spices. A recent study employed a VNIR hyperspectral system in the 400-900 nm range under UV illumination centered at 365 nm to differentiate peanut kernels inoculated with toxigenic (NRRL3357) and atoxigenic (SRRC-AF36) *A. flavus* [36]. Multivariate analysis of variance (MAV) used to analyze the spectral data demonstrated 100% accuracy in differentiating both inoculated groups from the controls, but a successful separation of the two inoculated groups was achieved with only 80% accuracy. An interesting observation was that the fluorescence intensity of the peanuts without skins was 10-fold that of the peanuts with skins, indicating an important role that intensity may play in fluorescence data analysis noted previously in corn kernels by Hruska *et al.* [47]. Fluorescence HSI was also applied to identify figs artificially and naturally contaminated with AFs. The naturally contaminated figs exhibited fluorescence emissions between 450 and 490 nm, but the artificially inoculated figs, by application of known aliquots of AF to the fig surface, did not fluoresce [38].

Fluorescence multispectral imaging, a variant of fluorescence HSI with fewer spectral bands used to speed up acquisition and processing and reduce the volume of data, was also applied for detecting and classification of AF-contaminated figs and hazelnuts [37]. The authors classified images of AF contaminated figs based on BGY fluorescence acquired with a fluorescence multispectral imaging (MSI) system (within 410-720 nm range) and found that the fluorescence emissions of contaminated figs were highly correlated with AF contamination, with an approximately 12% chance of misclassification. In a different study, the same authors used a multispectral system equipped with 12 filters for series of specific spectral bands in the 400-600 nm range, to image AF-contaminated and uncontaminated hazelnuts and red chili peppers [39]. The hazelnut images were classified with a two-dimensional local discriminant bases (LDB) algorithm and yielded a mean accuracy classification rate of 92.3%, effectively decreasing AF contamination from 608 ppb to 0.84 ppb after removing the contaminated nuts. The LDB algorithm was less successful in classifying the crushed chili peppers, achieving a moderate classification accuracy rate of 80% and marginally reducing contamination from 38.26 ppb to 22.85 ppb after removal of the contaminated peppers [39]. Lunadei *et al.* [33] exploited the BGYF signal to classify AF contaminated pistachios and cashews with an MSI system under UV-A illumination using two optimal wavebands for each nut variety selected with discriminant analysis, 480 and 520 nm for pistachios and 440 and 600 nm for cashews. Classification accuracy achieved was 92% and 82% for pistachios and cashews, respectively. The image data was highly correlated with reference chemical analysis.

2.1.2. Evaluation of Other Major Fungal Contaminants

Although OTAs exhibit intrinsic fluorescence with maximum emission peaks at 467 nm in 96% ethanol and 428 nm in absolute ethanol when excited at 340 nm [49], hyperspectral studies that took advantage of this attribute via fluorescence HSI or MSI for evaluating OTA-infected consumables, are sparse. As far as we know, only one study reported utilizing fluorescence HSI for detecting OTA in dried figs [38]. The study investigated the suitability of using HSI (460-720 nm) under UV-A illumination for detecting OTA in naturally and artificially inoculated figs. The authors reported a low-intensity fluorescence exhibited by the naturally contaminated figs with peaks detected at 443 and 477 nm, excited at 333 nm, with no evidence of fluorescence on the artificially contaminated figs, casting doubt on adopting fluorescence HSI for this particular application. There do not appear to be additional studies using the fluorescence HSI platform for evaluating any other major mycotoxins in current literature.

2.2. Application of Reflectance HSI

2.2.1. Evaluation of Aflatoxin and Related Fungal Contaminants

With the development of HSI system and image analysis algorithm, the reflectance HSI has been increasingly used for the evaluation of aflatoxin contamination and *Aspergillus* fungus infection in different varieties of agricultural products. Currently, corn is the most studied agricultural product in this aspect due to its high susceptibility to *Aspergillus* fungus infection and its importance as a world food crop. The corn research has mainly focused on the artificially surface-contaminated kernels or laboratory/field-inoculated ears. To the best of our knowledge, to date, no report is available where naturally contaminated corn kernels were studied using reflectance HSI. The artificially surface-contaminated kernels were generally prepared by dropping different concentrations of AFB_1 in an organic solvent (*e.g.*, methanol, acetonitrile) onto single corn kernels to achieve different contamination levels on the surface [50 - 53], or by emerging the healthy kernels into different concentrations of AFB_1 solution [54]. Even though encouraging positive prediction results were achieved in these reported studies using the reflectance HSI between different spectral ranges of 400-1000, 900-1700 and 1000-2500 nm [50 - 54], it needs to be noted that these artificially prepared samples, by surface contamination, were produced under controlled conditions. These samples lack the biological processes induced by fungal growth and metabolic activities, which may cause internal and/or external changes in kernels, and therefore cannot represent the actual condition of kernels containing aflatoxin due to fungal infection.

To further address the aflatoxin contamination of kernels due to fungal infection, several studies have been conducted by artificial inoculation of corn ears in the field using fungal spores. The potential of VNIR reflectance HSI over the spectral range of 400-900 nm has been investigated with such artificially field inoculated kernels [55]. The contaminated kernels were artificially created by inoculating corn ears in the field with toxigenic *A. flavus* spores at the early dough stage. Using with 20 ppb and 100 ppb as classification thresholds, the best overall prediction accuracy of 90.0% was achieved for both 2-class discriminant models using decision trees algorithm. Fig. (**3**) shows the mean spectra of the three groups of kernels with aflatoxin concentration above 100 ppb, 20-100 ppb and below 20 ppb. Spectral difference was observed near the 700-800 region among the three group mean spectra, *i.e.*, the slope of the mean reflectance from 700 to 800 nm was positive for the group with aflatoxin concentration above 100 ppb, and negative for the two other groups. Therefore, the band ratio image of 800 to 700 nm was calculated to build a simple detection algorithm, which achieved an overall accuracy of 80.0% based on the 100-ppb threshold. Though the accuracy obtained using the band ratio algorithm was lower than that using the full spectra, the two-band ratio analysis provided a simpler way for identifying aflatoxin-contaminated kernels with less computation time and a detection approach with significantly lower cost. Further, Zhu *et al.* [45] integrated fluorescence and reflectance VNIR hyperspectral images to identify aflatoxin-contaminated corn kernels. Both kernel side (endosperm, germ) data were collected and the classification thresholds of 20 and 100 ppb were used separately to build the SVM and *k*-nearest neighbor (KNN) models. The results of this study show that fluorescence and reflectance VNIR hyperspectral data generally achieved similar classification accuracy. The integrated analysis achieved better results than separate fluorescence or reflectance analysis on the germ side, and conspicuous improvement in the true positive rate (TPR) of the germ side was observed after integration. The best overall prediction accuracy of 95.3% was attained using the germ-side LS-SVM model in the integrated form and the 100-ppb classification threshold. More importantly, the mean aflatoxin concentration in the prediction samples could be reduced from 2662.01 ppb to 64.04, 87.33, and 7.59 ppb after removing samples that were classified as contaminated by fluorescence, reflectance, and integrated analysis, respectively, on the germ side.

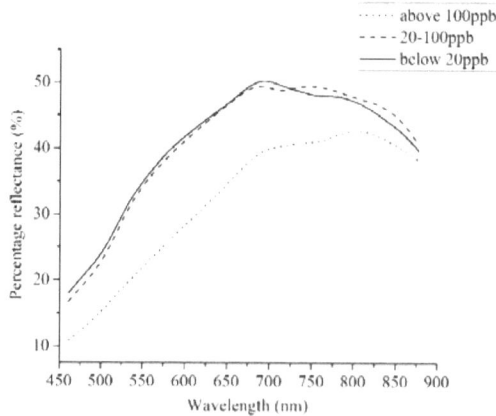

Fig. (3). Mean germ-side spectra of the three groups of kernels with aflatoxin concentration above 100 ppb, 20-100 ppb and below 20 ppb [55].

Similar to the above-mentioned studies using shorter VNIR spectral range, reflectance HSI over short-wave infrared (SWIR) range has also been investigated with field inoculated corn kernels. Wang *et al.* [56, 57] demonstrated the usefulness of the reflectance HSI combined with principal component analysis (PCA) and spectral angle mapper (SAM) classifier in identifying AFB_1-contaminated corn kernels of several different varieties. In the earlier study, which included two varieties of kernels, a minimum classification accuracy of 92.3% was obtained [56]. The latter study used the thresholds of 10 and 100 ppb to classify the single kernels into 3 classes of $AFB_1 < 10$ ppb, ppb $10 \leq AFB_1 < 100$ ppb and $AFB_1 \geq 100$ ppb [57]. For the 3 varieties of kernels studied, the classification accuracies obtained for the low and high content groups, *i.e.*, $AFB_1 < 10$ ppb and $AFB_1 \geq 100$ ppb, ranged between 80.0% and 96.2%, using either the germ-side or the endosperm-side data. However, the classification accuracy for the medium content group, *i.e.*, the group of ppb $10 \leq AFB_1 < 100$ ppb varied in the range of 50.0-85.71%. A further verification using the fourth variety of 30 kernels obtained the classification accuracies 83.3% and 88.9% for the low and high content groups, using germ-side and endosperm-side data, respectively. The classification accuracies for the medium content group only achieved 66.7% and 50.0% using the germ-side and endosperm-side data. Both of their studies indicate that the wavelengths of 1729 and 2344 nm are related to the existence of AFB_1 on corn kernels. Chu *et al.* [58] further investigated the reflectance HSI between the same spectral range as Wang *et al.* [56, 57] for the qualitative and quantitative determination of AFB_1 contamination on single corn kernels. Four varieties of kernels were used in their study and based upon the mean spectra of single kernels, PCA and SVM classifiers were employed to classify the kernels into 3 categories of $AFB_1 < 20$ ppb, ppb $20 \leq AFB_1 \leq 100$ ppb and $AFB_1 > 100$ ppb. Using a combined dataset of the four varieties of kernels, the SVM classifier obtained an overall validation accuracy of 82.5%. The support vector regression (SVR) models were established for quantitative analysis, and the authors obtained the highest coefficient of determination of 0.71 for validation (R_V^2), using the first eight PCs. Based upon the first three PCs, an R_V^2 of 0.70 was obtained for validation along with a large root mean square error of validation (RMSEV) of 524.4 ppb.

In addition to utilizing reflectance HSI for detecting aflatoxin contamination on corn kernels, its potential in discriminating corn kernels infected with different *A. flavus* species has also been reported. Fiore *et al.* [59] demonstrated the possibility of using HSI between 400 and 1000 nm to rapidly discriminate corn kernels artificially infected with toxigenic fungi from uninfected controls after 48 h inoculation with *A. flavus* or *A. niger*. More recently, Tao *et al.* [60] reported using reflectance HSI between 900 and 2500 nm to discriminate corn kernels artificially infected with aflatoxigenic and non-aflatoxigenic *A. flavus*. The aflatoxigenic AF13 and non-aflatoxigenic AF36 were used for artificial inoculation on surface-sterilized kernels, and the inoculated kernels were incubated at 30 °C for 3 and 8 days separately to increase the kernel diversity. Fig. (4) depicts the mean and standard deviation (SD) spectra of the four categories. Overall, the mean spectra of the four categories of corn kernels are analogous to each other, except for the difference existing in reflectance intensity. However, as the SD values (Fig. 4b) are relatively large for all categories of kernels, it is not appropriate to use the reflectance intensity difference to categorize the kernels. PCA combined with linear discriminant analysis (LDA) was therefore employed to discriminate different categories of kernels in this study. The four pairwise models, namely, the models established between each category (incubation day) of the AF13-inoculated kernels at days 3 and 8 and each category (incubation day) of the AF36-inoculated kernels at days 3 and 8, all obtained $\geq 98.0\%$ overall prediction accuracies. The common model that takes the AF13-inoculated kernels at days 3 and 8 as one class and the AF36-inoculated kernels at days 3 and 8 as the second class, achieved an overall accuracy of 99.0% for the prediction samples. The obtained results in this study, therefore, indicate great potential of reflectance HSI between 900 and 2500 nm in discriminating corn kernels infected by aflatoxigenic and non-aflatoxigenic *A. flavus*.

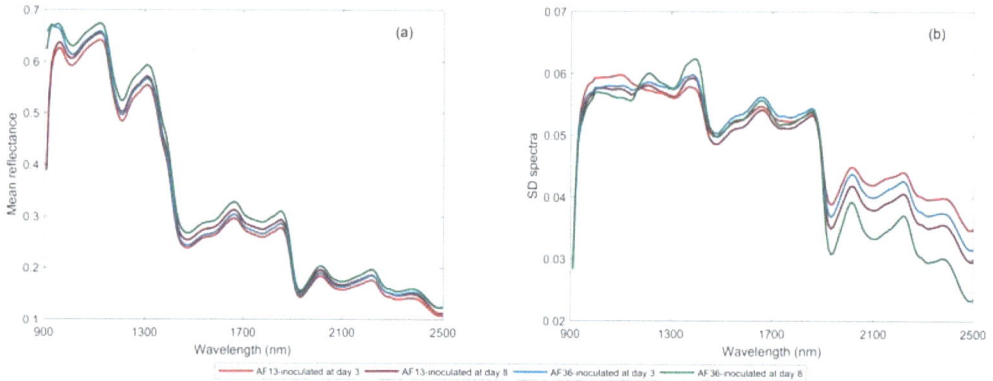

Fig. (4). The mean and SD spectra of different groups of corn kernels: **a)** mean spectra; **b)** SD spectra [60].

A few studies have been reported using reflectance HSI between 900 and 2500 nm to identify moldy peanuts from healthy peanuts [61 - 63]. In these reports, the moldy peanuts were generally artificially created by being placed in an incubator with appropriate temperature and relative humidity to facilitate the rapid growth of the existing fungus. The latter two reports [62, 63] further confirmed the existence of AFB_1 in the moldy peanuts by a reference method, while the first report [61] did not mention this procedure. All these reports show great capability of reflectance HSI combined with advanced chemometric algorithms in identifying moldy peanuts from healthy peanuts in terms of pixel-wise and kernel-level classifications. The classification accuracies obtained in these studies are all higher than 87.0%. Fig. (**5**) presents the mean spectra of healthy and moldy peanuts, where obvious spectral differences were observed between the two classes, particularly in the spectral region of 1000-1500 nm. The vertical dashed lines show the seven wavelengths of 1005, 1208, 1450, 1927, 2078, 2190 and 2251 nm that were determined by successive projection algorithm (SPA) as optimal bands for identification of moldy peanuts.

Fig. (**6**) presents the kernel-scale classification results with SVM classifier (for β = 0.15) using the optimal bands selected by SPA or wavelet features (WFs) for the Huayu variety of peanut kernels. From this figure, it was observed that for the training data, irrespective of whether the WFs or the optimal bands were used, the peanuts of both the healthy and moldy groups could all be correctly identified (Figs. **6a** and **b**). While, for the test data, 3 kernels were misclassified using the optimal bands and 1 kernel was misclassified using the WFS. The overall prediction accuracy using the optimal bands by SPA was 97.8% and 98.4% using WFs. Therefore, the WFs slightly outperformed the optimal bands with higher overall prediction accuracy.

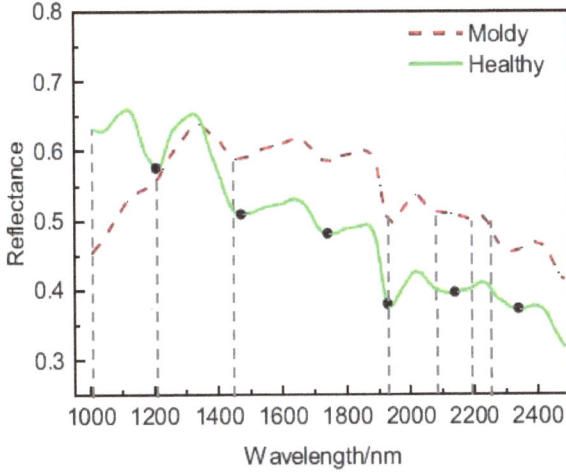

Fig. (5). Spectral response of healthy and moldy peanuts: the black dots indicate absorption peaks of healthy peanuts and the vertical dashed lines indicate the optimal wavelength variables selected by SPA [63].

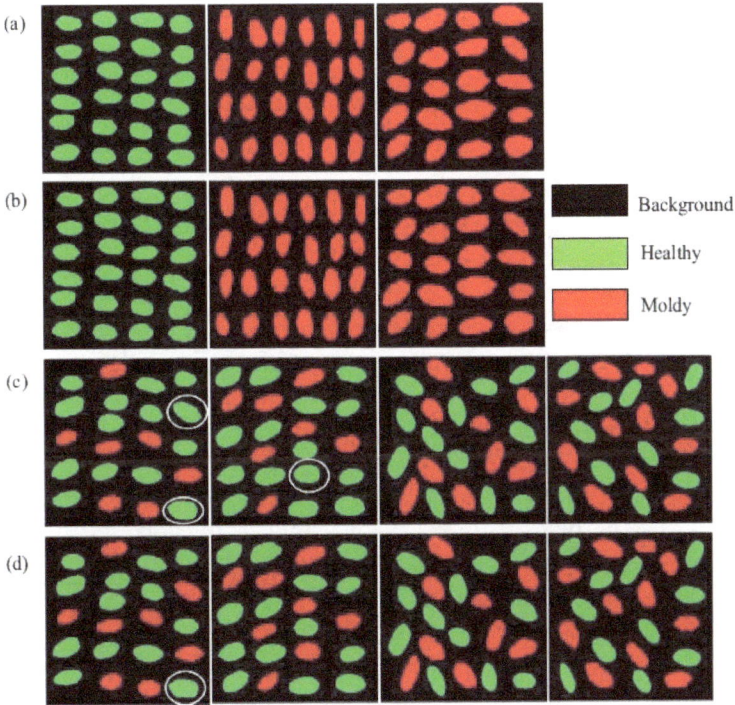

Fig. (6). Kernel-scale classification results with SVM classifier for the training datasets using **(a)** optimal bands and **(b)** WFs, and for the test dataset using **(c)** optimal bands and **(d)** WFs (The white circles indicate misclassified peanut kernels) [63].

In addition to corn and peanuts, studies using reflectance HSI have also been reported with pistachios and chili pepper. Wu and Xu [64] reported a study using VNIR reflectance HSI between 408 and 1007 nm to detect artificially surface-contaminated pistachio kernels with different concentrations of AFB_1 solution. A total of six contamination levels were included in this study, namely, 0 (control), 5, 10, 20, 30 and 50 ppb. Their results demonstrated that PCA score plot had an overall separation trend between the control and all contaminated kernels regardless of contamination level. The LDA classifier based upon the preprocessed spectra in the region of 694-988 nm obtained an overall validation accuracy of 91.0% for the 6-class classification. Additionally, the stepwise multiple linear regression (SMLR) model using 17 wavelengths between 408 and 1007 nm produced a moderate correlation coefficient for validation (R_V) of 0.91, along with the RMSEV of 5.5 ppb. Atas *et al.* [65] compared the capability of reflectance and fluorescence HSI ranging from 400 to 720 nm (10 nm spectral bandwidth) in identifying naturally contaminated ground red chili pepper flakes. The feature vectors of energy values of individual spectral bands, difference images of consecutive spectral bands and their quantized values were extracted and hierarchical bottleneck backward elimination (HBBE), Guyon's SVM-recursive feature elimination (SVM-RFE), classical Fisher discrimination power and PCA were employed to select the informative features. Both multilayer perceptrons (MLPs) and LDA were used as classifiers to discriminate the contaminated and healthy samples based on the threshold of 10 ppb. The obtained results showed that among all the models developed, the best classification accuracy of 83.3% was achieved under halogen illumination using the selected quantized absolute difference energy features by the HBBE method and the MLP classifier.

2.2.2. Evaluation of OTA and Related Fungal Contaminants

A few studies have demonstrated the capability of reflectance HSI between 1000 and 1700 nm in the identification of OTA and related fungal infection in stored wheat and barley [66 - 68]. In these studies, the fungus-infected or OTA-contaminated samples were artificially prepared by inoculation of specific fungal strains, before which they were first surface sterilized and autoclaved at 120 °C for 20 min. The healthy control samples were also surface sterilized and autoclaved. In the earlier study, Senthilkumar *et al.* [66] explored the potential of reflectance HSI in detecting different degrees of infection of *A. glaucus* and *Penicillium* spp. and OT contamination in wheat samples. The inoculated wheat kernels with *Aspergillus glaucus* and *Penicillium* spp. were incubated in special environmental chambers for 2, 4, 6, 8 and 10 weeks. The wheat samples used for different levels of OTA contamination detection were obtained through inoculation *P. verrucosum* and incubation for 18, 20, 22, 24 and 26 weeks. Three significant wavelengths of 1280, 1300, and 1350 nm corresponding to the highest factor loadings of the first PC for wheat samples inoculated with *A. glaucus* and *Penicillium* spp. were found regardless of the infection period. The first PC analysis of OTA contaminated samples provided 1300, 1350 and 1480 nm as significant wavelengths based on their highest factor loadings. The six statistical features (mean, median, standard deviation, variance, maximum, and minimum) and ten histogram features corresponding to the significant wavelengths were extracted and subjected to linear, quadratic and Mahalanobis discriminant classifiers, separately, for building pair-wise, two-way and six-way classification models. For the two-class discriminant models, including healthy *vs* the combined set of varying infection periods of *A. glaucus* inoculated wheat kernels, healthy *vs* the combined set of varying infection periods of *Penicillium* spp. inoculated wheat kernels, healthy *vs* the combined set of wheat kernels at five different OTA-contamination levels, healthy *vs* the combined set of different infection periods of *A. glaucus* and *Penicillium* spp. inoculated wheat kernels, the classification accuracy was all higher than 95.0%. For the six-way classifications of healthy and five infection periods (2, 4, 6, 8, and 10 weeks post-inoculation) of *A. glaucus* inoculated wheat kernels, healthy and five infection periods (2, 4, 6, 8, and 10 weeks post-inoculation) of *Penicillium* spp. inoculated wheat kernels, the accuracies were all higher than 88.0% except the accuracies obtained for the kernels at the 2-week post-inoculation by *A. glaucus*. For the six-way classification of healthy and OTA-contaminated wheat kernels at five different levels, the accuracies obtained were all over 98.0%. Further, Senthilkumar *et al.* [67] investigated the potential of reflectance HSI in discriminating wheat kernels infected with OTA-producing *P. verrucosum* strains and non-toxigenic *P. verrucosum* strains. The inoculated wheat kernels were incubated for 18, 20, 22, 24 and 26 weeks separately. PCA analysis on the combined samples of sterile, kernels inoculated with non-OTA producing *P. verrucosum* and OTA-contaminated kernels provided 1280, 1300, 1350, and 1480 nm as key wavelengths to detect kernels infected with non-OTA producing *P. verrucosum* and OTA-contaminated kernels. The results obtained in this study [67] were similar to their previous finding in Senthilkumar *et al.* [66]. By performing similar modeling procedures as in the aforementioned study, the 2-class classifications between the kernels infected with each of the two OTA-producing fungi with different infection periods and kernels infected with each of the two non-OTA producing fungi with different infection periods yielded over 98.0% accuracies. Moreover, the 2-class classification between kernels infected with the two OTA-producing fungi obtained higher than 99.0% classification accuracies. This indicated that the difference between the kernels infected with different strains of OTA-producing fungi may exist due to the specific fungal growth and metabolic activity of each fungal strain. Although this type of difference existed, the common 2-class classification between the kernels infected with OTA-producing and non-OTA producing fungi regardless of the infecting fungus and infection time, also obtained over 99.0% classification accuracies, demonstrating the possibility of establishing one common model to discriminate kernels infected with different strains of OTA-producing fungi from kernels infected with different strains of non-OTA producing fungi. However, the classification accuracies between kernels infected with the two strains of non-OTA producing fungi were all lower than 70.0%.

The study conducted on stored barley investigated the detection of fungal infection by *A. glaucus* and *Penicillium* spp. after 2, 4, 6, 8 and 10 weeks and discrimination of different contamination levels of OTA after inoculation with OTA producing *P. verrucosum* and non-OTA producing *P. verrucosum* for 18, 20, 22, 24 and 26 weeks [68]. Slightly different from the significant wavelengths determined on wheat kernels, the results obtained on barley samples demonstrated that the wavelengths of 1260, 1310, and 1360 nm are the most informative corresponding to *A. glaucus*, *Penicillium* spp., and non-OTA producing *P. verrucosum* infected barley kernels and the wavelengths 1310, 1360, and 1480 nm corresponding to OTA contaminated barley kernels based on the first PC loadings yielded from PCA analysis. Similarly, the extracted statistical features and histogram features at the determined significant wavelengths were used as inputs for linear, quadratic and Mahalanobis discriminant classifiers to build the pairwise, two-class, and six-class, discriminant models. The obtained results showed that the three classifiers differentiated sterile kernels with a classification accuracy of more than 94%, fungal infected kernels with more than 80% at initial periods of fungal infection, and attained 100% classification accuracy after four weeks of fungal infection. OTA contaminated kernels could be differentiated from sterile kernels with a classification accuracy of 100%. Different periods of fungal infection and different levels of OTA contamination were discriminated with a classification accuracy of more than 82%.

2.2.3. Evaluation of DON and Related Fungal Contaminants

Optical detection of DON in contaminated wheat using reflectance HSI has been investigated by several authors. Liang *et al.* [69] studied the potential of reflectance HSI between 400 and 1000 nm in discriminating different DON contamination levels in bulk wheat kernels. In this study, approximately 70 wheat kernels measuring approximately 30 mm in diameter and 5 mm in thickness were treated as one sample. A total of 180 samples were imaged, with 60 samples included in each of the three different contamination levels. By a series of processing including extraction of mean reflectance spectra, spectral preprocessing by standard normal variate (SNV) transformation and multiplicative scatter correction (MSC), selection of the most informative wavelengths using successive projections algorithm (SPA) and random frog and 3-class discriminant model establishment with SVM and PLS-DA classifiers, the authors obtained the best model with the combination of MSC, SPA and SVM. This 3-class model achieved the classification accuracies of 100.0% and 97.9% for the training and prediction sets using the classification thresholds of 250, 1162 and 2655 µg/kg. The selected wavelengths constructing the best discriminant model were 407, 435, 636, 826, 836, 837, 847, 866, 915, 959, 973, 999 nm. Further, Liang *et al.* [70] compared the performance of VNIR HSI ranging from 400 to 1000 nm and SWIR HSI ranging from 1000 to 2000 nm in the identification of DON contaminated wheat kernels and wheat flour samples. Similar to the above-mentioned study, each wheat kernel sample containing approximately 25 g of kernels was placed flat on a black background plate for imaging. After imaging, the kernel samples were ground and placed into glass Petri dishes for re-imaging as wheat flour samples. The DON threshold of 1000 ppb (the maximum allowed limit in China) was used for establishing 2-class discriminant models. By comparing all the established discriminant models, the authors found that the VNIR information was more useful for detecting DON contamination in wheat kernels compared to the SWIR range, whereas the SWIR system was more accurate than the VNIR for DON detection in flour. Among all the established models, the best prediction accuracies of 100.0% and 96.4% were achieved for wheat kernel and flour samples, respectively.

DON is produced mainly by *Fusarium* species such as *F. graminearum* and *F. culmorum*. Both *F. graminearum* and *F. culmorum* are listed as pathogens for wheat that cause a disease named Fusarium head blight (FHB), which is also identified as scab [71]. The occurrence of FHB may indicate contamination of wheat with DON. Therefore, researchers investigated using reflectance HSI for detecting FHB as an indirect indication of DON contamination in wheat. In a preliminary study, Delwiche *et al.* [72] demonstrated the potential of VNIR (400-1000 nm) and NIR (900-1700 nm) reflectance HSI in separating FHB damaged wheat kernels from sound kernels, and located the informative pairs of wavelengths, 502, 678 nm and 1198, 1496 nm over both spectral ranges, as for that purpose. More recently, Delwiche *et al.* [73] estimated the percentages of FHB damaged hard wheat kernels using reflectance NIR HSI. In this experiment, the authors expanded the number of wheat kernels in each sample to approximately 200 kernels, compared to 32 kernels per sample in the earlier study. LDA analysis of mean reflectance values of the interior pixels of kernels scanned by NIR HSI identified four wavelengths (1000, 1197, 1308 and 1394 nm) as sufficient for differentiating sound and FHB damaged kernels. Cross-validation accuracies were high (> 95%) on individual kernel level for kernels with high visual contrast between sound and FHB damaged conditions, while true model accuracy, when evaluated on external samples, decreased because of the ambiguous appearance of some kernels. Using the median reflectance value at the four determined wavelengths, a standard error of prediction (SEP) value of 4.87% was obtained using 82 external samples (each sample containing 198-250 kernels).

From a different perspective, Ropelewska and Zapotoczny [74] reported using textural features extracted from reflectance hyperspectral images to classify *Fusarium*-infected and healthy wheat kernels. Two varieties of wheat kernels were investigated in this study, and kernel images from both ventral and dorsal sides were collected between 400 and 1100 nm. The textural features, including histogram-based textural features, gradient-based features, co-occurrence matrix-based features, run-length matrix-based features and textural parameters based on an autoregressive model and Haar wavelet transform, were extracted at the 550, 710 and 850 nm wavelengths. The extracted textural features at each selected wavelength were used as inputs for the different classifiers. The classification accuracy of kernels positioned on the ventral side was determined at 78-100% and 78-98% when positioned on the dorsal side. When combining textural parameters from both kernel sides, the classification accuracies obtained were between 76% and 98%. Barbedo *et al.* [75] presented a hyperspectral image processing-based algorithm for automatic detection of FHB in wheat kernels using HSI. The algorithm is mostly based on widely employed morphological mathematical operations and spectral band manipulations, making it easy to implement and is computationally fast. The *Fusarium* index (FI) yielded by the algorithm can be used as an indicator of the likelihood that a kernel is diseased rather than as a categorical assessment. With a classification accuracy above 91%, the developed algorithm was considered to be robust to factors such as shape, orientation, shadowing and clustering of kernels. The measured correlation between FI values and DON concentrations of kernels included within each image (each image containing a mixture of 25-50 kernels of four varieties) as a whole, was 84%. Alisaac *et al.* [76] reported a comprehensive study using reflectance HSI over both 400-1000 nm and 1000-2500 nm ranges to monitor the infection of wheat kernels and flour with three *Fusarium* species separately. Two varieties of spring wheat differing in their susceptibility to FHB were grown under greenhouse conditions and inoculated at anthesis with *F. graminearum*, *F. culmorum* and *F. poae* at five different spore concentrations. It was found that the mean spectral reflectance of *Fusarium* infected wheat kernels was significantly higher than of the un-inoculated kernels. Interestingly, for wheat flour, the spectral reflectance of flour from un-inoculated samples was higher than that of the *Fusarium* infected samples in the Vis and lower in the NIR and SWIR ranges. Spectral reflectance of kernels was positively correlated with fungal DNA and DON contents. For wheat flour, the correlation coefficient exceeded -0.80 in the Vis range and remarkably, peaks of correlation appeared at 1193, 1231, 1446 to 1465, and 1742 to 2500 nm in the SWIR range. However, no discriminant modeling was conducted in this study.

In addition to solely discriminating the *Fusarium* infected wheat kernels from healthy kernels, a few studies also reported using VNIR HSI reflectance to classify the severity levels of *Fusarium* infection on wheat kernels [77, 78]. Shahin and Symons [77] investigated using the reflectance VNIR HSI between 400 and 1000 nm to classify the Canada Western Red Spring wheat kernels into 3 categories of sound, mildly damaged and severely damaged with a two-step approach. The two-step approach refers to first classifying the kernels into 2 categories of sound and *Fusarium* damaged, and then discriminating the *Fusarium* damaged kernels into another 2 categories of mildly and severely damaged. Based on either the full spectra between 450 and 950 nm or the 6 selected wavelengths of 484, 567 684, 817, 900 and 950 nm, the PCA combined with LDA classifier was employed to build the discriminant models. The results in this study showed that the models established using the 6 selected wavelengths obtained comparable accuracies in both, classifying the sound and *Fusarium* damaged wheat kernels, and the mildly and severely damaged kernels. The first-step accuracies obtained in classifying the sound and *Fusarium* damaged wheat kernels were 92.0% and 92.3% using the full spectra and the 6 selected wavelengths, respectively. The second-step accuracies in discriminating the mildly and severely damaged wheat kernels were 86.0% and 84.0% using the full spectra and the 6 selected wavelengths. More recently, Zhang *et al.* [78] also presented a study utilizing the reflectance HSI over the same spectral range of 400-1000 nm to classify wheat kernels into 3 classes of healthy, moderately and severely damaged kernels. Fig. (7) presents the original mean spectra of the 3 classes of wheat kernels, which showed a generally similar spectral trend. Three optimal selection methods, including PCA, SPA and random forest (RF) were employed to select the most informative wavelengths. The image features were extracted at the selected wavelengths by gray level co-occurrence matrix (GLCM), which is a common method to describe texture by studying the spatial correlation of gray scale. Both the spectral features and the fused features of spectral and textural data were used as inputs for the classifiers of SVM, RF and naive Bayes (NB) to build discriminant models. Among all the developed models, the RF model using the selected fused features by SPA achieved the best prediction accuracy of 96.4%.

Beside the major studies focused on wheat, one report is available on oat kernels. Tekle *et al.* [79]. investigated using the reflectance HSI between 1000 and 2500 nm to detect DON content and *Fusarium* damage in single oats (*Avena sativa* L.). Four categories of oat kernels were included in this study, namely, asymptomatic, mildly damaged and severely damaged from a *Fusarium* inoculation trial farm, and un-inoculated kernels of the same cultivar as controls. The mean DON contents of 31 kernels in the calibration set were 0.09 ± 0.05, 1.93 ± 4.49, 25.31 ± 53.94, 136.34 ± 123.04 ppm, for the un-inoculated control, asymptomatic, mildly damaged and severely damaged kernels, respectively. Pixel-wise discriminant models showed that percentages of the damaged pixels of the un-inoculated control, asymptomatic, mildly damaged, severely damaged kernels of the validation set were 26.5 ± 7.4, 29.3 ± 7.1, 46.9 ± 18.4 and $73.3 \pm 16.3\%$, indicating the potential of NIR HSI in identifying *Fusarium* damage on oat kernels. The correlation coefficients between the measured and predicted DON content (in the transformed type of $[\log(\mathrm{DON})]^3$) of the validated kernels were 0.81 and 0.79 using the established PLSR and PLS-LDA models. Scanning electron micrographs were also taken of different categories of oat kernels, which showed the un-inoculated and asymptomatic kernels to be plump and free of any fungal mycelia, while the severely damaged kernels were shriveled and heavily colonized with *F. graminearum*.

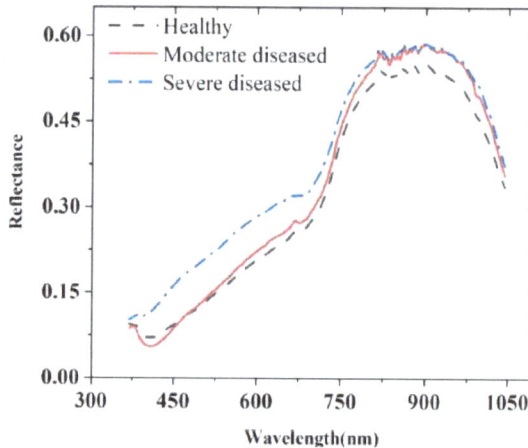

Fig. (7). The mean spectra of healthy, moderately and severely infected wheat kernels [78].

Although most of the studies reported encouraging and positive results regarding using reflectance HSI for detection of DON and related fungal contamination on wheat and oats, it needs to be noted that a few reports did not show as satisfactory results as in the above-mentioned studies. Barbedo *et al.* [80] reported using reflectance HSI over the spectral range of 528-1785 nm to classify naturally infected wheat kernels into 3-classes, class 1 - low level (≤ 0.5 mg/kg), class 2 - medium level (> 0.5 mg/kg and ≤ 1.25 mg/kg), class 3 - high level (> 1.25 mg/kg) according to the European limits, and 2 additional classes, (class 1 + class 2) and class 3. The wheat kernels included in this study were from 33 different wheat cultivars and each sample image contained 30-50 kernels. However, the three- and two-class classification accuracies only achieved 72% and 81%, respectively. Another low prediction result was reported by Femenias *et al.* [81]. Using reflectance HSI between 900 and 1700 nm, a 2-class validation accuracy only yielded 62.7% depending on the EU maximum level (1250 µg/kg). Additionally, the developed PLSR obtained a root mean square error of cross-validation (RMSECV) and a root mean square error of independent validation set (RMSEV) 405 µg/kg and 1174 µg/kg, respectively.

2.2.4. Evaluation of Other Mycotoxins and Fungal Contaminants

Corn plant infection with *F. verticillioides* generally indicates a significant risk of contamination with FMs. In many cases, the presence of *F. verticillioides* is undetected because it does not cause visible damage. A technique for detecting the invisible damage is thus desired. Williams *et al.* [82] reported a study employing SWIR HSI between 1000 and 2500 nm to detect corn kernels infected with *F. verticillioides*. Field inoculated kernels at the mid-silk stage were used in this study. PCA analysis demonstrated a distinct difference between infected and sound kernels along the first PC and two distinguishable clusters were found. The loading plot of the first PC showed important absorbance peaks for the two classes, *i.e.*, 1960 and 2100 nm for the infected class and 1450, 2300 and 2350 nm for the non-infected class. The pixel-wise PLS-DA model yielded a root mean square error of prediction (RMSEP) of 0.23 in classifying the infected and non-infected regions.

CONCLUSION AND FUTURE OUTLOOK

This chapter summarized recent research advances of HSI in both fluorescence and reflectance modes for detection of the main mycotoxins and related fungal contamination in varieties of agricultural products. Fluorescence HSI has been mainly focused on AF contamination detection in corn, and promising results have been achieved. Reports using fluorescence HSI for the detection of other mycotoxins are still rare. In comparison, reflectance HSI has been explored extensively for the detection of several main mycotoxins, including AFs, OTA and DON and their related fungal infections, and its great potential has been demonstrated in available reports. Regarding both fluorescence and reflectance HSI, most studies concentrate on discriminant analysis based upon the maximum allowed limits in different regions. Only a few studies have reported quantitative results of HSI for evaluation of mycotoxin contamination, and the results indicate that HSI requires additional refinement for potential quantification of mycotoxins in agricultural products.

Importantly, the most informative spectral bands have been extracted and reported for both fluorescence and reflectance HSI technologies with different detection objectives. Thus, the development of more cost-effective and user-friendly HSI-based, MSI instruments appears to be a logical and practical transition of this technology leading to broader utilization by various sectors of the industry. However, it needs to be pointed out that relatively small sample sizes were generally used in most studies. The efficiency of HSI system in high-throughput analysis of bulk samples has not been established so far. As such, a much larger sample size involving more diverse sample conditions has to be considered to minimize the gap between laboratory and industrial applications. In addition, compared to conventional wet chemical and microbiological methods, the HSI-based methods still appear to have low detection sensitivity and specificity. This can be caused by many factors which are often inevitable in nondestructive detection, including the background components of food matrices and other types of mycotoxins coexisting in the test samples. Apart from the continuous development of hardware systems, exploration of more advanced imaging processing and chemometric techniques may also yield more robust prediction models and improve predictive accuracies.

CONSENT FOR PUBLICATION

Not applicable.

CONFLICT OF INTEREST

The authors confirm that this chapter contents have no conflict of interest.

ACKNOWLEDGEMENTS

This publication is a contribution of the Mississippi Agricultural and Forestry Experiment Station (Starkville, MS). This material is based upon work supported by the U.S. Department of Agriculture, Agricultural Research Service, under agreement No. 58-6-54-8-009, and by the National Institute of Food and Agriculture, U.S. Department of Agriculture, Hatch multistate project under accession number 1018484. Any opinions, findings, conclusion, or recommend-ations expressed in this publication are those of the author(s) and do not necessarily reflect the view of the U.S. Department of Agriculture.

REFERENCES

[1]   J. Selvaraj, "Mycotoxin detection-Recent trends at global level", *J. Integr. Agric.,* vol. 14, no. 11, pp. 2265-2281, 2015.
[http://dx.doi.org/10.1016/S2095-3119(15)61120-0]

[2]   V. Pereira, J. Fernandes, and S. Cunha, "Mycotoxins in cereals and related foodstuffs: A review on occurrence and recent methods of analysis", *Trends Food Sci. Technol.,* vol. 36, no. 2, pp. 96-136, 2014.
[http://dx.doi.org/10.1016/j.tifs.2014.01.005]

[3]   World Health Organisation; "International Agency for Research on Cancer (WHO-IARC), "Tosins derived from *Fusarium moniliforme*: fumonisins B_1 and B_2, and fusarin C.", *IARC Monogr. Eval. Carcinog. Risks Hum.*, vol. 56, pp. 445-462, 1993.

[4]   S. Agriopoulou, E. Stamatelopoulou, and T. Varzakas, "Advances in occurrence, importance, and mycotoxin control strategies: prevention and detoxification in foods", *Foods*, vol. 9, no. 2, p. 137, 2020.
[http://dx.doi.org/10.3390/foods9020137] [PMID: 32012820]

[5]   F. Tao, H. Yao, Z. Hruska, L. Burger, K. Rajasekaran, and D. Bhatnagar, "Recent development of optical methods in rapid and non-destructive detection of aflatoxin and fungal contamination in agricultural products", *Trends Analyt. Chem.*, vol. 100, pp. 65-81, 2018.
[http://dx.doi.org/10.1016/j.trac.2017.12.017]

[6]   E.G. Lizárraga-Paulín, E. Moreno-Martínez, and S.P. Miranda-Castro, "Aflatoxins and their impact on human and animal health: an emerging problem", In: *Aflatoxins-Biochemistry and Molecular Biology.*, R.G. Guevara-González, Ed., InTech, 2011, pp. 255-282.

[7]   H.L. Trenk, M.E. Butz, and F.S. Chu, "Production of ochratoxins in different cereal products by *Aspergillus ochraceus*", *Appl. Microbiol.*, vol. 21, no. 6, pp. 1032-1035, 1971.
[http://dx.doi.org/10.1128/AEM.21.6.1032-1035.1971] [PMID: 5564676]

[8]   A-M. Domijan, and M. Peraica, "14.07 - Carcinogenic Mycotoxins", In: *Comprehensive Toxicology,* McQueen C.A., Ed., vol. 14. Second Edition. , 2010, pp. 125-137.

[9]   P. Sobrova, V. Adam, A. Vasatkova, M. Beklova, L. Zeman, and R. Kizek, "Deoxynivalenol and its toxicity", *Interdiscip. Toxicol.*, vol. 3, no. 3, pp. 94-99, 2010.
[http://dx.doi.org/10.2478/v10102-010-0019-x] [PMID: 21217881]

[10]   J.P. Rheeder, W.F. Marasas, and H.F. Vismer, "Production of fumonisin analogs by *Fusarium* species", *Appl. Environ. Microbiol.*, vol. 68, no. 5, pp. 2101-2105, 2002.
[http://dx.doi.org/10.1128/AEM.68.5.2101-2105.2002] [PMID: 11976077]

[11]   A. Astoreca, C. Magnoli, C. Barberis, S.M. Chiacchiera, M. Combina, and A. Dalcero, "Ochratoxin A production in relation to ecophysiological factors by *Aspergillus* section Nigri strains isolated from different substrates in Argentina", *Sci. Total Environ.*, vol. 388, no. 1-3, pp. 16-23, 2007.
[http://dx.doi.org/10.1016/j.scitotenv.2007.07.028] [PMID: 17920659]

[12]   A. Astoreca, C. Magnoli, M.L. Ramirez, M. Combina, and A. Dalcero, "Water activity and temperature effects on growth of *Aspergillus niger, A. awamori* and *A. carbonarius* isolated from different substrates in Argentina", *Int. J. Food Microbiol.*, vol. 119, no. 3, pp. 314-318, 2007.
[http://dx.doi.org/10.1016/j.ijfoodmicro.2007.08.027] [PMID: 17897746]

[13]   J.C. Frisvad, J. Smedsgaard, R.A. Samson, T.O. Larsen, and U. Thrane, "Fumonisin B_2 production by *Aspergillus niger*", *J. Agric. Food Chem.*, vol. 55, no. 23, pp. 9727-9732, 2007.
[http://dx.doi.org/10.1021/jf0718906] [PMID: 17929891]

[14]   J.M. Mogensen, J.C. Frisvad, U. Thrane, and K.F. Nielsen, "Production of Fumonisin B_2 and B_4 by *Aspergillus niger* on grapes and raisins", *J. Agric. Food Chem.*, vol. 58, no. 2, pp. 954-958, 2010.
[http://dx.doi.org/10.1021/jf903116q] [PMID: 20014861]

[15]   P. Kumar, D.K. Mahato, M. Kamle, T.K. Mohanta, and S.G. Kang, "Aflatoxins: a global concern for food safety, human health and their management", *Front. Microbiol.*, vol. 7, p. 2170, 2017.
[http://dx.doi.org/10.3389/fmicb.2016.02170] [PMID: 28144235]

[16]   C. Dall'Asta, and P. Battilani, "Fumonisins and their modified forms, a matter of concern in future scenario?", *World Mycotoxin J.*, vol. 9, no. 5, pp. 727-739, 2016.
[http://dx.doi.org/10.3920/WMJ2016.2058]

[17]   E. Cendoya, M.L. Chiotta, V. Zachetti, S.N. Chulze, and M.L. Ramirez, "Fumonisins and fumonisin-producing *Fusarium* occurrence in wheat and wheat by products: A review", *J. Cereal Sci.*, vol. 80, pp. 158-166, 2018.

[http://dx.doi.org/10.1016/j.jcs.2018.02.010]

[18]   M.S. Braun, and M. Wink, "Exposure, occurrence, and chemistry of fumonisins and their cryptic derivatives", *Compr. Rev. Food Sci. Food Saf.,* vol. 17, no. 3, pp. 769-791, 2018.
[http://dx.doi.org/10.1111/1541-4337.12334]

[19]   W.F. Marasas, *Fumonisins in Food.* Springer: Berlin/Heidelberg, Germany, 1996.

[20]   C. Juan, A. Ritieni, and J. Mañes, "Determination of trichothecenes and zearalenones in grain cereal, flour and bread by liquid chromatography tandem mass spectrometry", *Food Chem.,* vol. 134, no. 4, pp. 2389-2397, 2012.
[http://dx.doi.org/10.1016/j.foodchem.2012.04.051] [PMID: 23442700]

[21]   M.Z. Zheng, J.L. Richard, and J. Binder, "A review of rapid methods for the analysis of mycotoxins", *Mycopathologia,* vol. 161, no. 5, pp. 261-273, 2006.
[http://dx.doi.org/10.1007/s11046-006-0215-6] [PMID: 16649076]

[22]   N.W. Turner, S. Subrahmanyam, and S.A. Piletsky, "Analytical methods for determination of mycotoxins: a review", *Anal. Chim. Acta,* vol. 632, no. 2, pp. 168-180, 2009.
[http://dx.doi.org/10.1016/j.aca.2008.11.010] [PMID: 19110091]

[23]   J.A. Richards, and X. Jia, *Remote sensing digital image analysis: an introduction.* Springer: Berlin, 1999.
[http://dx.doi.org/10.1007/978-3-662-03978-6]

[24]   H. Yao, Z. Hruska, R. Kincaid, R.L. Brown, D. Bhatnagar, and T.E. Cleveland, "Detecting maize inoculated with toxigenic and atoxigenic fungal strains with fluorescence hyperspectral imagery", *Biosyst. Eng.,* vol. 115, pp. 125-135, 2013.
[http://dx.doi.org/10.1016/j.biosystemseng.2013.03.006]

[25]   C. Mao, *"Focal plane scanner with reciprocating spatial window",* USA Patent 6,166,373, 2000.

[26]   B. Park, S. Lee, S-C. Yoon, J. Sundaram, W.R. Windham, and J.A. Hinton, "AOTF hyperspectral microscopic imaging for foodborne pathogenic bacteria detection", *SPIE Proceedings (Sensing for Agriculture and Food Quality and Safety III).,* 2011p. 802707
[http://dx.doi.org/10.1117/12.884012]

[27]   Y. Peng, and R. Lu, "An LCTF-based multispectral imaging system for estimation of apple fruit firmness: Part I. Acquisition and characterization of scattering images", *Trans. ASAE,* vol. 49, pp. 259-267, 2006.
[http://dx.doi.org/10.13031/2013.20225]

[28]   H. Yao, and D. Lewis, "Spectral pre-processing and calibration techniques", In: *Hyperspectral Imaging for Food Quality Analysis and Control,* Sun D.W, Ed., Elsevier, 2010.

[29]   R. Karoui, and C. Blecker, "Fluorescence spectroscopy measurement for quality assessment of food systems-a review", *Food Bioprocess Technol.,* vol. 4, no. 3, pp. 364-386, 2010.
[http://dx.doi.org/10.1007/s11947-010-0370-0]

[30]   M. Zude, *Optical Methods for Monitoring Fresh and Processed Food-Basics and Applications for a Better Understanding of Non-destructive Sensing.* Taylor & Francis: Boca Raton, 2008.

[31]   M.S. Kim, Y-R. Chen, and P.M. Mehl, "Hyperspectral reflectance and fluorescence imaging system for food quality and safety", *Trans. ASAE,* vol. 44, pp. 721-729, 2001.

[32]   A.P. Wacoo, D. Wendiro, P.C. Vuzi, and J.F. Hawumba, "Methods for detection of aflatoxins in agricultural food crops", *Journal of Applied Chemistry,* vol. 2014, pp. 1-15, 2014.
[http://dx.doi.org/10.1155/2014/706291]

[33]   L. Lunadei, L. Ruiz-Garcia, L. Bodria, and R. Guidetti, "Image-based screening for the identification of Bright Greenish Yellow fluorescence on pistachio nuts and cashews", *Food Bioprocess Technol.,* vol. 6, no. 5, pp. 1261-1268, 2013.
[http://dx.doi.org/10.1007/s11947-012-0815-8]

[34] Z. Hruska, H. Yao, R. Kincaid, R. Brown, T. Cleveland, and D. Bhatnagar, "Fluorescence excitation–emission features of aflatoxin and related secondary metabolites and their application for rapid detection of mycotoxins", *Food Bioprocess Technol.,* vol. 7, no. 4, pp. 1195-1201, 2014. [http://dx.doi.org/10.1007/s11947-014-1265-2]

[35] D.T. Wicklow, "Influence of *Aspergillus flavus* strains on aflatoxin and Bright Greenish Yellow fluorescence of corn kernels", *Plant Dis.,* vol. 83, no. 12, pp. 1146-1148, 1999. [http://dx.doi.org/10.1094/PDIS.1999.83.12.1146] [PMID: 30841140]

[36] F. Xing, "Detecting peanuts inoculated with toxigenic and atoxienic *Aspergillus flavus* strains with fluorescence hyperspectral imagery, Proc. SPIE 10217", *Sensing for Agriculture and Food Quality and Safety,* vol. IX, 2017.102170I

[37] H. Kalkan, A. Güneş, E. Durmuş, and A. Kuşçu, "Non-invasive detection of aflatoxin-contaminated figs using fluorescence and multispectral imaging", *Food Addit. Contam. Part A Chem. Anal. Control Expo. Risk Assess.,* vol. 31, no. 8, pp. 1414-1421, 2014. [http://dx.doi.org/10.1080/19440049.2014.926398] [PMID: 24848335]

[38] S. Benalia, B. Bernardi, S. Cubero, A. Leuzzi, M. Larizza, and J. Blasco, "Preliminary trials on hyperspectral imaging implementation to detect mycotoxins in dried figs", *Chem. Eng. Trans.,* vol. 44, pp. 157-162, 2015.

[39] H. Kalkan, P. Beriat, Y. Yardimci, and T.C. Pearson, "Detection of contaminated hazelnuts and ground red chili pepper flakes by multispectral imaging", *Comput. Electron. Agric.,* vol. 77, no. 1, pp. 28-34, 2011. [http://dx.doi.org/10.1016/j.compag.2011.03.005]

[40] H. Yao, Z. Hruska, R. Kincaid, R. Brown, T. Cleveland, and D. Bhatnagar, "Correlation and classification of single kernel fluorescence hyperspectral data with aflatoxin concentration in corn kernels inoculated with *Aspergillus flavus* spores", *Food Addit. Contam. Part A Chem. Anal. Control Expo. Risk Assess.,* vol. 27, no. 5, pp. 701-709, 2010. [http://dx.doi.org/10.1080/19440040903527368] [PMID: 20221935]

[41] H. Yao, "Selective principal component regression analysis of fluorescence image to assess aflatoxin contamination in corn", *2011 3rd Workshop on Hyperspectral Image and Signal Processing: Evolution in Remote Sensing (WHISPERS).,* pp. 1-4, 2011. [http://dx.doi.org/10.1109/WHISPERS.2011.6080970]

[42] H. Yao, Z. Hruska, R. Kincaid, R.L. Brown, D. Bhatnagar, and T.E. Cleveland, "SVM-based feature extraction and classification of aflatoxin contaminated corn using fluorescence hyperspectral data", *Proceedings of the 4th IEEE Workshop on Hyperspectral Image and Signal Processing: Evolution in Remote Sensing.,* 2012 Shanghai, China. Paper no. 112 [http://dx.doi.org/10.1109/WHISPERS.2012.6874234]

[43] H. Yao, Z. Hruska, R. Kincaid, R.L. Brown, D. Bhatnagar, and T.E. Cleveland, "Hyperspectral image classification and development of fluorescence index for single corn kernels infected with *Aspergillus flavus*", *Trans. ASABE,* vol. 56, no. 5, pp. 1977-1988, 2013.

[44] D. Han, H. Yao, Z. Hruska, R. Kincaid, K. Rajasekaran, and D. Bhatnagar, "Development of high-speed dual-camera system for batch screening of aflatoxin contamination of corn using multispectral fluorescence imaging", *Trans. ASABE,* vol. 62, no. 2, pp. 381-391, 2019. [http://dx.doi.org/10.13031/trans.13125]

[45] F. Zhu, H. Yao, Z. Hruska, R. Kincaid, R. Brown, D. Bhatnagar, and T.E. Cleveland, "Integration of fluorescence and reflectance visible near-infrared (VNIR) hyperspectral images for detection of aflatoxins in corn kernels", *Trans. ASABE,* vol. 59, no. 3, pp. 785-794, 2016. [http://dx.doi.org/10.13031/trans.59.11365]

[46] Z. Hruska, H. Yao, R. Kincaid, D. Darlington, R.L. Brown, D. Bhatnagar, and T.E. Cleveland, "Fluorescence imaging spectroscopy (FIS) for comparing spectra from corn ears naturally and artificially infected with aflatoxin producing fungus", *J. Food Sci.,* vol. 78, no. 8, pp. T1313-T1320,

2013.
[http://dx.doi.org/10.1111/1750-3841.12202] [PMID: 23957423]

[47]   Z. Hruska, H. Yao, R. Kincaid, R.L. Brown, D. Bhatnagar, and T.E. Cleveland, "Temporal effects on
       internal fluorescence emissions associated with aflatoxin contamination from corn kernel cross-
       sections inoculated with toxigenic and atoxigenic *Aspergillus flavus*", *Front. Microbiol.*, vol. 8, p.
       1718, 2017.
       [http://dx.doi.org/10.3389/fmicb.2017.01718] [PMID: 28966606]

[48]   Z. Hruska, H. Yao, R. Kincaid, F. Tao, R.L. Brown, T.E. Cleveland, K. Rajasekaran, and D.
       Bhatnagar, "Spectral-based screening approach evaluating two specific maize lines with divergent
       resistance to invasion by aflatoxigenic fungi", *Front. Microbiol.*, vol. 10, p. 3152, 2020.
       [http://dx.doi.org/10.3389/fmicb.2019.03152] [PMID: 32038584]

[49]   M. Bougrini, "Development of a novel capacitance electrochemical biosensor based on silicon nitride
       for ochratoxin A detection", *Sens. Actuators B Chem.*, vol. 234, pp. 446-452, 2016.
       [http://dx.doi.org/10.1016/j.snb.2016.03.166]

[50]   W. Wang, G.W. Heitschmidt, X. Ni, W.R. Windham, S. Hawkins, and X. Chu, "Identification of
       aflatoxin B1 on maize kernel surfaces using hyperspectral imaging", *Food Control*, vol. 42, pp. 78-86,
       2014.
       [http://dx.doi.org/10.1016/j.foodcont.2014.01.038]

[51]   W. Wang, G.W. Heitschmidt, W.R. Windham, P. Feldner, X. Ni, and X. Chu, "Feasibility of detecting
       aflatoxin B1 on inoculated maize kernels surface using Vis/NIR hyperspectral imaging", *J. Food Sci.*,
       vol. 80, no. 1, pp. M116-M122, 2015.
       [http://dx.doi.org/10.1111/1750-3841.12728] [PMID: 25495222]

[52]   D. Kimuli, W. Wang, W. Wang, H. Jiang, X. Zhao, and X. Chu, "Application of SWIR hyperspectral
       imaging and chemometrics for identification of aflatoxin B1 contaminated maize kernels", *Infrared
       Phys. Technol.*, vol. 89, pp. 351-362, 2018.
       [http://dx.doi.org/10.1016/j.infrared.2018.01.026]

[53]   D. Kimuli, W. Wang, K.C. Lawrence, S-C. Yoon, X. Ni, and G.W. Heitschmidt, "Utilisation of
       visible/near-infrared hyperspectral images to classify aflatoxin B1 contaminated maize kernels",
       Biosyst. Eng., vol. 166, pp. 150-160, 2018.
       [http://dx.doi.org/10.1016/j.biosystemseng.2017.11.018]

[54]   L.M. Kandpal, S. Lee, M.S. Kim, H. Bae, and B-K. Cho, "Short wave infrared (SWIR) hyperspectral
       imaging technique for examination of aflatoxin B1 (AFB1) on corn kernels", *Food Control*, vol. 51,
       pp. 171-176, 2015.
       [http://dx.doi.org/10.1016/j.foodcont.2014.11.020]

[55]   F. Zhu, "Visible near-infrared (VNIR) reflectance hyperspectral imagery for identifying aflatoxin-
       contaminated corn kernels", *2015 ASABE Annual International Meeting*, 2015 Paper number,
       152189995

[56]   W. Wang, K.C. Lawrence, X. Ni, S-C. Yoon, G.W. Heitschmidt, and P. Feldner, "Near-infrared
       hyperspectral imaging for detecting Aflatoxin B_1 of maize kernels", *Food Control*, vol. 51, pp. 347-
       355, 2015.
       [http://dx.doi.org/10.1016/j.foodcont.2014.11.047]

[57]   W. Wang, X. Ni, K.C. Lawrence, S-C. Yoon, G.W. Heitschmidt, and P. Feldner, "Feasibility of
       detecting aflatoxin B_1 in single maize kernels using hyperspectral imaging", *J. Food Eng.*, vol. 166,
       pp. 182-192, 2015.
       [http://dx.doi.org/10.1016/j.jfoodeng.2015.06.009]

[58]   X. Chu, W. Wang, S-C. Yoon, X. Ni, and G.W. Heitschmidt, "Detection of aflatoxin B_1 (AFB_1) in
       individual maize kernels using short wave infrared (SWIR) hyperspectral imaging", *Biosyst. Eng.*, vol.
       157, pp. 13-23, 2017.
       [http://dx.doi.org/10.1016/j.biosystemseng.2017.02.005]

[59] A. Del Fiore, M. Reverberi, A. Ricelli, F. Pinzari, S. Serranti, A.A. Fabbri, G. Bonifazi, and C. Fanelli, "Early detection of toxigenic fungi on maize by hyperspectral imaging analysis", *Int. J. Food Microbiol.,* vol. 144, no. 1, pp. 64-71, 2010.
[http://dx.doi.org/10.1016/j.ijfoodmicro.2010.08.001] [PMID: 20869132]

[60] F. Tao, H. Yao, Z. Hruska, R. Kincaid, K. Rajasekaran, and D. Bhatnagar, "Potential of near-infrared hyperspectral imaging in discriminating corn kernels infected with aflatoxigenic and non-aflatoxigenic *Aspergillus flavus,* Proc. SPIE 11016", *Sensing for Agriculture and Food Quality and Safety,* vol. XI, 2019.1101603

[61] X. Qiao, J. Jiang, X. Qi, H. Guo, and D. Yuan, "Utilization of spectral-spatial characteristics in shortwave infrared hyperspectral images to classify and identify fungi-contaminated peanuts", *Food Chem.,* vol. 220, pp. 393-399, 2017.
[http://dx.doi.org/10.1016/j.foodchem.2016.09.119] [PMID: 27855916]

[62] X. Qi, J. Jiang, X. Cui, and D. Yuan, "Identification of fungi-contaminated peanuts using hyperspectral imaging technology and joint sparse representation model", *J. Food Sci. Technol.,* vol. 56, no. 7, pp. 3195-3204, 2019.
[http://dx.doi.org/10.1007/s13197-019-03745-2] [PMID: 31274887]

[63] X. Qi, J. Jiang, X. Cui, and D. Yuan, "Moldy peanut kernel identification using wavelet spectral features extracted from hyperspectral images", *Food Anal. Methods,* vol. 13, pp. 445-456, 2020.
[http://dx.doi.org/10.1007/s12161-019-01670-w]

[64] Q. Wu, and H. Xu, "Detection of aflatoxin B_1 in pistachio kernels using visible/near-infrared hyperspectral imaging", *Trans. ASABE,* vol. 62, no. 5, pp. 1065-1074, 2019.
[http://dx.doi.org/10.13031/trans.13161]

[65] M. Atas, Y. Yardimci, and A. Temizel, "A new approach to aflatoxin detection in chili pepper by machine vision", *Comput. Electron. Agric.,* vol. 87, pp. 129-141, 2012.
[http://dx.doi.org/10.1016/j.compag.2012.06.001]

[66] T. Senthilkumar, D.S. Jayas, N.D.G. White, P.G. Fields, and T. Gräfenhan, "Detection of fungal infection and ochratoxin A contamination in stored wheat using near-infrared hyperspectral imaging", *J. Stored Prod. Res.,* vol. 65, pp. 30-39, 2016.
[http://dx.doi.org/10.1016/j.jspr.2015.11.004]

[67] T. Senthilkumar, D.S. Jayas, N.D.G. White, P.G. Fields, and T. Gräfenhan, "Detection of ochratoxin A contamination in stored wheat using near-infrared hyperspectral imaging", *Infrared Phys. Technol.,* vol. 81, pp. 228-235, 2017.
[http://dx.doi.org/10.1016/j.infrared.2017.01.015]

[68] T. Senthilkumar, D.S. Jayas, N.D.G. White, P.G. Fields, and T. Gräfenhan, "Detection of fungal infection and ochratoxin A contamination in stored barley using near-infrared hyperspectral imaging", *Biosyst. Eng.,* vol. 147, pp. 162-173, 2016.
[http://dx.doi.org/10.1016/j.biosystemseng.2016.03.010]

[69] K. Liang, Q.X. Liu, J.H. Xu, Y.Q. Wang, C.S. Okinda, and M.X. Shen, "Determination and visualization of different levels of deoxynivalenol in bulk wheat kernels by hyperspectral imaging", *J. Appl. Spectrosc.,* vol. 85, pp. 953-961, 2018.
[http://dx.doi.org/10.1007/s10812-018-0745-y]

[70] K. Liang, J. Huang, R. He, Q. Wang, Y. Chai, and M. Shen, "Comparison of Vis-NIR and SWIR hyperspectral imaging for the nondestructive detection of DON levels in Fusarium head blight wheat kernels and wheat flour", *Infrared Phys. Technol.,* vol. 106, 2020.103281
[http://dx.doi.org/10.1016/j.infrared.2020.103281]

[71] A.M. Khaneghah, L.M. Martins, A.M. von Hertwig, R. Bertoldo, and A.S. Sant'Ana, "Deoxynivalenol and its masked forms: characteristics, incidence, control and fate during wheat and wheat based products processing - a review", *Trends Food Sci. Technol.,* vol. 71, pp. 13-24, 2018.
[http://dx.doi.org/10.1016/j.tifs.2017.10.012]

[72]   S.R. Delwiche, M.S. Kim, and Y. Dong, "*Fusarium* damage assessment in wheat kernels by Vis/NIR hyperspectral imaging", *Sens. & Instrumen. Food Qual.,* vol. 5, pp. 63-71, 2011.
[http://dx.doi.org/10.1007/s11694-011-9112-x]

[73]   S.R. Delwiche, I.T. Rodriguez, S.R. Rausch, and R.A. Graybosch, "Estimating percentages of *fusarium*-damaged kernels in hard wheat by near-infrared hyperspectral imaging", *J. Cereal Sci.,* vol. 87, pp. 18-24, 2019.
[http://dx.doi.org/10.1016/j.jcs.2019.02.008]

[74]   E. Ropelewska, and P. Zapotoczny, "Classification of *Fusarium*-infected and healthy wheat kernels based on features from hyperspectral images and flatbed scanner images: a comparative analysis", *Eur. Food Res. Technol.,* vol. 244, pp. 1453-1462, 2018.
[http://dx.doi.org/10.1007/s00217-018-3059-7]

[75]   J.G.A. Barbedo, C.S. Tibola, and J.M.C. Fernandes, "Detecting *Fusarium* head blight in wheat kernels using hyperspectral imaging", *Biosyst. Eng.,* vol. 131, pp. 65-76, 2015.
[http://dx.doi.org/10.1016/j.biosystemseng.2015.01.003]

[76]   E. Alisaac, J. Behmann, A. Rathgeb, P. Karlovsky, H-W. Dehne, and A-K. Mahlein, "Assessment of *Fusarium* infection and mycotoxin contamination of wheat kernels and flour using hyperspectral imaging", *Toxins (Basel),* vol. 11, no. 10, p. 556, 2019.
[http://dx.doi.org/10.3390/toxins11100556] [PMID: 31546581]

[77]   M.A. Shahin, and S.J. Symons, "Detection of *Fusarium* damaged kernels in Canada Western Red Spring wheat using visible/near-infrared hyperspectral imaging and principal component analysis", *Comput. Electron. Agric.,* vol. 75, pp. 107-112, 2011.
[http://dx.doi.org/10.1016/j.compag.2010.10.004]

[78]   D. Zhang, G. Chen, H. Zhang, N. Jin, C. Gu, S. Weng, Q. Wang, and Y. Chen, "Integration of spectroscopy and image for identifying *fusarium* damage in wheat kernels", *Spectrochim. Acta A Mol. Biomol. Spectrosc.,* vol. 236, no. 118344, 2020.118344
[http://dx.doi.org/10.1016/j.saa.2020.118344] [PMID: 32330824]

[79]   S. Tekle, I. Måge, V.H. Segtnan, and Å. Bjørnstad, "Near-infrared hyperspectral imaging of *Fusarium*-damaged oats (*Avena sativa* L.)", *Cereal Chem.,* vol. 92, no. 1, pp. 73-80, 2015.
[http://dx.doi.org/10.1094/CCHEM-04-14-0074-R]

[80]   J.G.A. Barbedo, C.S. Tibola, and M.I.P. Lima, "Deoxynivalenol screening in wheat kernels using hyperspectral imaging", *Biosyst. Eng.,* vol. 155, pp. 24-32, 2017.
[http://dx.doi.org/10.1016/j.biosystemseng.2016.12.004]

[81]   A. Femenias, F. Gatius, A.J. Ramos, V. Sanchis, and S. Marín, "Standardisation of near infrared hyperspectral imaging for quantification and classification of DON contaminated wheat samples", *Food Control,* vol. 111, no. 107074, 2020.
[http://dx.doi.org/10.1016/j.foodcont.2019.107074]

[82]   P. Williams, M. Manley, G. Fox, and P. Geladi, "Indirect detection of *Fusarium verticillioides* in maize (*Zea mays* L.) kernels by near infrared hyperspectral imaging"., *J. Near Infrared Spectrosc.,* vol. 18, pp. 49-58, 2010.
[http://dx.doi.org/10.1255/jnirs.858]

CHAPTER 7

Intelligent Sensing Technology for Processing of Agro-products

Zhiming Guo[1,2,*]

[1] *School of Food and Biological Engineering, Jiangsu University, Zhenjiang 212013, China*

[2] *International Research Center for Food Nutrition and Safety, Jiangsu University, Zhenjiang 212013, China*

Abstract: Intelligent sensing technology of agricultural products can effectively guarantee food quality and safety, and is the key technical support to promote the rapid development of the world's agricultural products processing industry, bringing more opportunities and development space to the emerging agricultural products processing industry. Intelligent sensing technology for agricultural products is a multidisciplinary research field, which has the advantages of fast detection speed, convenient operation, and easy online detection. This work reviews the research of optical, acoustic, electrical, magnetic, and bionic sensing technologies in the processing of agricultural products, expounds the principle, structure, and typical applications of each sensing technology, and summarizes the problems and trends in the development of each sensing technology. Intelligent sensing technology for agricultural product quality and safety is developing towards the direction of high sensitivity, automation, networking, intelligence, and multi-function, and has gradually become an indispensable and important technical means for agricultural product quality and safety inspection. The intelligent sensing technology of agricultural products is developing synchronously with the integration of the Internet of things, big data, and cloud computing, which can realize the standardization, refinement, and intelligent management of the agricultural products processing process.

Keywords: Acoustic sensor, Agricultural processing, Analog sensory, Electrical sensor, Intelligent sensing, Magnetic sensor, Optical sensor.

1. INTRODUCTION

The processing of agricultural products is an important link of agricultural industrialization, which improves the rate of conversion of agricultural products, the key to realize value-added processing, to accelerate the development of modern agriculture in the world. Agricultural and food products with high quality

---

[*] **Corresponding author Zhiming Guo:** School of Food and Biological Engineering, Jiangsu University, Zhenjiang 212013, China; & International Research Center for Food Nutrition and Safety, Jiangsu University, Zhenjiang 212013, China; Tel: +86 18260628702; Fax: +86 51188780201; E-mail: guozhiming@ujs.edu.cn

Jiangbo Li & Zhao Zhang (Eds.)

and safety are essential parameters for the consumers, and it is important to develop a compulsory examination of agricultural products and processed products [1 - 3]. Intelligent sensing technologies of the processing of agricultural products are the critical support for the healthy and rapid development of the world agricultural industry in the new era. A new round of scientific and technological revolution will bring subversive changes to the agricultural products processing and manufacturing industry, and will also bring more opportunities and development space to the emerging agricultural products industry [4, 5]. Some of the technological advancements being employed and currently in progress in the field of agricultural products processing and manufacturing include high-end processing and manufacturing equipments, sensors that simulate human senses like electronic sensors for collection of internal and external information; intelligent sensors that mimic the human brain and nervous system functions from external feeling to transmission of signals to the brain, and also give connotation of multiple sensing levels and degrees; and smart sensors used for monitoring agricultural products. Intelligent sensing technologies of agricultural products can effectively guarantee the quality and safety of food and will become the mainstream of the future development of the agricultural products processing industry.

Food quality and safety is a complex issue in the context of the internationalization of agricultural production and economic globalization [6]. Researchers focus on the research and development of agricultural product intelligent detection and processing technology and equipment, study the cutting-edge scientific issues of agricultural product processing technology, and respond to the new challenges of the scientific and technological revolution of the agricultural product processing industry by carrying out high-level, substantive and sustainable scientific and technological breakthroughs.

Intelligent sensing technology for agricultural products is multidisciplinary, involving computer technology, information technology, sensor technology, image processing, spectral technology, applied mathematics, pattern recognition, and other knowledge in a number of disciplines [7 - 10]. Agricultural intelligent sensing technologies which utilize sound, light, and magnetic field in their operation can acquire a lot of information that reflects the kind of investigated properties of the product under analysis. These sensing technologies have the advantages of rapidness, ease of operation and online testing [11].

It is one of the hotspots in the research of the current agricultural products processing. The traditional wet chemical method is generally handling destructive test samples, although the detection result is of high precision, but the method is high consumption, excessive complexity, and time delay. Meanwhile, the analytical process using chemical agents will produce waste gas and liquid leading to environmental pollution. Compared with the wet chemical analysis method, the intelligent sensing technology can be efficiently used as a fast, accurate and cost-effective way to indicate the quality and safety [12]. Intelligent sensing technology is developing towards the direction of high sensitivity, automation, network, intelligence and multi-function, and gradually becomes an indispensable and important technical means of agricultural product quality and safety detection, which is complementary to the detection of large and precise physical and chemical analysis instruments.

Agricultural products intelligent sensing technology is developed synchronously with the integration of the internet of things, big data and cloud computing, which can realize the standardization, refinement and intelligent management of agricultural products processing. This chapter summarizes the principles and characteristics of optical, acoustic, electrical, magnetic and analog sensory sensing technology, analyses the research status of sensing technology in agricultural product processing, and introduces typical application cases, points out the problems of sensing technology in agricultural product processing, and its application prospect.

2. OPTICAL SENSING TECHNOLOGY IN AGRICULTURAL PRODUCTS PROCESSING

2.1. Principle of Optical Sensing Technology

Optical sensor technology is mainly based on the principle of interaction between light and matter. When light is incident from one medium to another, due to the different refractive index of the medium, the propagation speed of light will change, which causes the change of the propagation direction of light, which is the refraction phenomenon of light. Fig. (**1**) shows the electromagnetic spectrum range used in agricultural processing.

The biological tissue of agricultural products is composed of cells of different sizes, densities and components, which are opaque, turbid and highly scattering in microcosmic. When light is transmitted in tissue, it interacts with the tissue in a variety of ways, as shown in Fig. (**2**). Wherein absorption and scattering will occur simultaneously, and multiple scattering plays a leading role. Photons are usually converted into another form of energy (such as heat energy) after being absorbed, and the transmission direction of photons changes after being scattered, but they will continue to transmit until they are absorbed by the tissue or escape from the surface of the medium. The absorption of light is generated by the transition of molecules from the ground state to high energy level, which is mainly related to the chemical composition of tissues, such as water, sugar, *etc*.

The scattering of light is caused by the change of the refractive index, which is mainly related to the physical properties of tissues, such as hardness, texture, *etc.* Therefore, the change of biological tissue properties will lead to the change of corresponding absorption or scattering characteristics, and then the change of detected signal (reflection spectrum, transmission spectrum, *etc.*). All of these methods are used to determine the quality of agricultural products by the characteristics of light absorption, scattering, reflection and transmission.

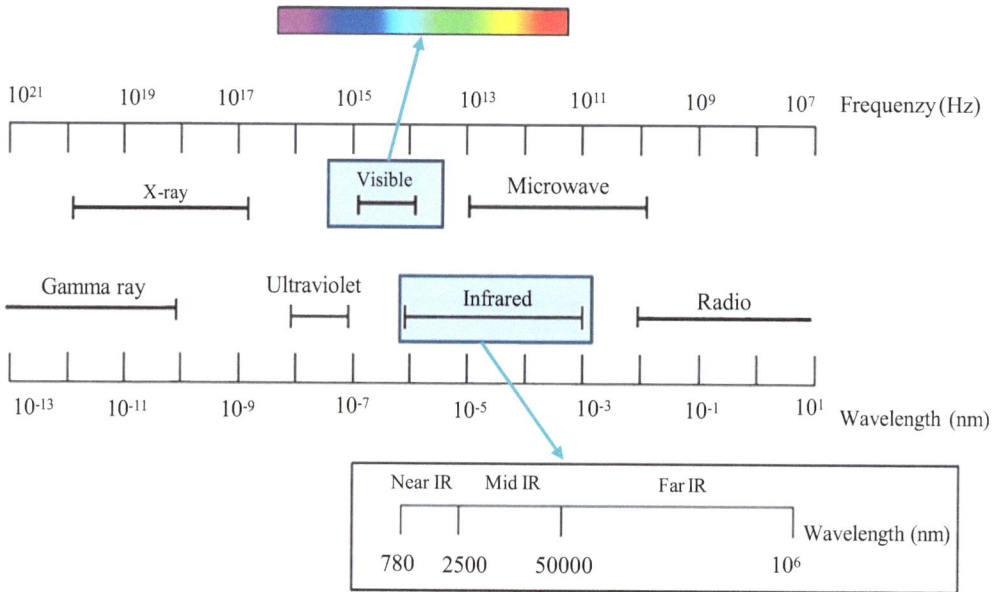

Fig. (1). The electromagnetic spectrum range used in the agricultural processing.

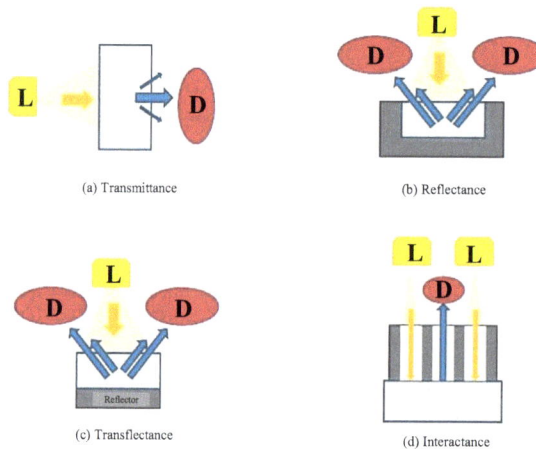

Fig. (2). Modes for the acquisition of spectra. L: light source, D: detector.

2.2. Advances in Optical Sensor Technology

In recent years, with the improvement of living standards, people pay more attention to the quality and safety of agricultural products. For meeting the needs of agricultural product quality and safety rapid detection, the corresponding rapid detection technology of agricultural product quality, especially non-destructive detection technology, has developed rapidly in recent years. Nondestructive detection technologies, as emerging detection methods, can achieve the quality detection of agricultural products without destroying the original state and chemical properties of the substance. Optical sensor technology is one of the most widely used nondestructive detection technologies due to the advantages of fast, simple, nondestructive. This chapter overviews examples of optical sensor technologies used in agricultural products processing. Table **2** summarizes the application of various kinds of optical sensor technologies in agricultural products quality and safety detection.

2.2.1. Application of Optical Sensor Technology in Agricultural Products Quality

In the processing of agricultural products, the detecting and grading of agricultural products quality can improve the utilization rate of agricultural products, and develop the commercial value of agricultural products. Agricultural products quality indexes mainly include color, shape, size, texture, taste, nutrition and defect. María-Teresa Sánchez *et al.* [13] in 2018 used near-infrared spectroscopy combined with modified partial least square regression models to detect the color (a* and b*), texture including maximum fracture force, toughness, stiffness and displacement, and dry matter of spinach leaves. The prediction performance of the models was good with $r^2_{cv} = 0.74$ for dry matter content and $r^2_{cv} > 0.6$ for textural parameters, which showed that optical sensor technology can be used in the routine, rapid and non-destructive analysis of spinach. Samuel Verdú *et al.* [14] in 2019 used the image analysis of diffraction patterns generated from laser-milk interaction to monitor the fermentation process of yoghurt. In the fermentation process, raw materials had different microbial stability and multi-phase components, which caused slight differences in the fermentation process. Therefore, according to the changes of the optical characteristics of agricultural products produced in their processing, optical sensor technology can be used to monitor the processing and evaluate the status of the processed products in time, finally providing valuable information for new developments to be applied to fermentation systems, product formulas and raw material adaptations. The regression models established in this research had satisfactory quantification performance with $R^2 = 0.82$ to 0.99 for each physicochemical parameter, which confirmed that the optical sensor technology can be used to monitor the processing of agricultural products. Aichen Wang *et al.* in 2014 used the NIR spectroscopy combined with multivariate analysis to detect the SSC of citrus fruit. For reducing the overlapping of spectra, simplifying spectral data and improving the operation speed, a variety of variable selection methods were applied to select the effective wavelength. The GA–SPA–MLR model achieved the best detection results with r_p and RMSEP of 0.893 and 0.436 °Brix, respectively, which proved that optical sensing technology had the potential to be used for online SSC determination of citrus fruits. Optical sensor technology has great application potential in the quality detection of agricultural products. Therefore, it is of great significance to study the optical properties of agricultural products better by employing optical sensor technology. Mengyun Zhang *et al.* [15] in 2019 researched optical properties of blueberry flesh and skin and obtained the absorption coefficient, reduced scattering coefficient and scattering anisotropy of blueberry flesh and skin in the spectral regions of 500–800 nm and 930–1400 nm. The light propagation model of blueberries using Monte Carlo multi-layered (MCML) simulation was established, which guided to develop non-destructive detection methods based on optical characteristics for blueberry internal bruising detection.

2.2.2. Application of Optical Sensor Technology in Agricultural Products Safety

Crop can absorb harmful substances such as heavy metals from the soil during their growth period. In addition, in order to prevent the effects of pests and weeds on crops, pesticides are inevitably sprayed. Microorganisms infection often happens during the growth and storage of agricultural products, which not only leads to the loss of nutritional value but also produce mycotoxins in agricultural products. If these substances cannot be removed during the processing of agricultural products, they would be absorbed by the human body and harm people's health. Therefore, agricultural products safety detection in the processing of agricultural products is of great significance to ensure good health. Agricultural products safety mainly includes the absence or reduced levels of toxins, heavy metals, agricultural residues, *etc.* Gaoqiang Lv *et al.* [16] in 2018 used depth-profiling Fourier transform infrared photoacoustic spectroscopy (FTIR-PAS) combined with PCA algorithm to detect tricyclazole residues on ripe rice husks. During detection, the organonitrogen pesticides molecules absorbed electromagnetic radiation, and the functional groups carried generated pressure fluctuations, which were captured by the spectrometer and finally transformed into spectral signals. The PCA results showed that the majority of pesticide residue samples can be distinguished from the samples without pesticide residue. Which confirmed that FTIR-PAS can be considered as an effective method for detecting organonitrogen pesticide residue in agriculture products. Pei Ma *et al.*

[17] in 2019 used SERS to quantitatively detect the chlorpyrifos pesticide residues in tomatoes. The SERS spectra of chlorpyrifos pesticide residues in tomato slices and tomato juice samples were collected respectively, and the quantitative detection model of chlorpyrifos was established at 678 cm^{-1} by combining with linear discrimination algorithm. The detection limit was as low as 10^{-9}mol/L, which showed that optical sensing technology SERS can be seen as a promising method in the quantitative detection of pesticide residues in agricultural products. Felix Y.H. Kutsanedzie *et al.* [18] in 2020 used SERS coupled with AgNPs to detect ochratoxin A and aflatoxin B1 in cocoa beans. The established quantitative detection model achieved the best detection results with Rc=0.9924 for aflatoxin B1 and Rc=0972 for ochratoxin A. The limit of detection achieved was 4.15 pg/mL for AFT B1 and 2.63 pg/mL for ochratoxin A, which were extremely lower than that of the current maximum permissible limit, indicating the potential of optical sensing technology to detect OTA and AFT B1 levels in cocoa beans and other agriculture products.

2.3. Typical Application of Optical Sensor Technology

2.3.1. An Effective Method to Inspect and Classify the Bruising Degree of Apples Based on the Optical Properties

Optical properties can provide a lot of information about the internal state of agricultural products, which has a broad application prospect in the detection and classification of fruit bruises. When light interacts with turbid biomaterials, the complex phenomenon of absorption and scattering will occur. The characteristics of light absorption and scattering are related to the composition of agricultural products tissue, such as sugar, chlorophyll, soluble solids and water, and microstructure, such as density, particle size, refractive index and cell structure. Therefore, the change of physiological, pathological and metabolic processes of fruit will lead to the change of fruit optical properties. The degree of small bruise of fruit can be quantitatively detected by studying the changes of the two basic optical properties of absorption coefficient (μ_a) and reduced scattering coefficient (μ_s), which can provide a more accurate basis for predicting and classifying the degree of small bruise of apple by using its optical properties.

Zhang *et al.* [19] in 2017 investigated the optical properties of apple tissues with different bruising degree, discriminated and classified the bruised apples based on the optical properties combined with principal component analysis and support vector machine. Fig. (3) shows a single integrating sphere measuring system for apple optical characteristics.

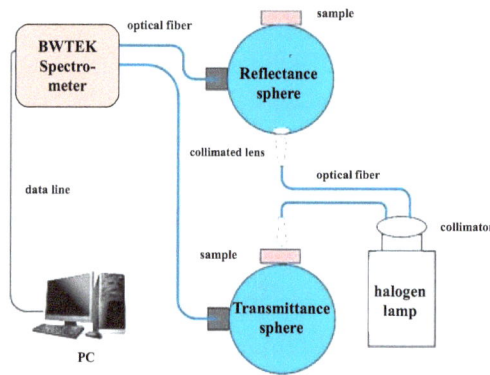

Fig. (3). The system of single integrating spheres.

First, the apples without bruises fall freely from different heights to perform impact damage experiments, which produced apple samples with different degrees of browning. Then the sample flesh of the marked area was cut off by a 3 mm thickness and 2 cm diameter slice with a microtome and sandwiched between two glass slides with saline solution. Finally, the samples were measured in diffuse reflectance and diffuse transmittance mode with the integrating sphere system. Fig. (**4**) presents the measurement procedure of integrating sphere system for the diffuse reflectance and diffuse transmittance acquisition of apple sample spectra.

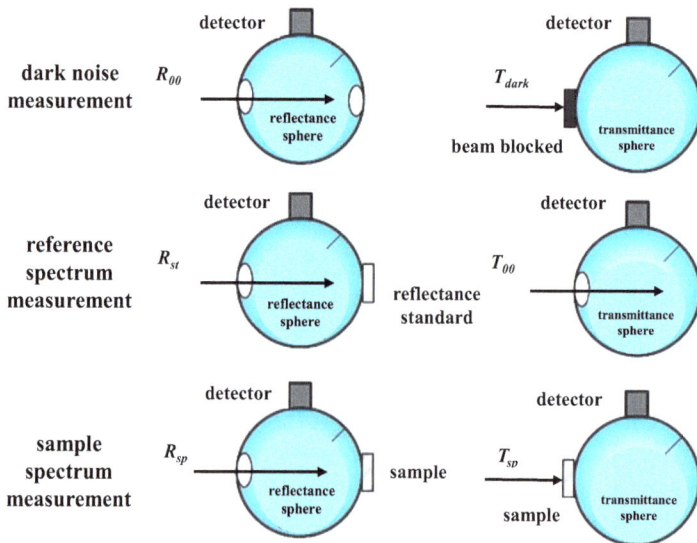

Fig. (4). Measurement of diffuse reflectance and diffuse transmittance.

Fig. (**5**) showed the absorption coefficient and reduced scattering coefficient of all apple samples in the wavelength range of 400–1050 nm. It can be seen that the absorption coefficient increased with the increase in browning degree, while reduced scattering coefficient showed the opposite trend, which may be due to cell wall depolymerization and an increase in the solubility of the middle lamella, which leads to cell separation. Fig. (**6**) showed the microstructure between different apple flesh. It can be seen that the intercellular space of the bruised apple flesh was bigger that of intact apple flesh, which showed that there was obvious cell wall depolymerization in the bruised apple flesh.

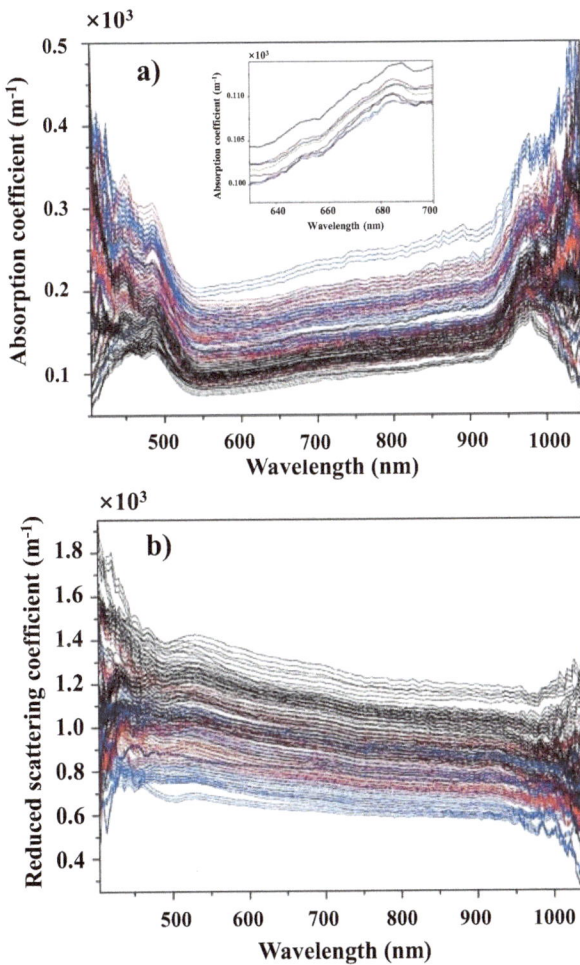

Fig. (**5**). (**a**) Absorption coefficient and (**b**) reduced scattering coefficient of all the samples.

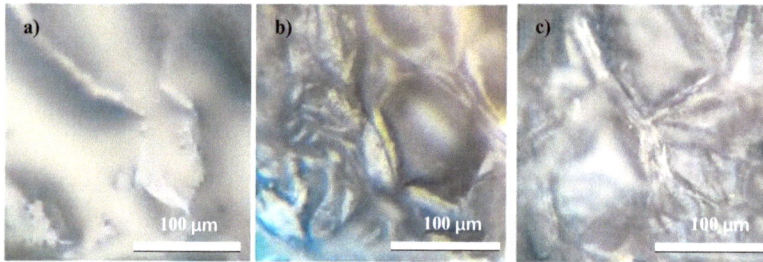

Fig. (6). The image of **(a)** intact apple flesh **(b)** bruised apple flesh **(c)** apple flesh after storage for 10 day.

According to the relationship between the optical properties and apple tissues, the discriminant models of different apple samples including 'intact', 'mild' bruised and 'severe' bruising was established by SVM. The discrimination accuracy is shown in Table **1**. It can be seen that the discriminant accuracy of only the 'mild' bruised samples and the 'severe' bruised samples was not perfect with an accuracy of 80% and 90% respectively. However, the discrimination accuracy of bruised and intact samples achieved 100% accuracy, which shows that the optical sensing technology based on the optical characteristics of agricultural products has practical application value in the future.

Table 1. The discrimination results of external validation of model based on reduced scattering coefficient.

Sample	Total Sample Number	Correct Sample Number	Misjudged Sample Number	Accuracy	Total Accuracy
No bruising	20	20	0	100%	92.5%
'Mild' bruising	10	9	1	90%	
'Severe' bruising	10	8	2	80%	

2.3.2. Assessing Firmness and Ssc of Pears Based on Absorption and Scattering Properties using an Automatic Integrating Sphere System

The transmission or attenuation of light in agricultural products is mainly in the form of absorption and scattering. The absorption in visible and near-infrared bands is related to important chemical quality indexes such as soluble solids content (SSC) of fruits, and the scattering reflect the microstructure properties such as firmness. Therefore, the quality indexes of fruit can be detected by measuring the optical properties of the fruit. He *et al.* in 2016 used an automatic integrating sphere system with a wavelength range of 400-1150 nm to detect the firmness and SSC of pears [20]. Fig. (**7**) shows the schematic diagram of automatic integrating sphere system for the detection of fruit tissue optical properties.

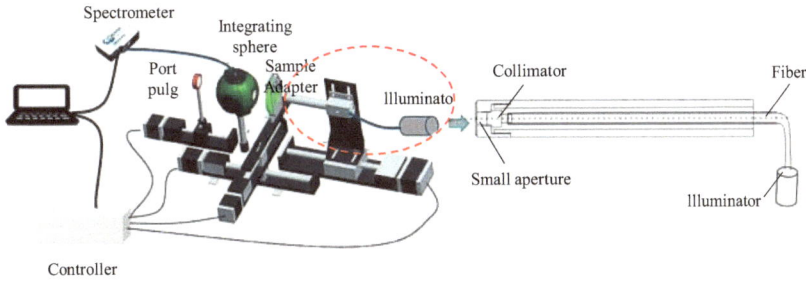

Fig. (7). Schematic diagram of the automatic integrating sphere system for measuring optical properties of fruit tissues in 400–1150 nm.

The measurement positions of firmness, SSC and optical characteristics were presented in Fig. (**8**). First, the pears were cut into two halves, then two pieces of pear tissue about 10 mm thick were-taken from the intact pears on opposite sides. Then, firmness was measured using four penetration tests on each side by TA.XT. Plus Text Analyzer. Reflectance and transmittance spectra were collected at three points located in the center of the pear slices, and the optical properties including absorption coefficient and reduced scattering coefficient were estimated by the inverse adding-doubling method based on the measured spectra and anisotropy coefficient (g). SSC was assessed by dropping the pear juice obtained by a manual fruit squeezer into a digital refractometer.

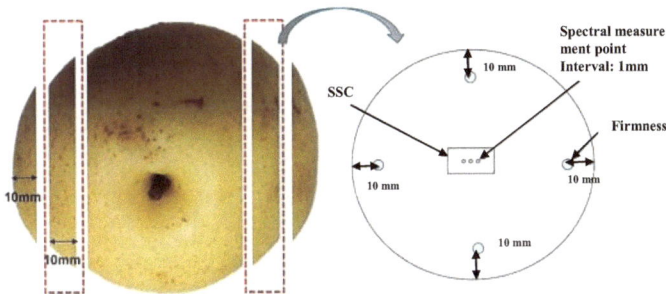

Fig. (8). Positions for penetrometer firmness, SSC, and optical measurements on pears.

Figs. (**9a** and **c**) show the optical properties μ_a and μ'_s of pear tissues and their shelf-life in the spectral range of 400–1150 nm. While Figs. (**9b** and **d**) show the changing trend of the optical properties of pears at 980 nm. It can be seen that the μ_a decreased with increase in shelf-life , while the μ'_s showed the opposite trend. Both of them had a good linear relationship with the shelf-life and recorded determination coefficients (R^2) of 0.50 and 0.80, respectively.

(a)

(b)

(c)

(d)

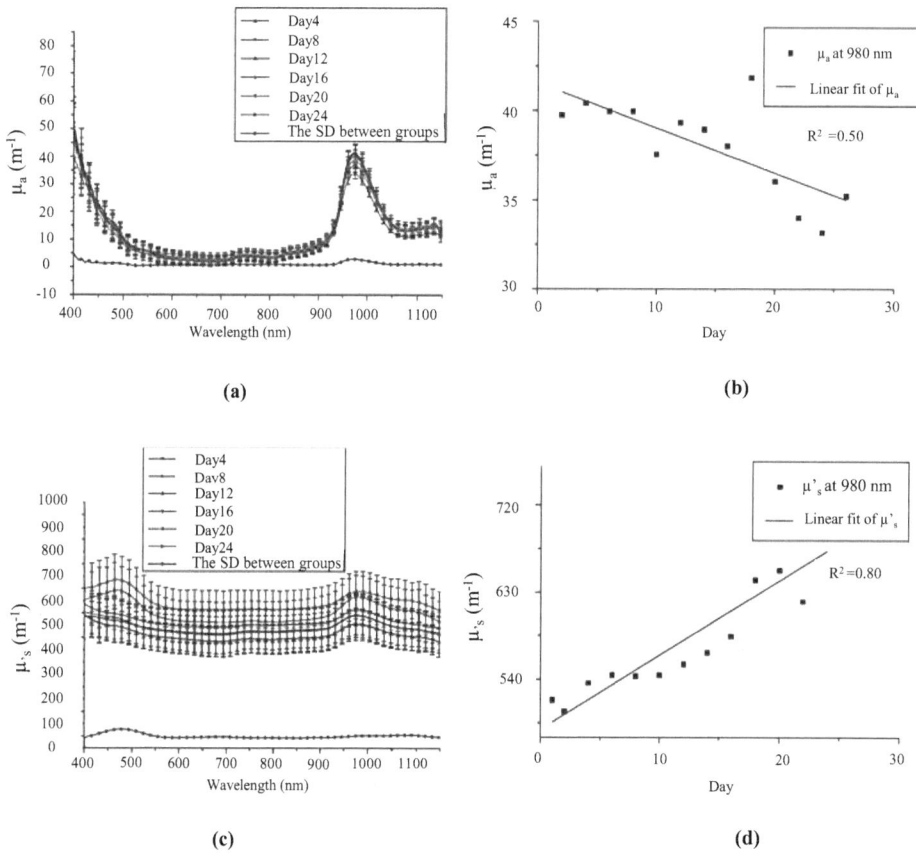

Fig. (9). (a) μ_a and (c) μ'_s of pear tissues in the region of 400–1150 nm respectively for the 6 batches out of total 13 batches and the averaged values of (b) μ_a and (d) μ'_s at 980 nm for all 13 batches and R^2 = 0.50, 0.80 respectively for (b) and (d). (SD: standard deviation).

The absorption coefficient, μ_a in the wavelength range of 900–1050 nm was used to calculate the SSC since it contained the absorption peak attributed to SSC, and the reduced scattering coefficient μ'_s in the spectral range of 500-1050nm was used to establish the detection model of firmness as it is related to the scattering properties resulting from the microstructure of the fruit. Figs. (**10a** and **b**) show the PLSR models for SSC and firmness prediction of pear based on the optical coefficients. It can be seen that the models achieved good prediction results for SSC and firmness with R^2_c=0.83, and 0.67 respectively. This research about the relationship between the optical properties of pear tissues and the corresponding SSC and firmness has provided references for future non-destructive detection based on optical sensing technology.

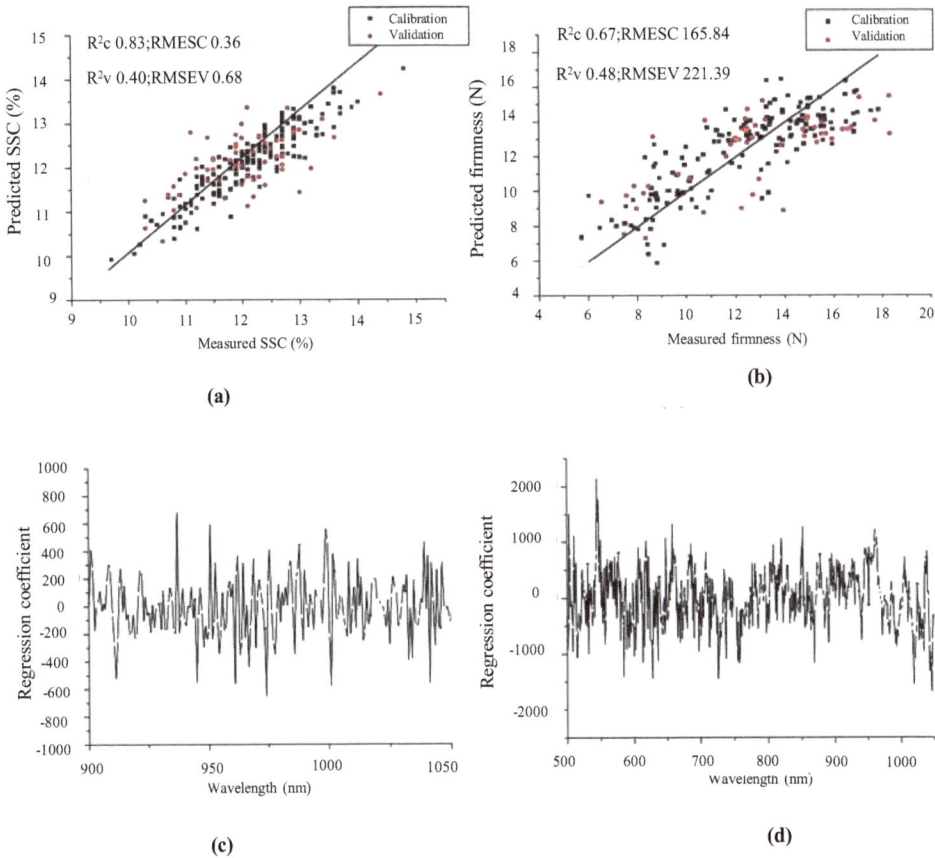

Fig. (10). (a) The PLSR model of pear's SSC prediction based on μ_a spectra in the range of 900–1050 nm; **(b)** The PLSR model of pear's firmness based on μ'_s in the range of 500–1050 nm; The regression coefficient plots for the PLSR models for predicting **(c)** SSC and **(d)** firmness.

2.4. The Future Trends of Optical Sensor Technology

Optical sensor technology has the advantages of low price, fast and no damage, and has a broad application prospect in agricultural products processing. However, presently, there is not enough knowledge about the transmission law of light in agricultural product tissues. The detection model based on optical coefficient is not stable and easy to be affected by environmental conditions. It usually needs a large number of samples to model and analyze the acquired signal and detection index based on the chemometrics method, the specific modeling process is not visible and lacks strict theoretical basis. The prediction results are usually affected by testing equipment, testing objects, testing environment, modeling methods and other factors, which eventually lead to the instability of the prediction results, the poor universality of the model, and the inability to further improve the analysis results and other problems. In addition, optical sensing technology cannot distinguish the absorption and scattering of light generated from tissue. The measured optical signal is a combination of absorption and scattering, which could easily lead to the loss of effective information. With the development of agricultural products processing technology, rapid non-destructive testing technology is urgently needed to meet the needs of online production. In the future, more attention is suggested to be focused on the research about the optical properties of agricultural products tissue, and establishing stable light transmission models, which would provide the reliable theoretical support for the application of optical sensing technology. In addition, more researches are also aimed at simplifying models, reducing the runtime, and meeting the needs of online applications.

Table 2. Summary of the application of optical sensor technology in agricultural products quality and safety detection.

Optical Sensor Technology	Sample	Parameters	References
NIR	Korean Hulled Barley	fusarium infection	[21]
	onion	internal rot	[22]
	apple	moldy core	[23]
	tangerine fruit	granulation	[24]
	citrus fruit	SSC	[25]
	cherry tomato	soluble solids and lycopene contents	[26]
	tomato	soluble solids content, pH	[27]
	Fish Muscle	moisture content, lipid content, protein content, pH, and freshness indicators	[28]
	Rice surface	ricyclazole pesticide	[16]
	lettuce leaves	fen valerate and chlorpyrifos	[29]
	lettuce leaves	organophosphorus pesticides	[30]
	rice	cadmium	[31]
	Molasses	sugar content	[32]
	Black tea	taste-related compounds content	[33]
	'Hami' melons	SSC	[34]

(Table 2) cont.....

Optical Sensor Technology	Sample	Parameters	References
HSI	apple	bruise region detection	[35]
	Oil Chestnuts	blue mold infection	[36]
	citrus fruits	decay lesions	[37]
	oranges	decay	[38]
	pear	SSC and firmness	[39]
	Peaches	fungus infection	[40]
	peach	decay	[41]
	Strawberry	decay	[42]
	nectarines	split pit and ripeness monitoring	[43]
	Chicken meat	bacterial contamination	[44]
	potatoes	blackspot	[45]
	peanut	Aflatoxin	[46]
	Flos Lonicer	total phenolic content	[47]
	sugar beet	fungal pathogen	[48]
Raman	Citrus fruits	freshness assessment	[49]
	tomato	carotenoids	[50]
	Wheat and Grain	fungal Infections	[51]
	tomato	chlorpyrifos pesticide residues	[17]
	rice	mold colonies	[52]
image processing	rice	quality and defect	[53]
laser-induced breakdown spectroscopy	tea	Pb	[54]
multispectral imaging	Potato	defects	[55]
frequency domain diffuse optical tomography	apple	underlying moldy lesions	[56]

3. ACOUSTIC SENSING TECHNOLOGY IN AGRICULTURAL PRODUCTS PROCESSING

3.1. Principle of Acoustic Sensing Technology

Sound wave is a kind of mechanical wave produced by the vibration of an object (sound source). In recent years, the acoustic characteristics of agricultural products for their quality non-destructive classification and testing have begun to develop and progress. The acoustic signal is converted into an electrical signal by the method of priority, and the converted electrical signal is stored and processed by the computer. At present, the acoustic sensing technology commonly used in non-destructive testing of agricultural products mainly includes three models: spectrum, ultrasonic and ultrasonic imaging.

Frequency spectrum is the abbreviation of frequency spectrum density and frequency distribution curve. Acoustic Spectrum is a tool used to describe the distribution of acoustic energy in sound and its components. The sound spectrum can be thought of as a "picture of sound". Static spectrum can reflect the relationship between frequency and amplitude of sound at a certain moment in the process of sound production. Its x-coordinate is the frequency and its y-coordinate is the amplitude. Dynamic sound spectrum can reflect the change of sound intensity and pitch with time in a certain period of time. Its abscissa is time and its ordinate is force or pitch.

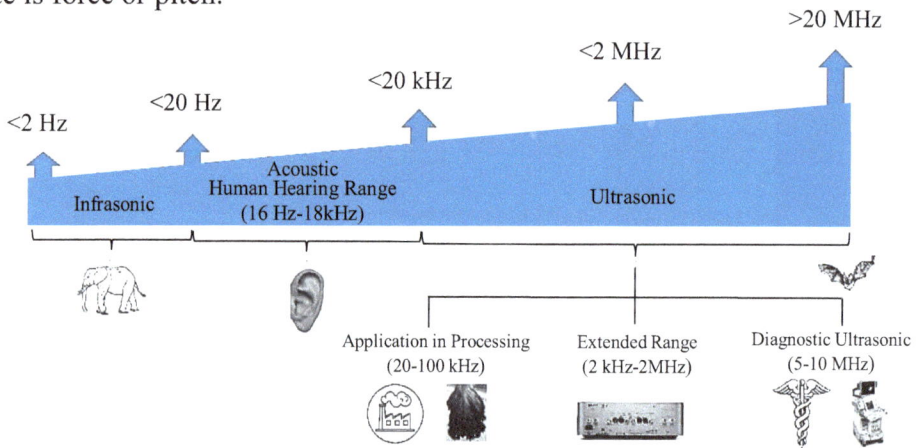

Fig. (11). Frequency range of Ultrasound used in agricultural processing.

Fig. (**11**) shows the frequency range of ultrasonic waves used in agricultural production. The ultrasonic frequency is higher than the limit of human perception, its spectrum is within the frequency range between 20kHz and 1GHz. The frequency utilized by ultrasonic equipment ranges from 20 kHz to a few GHz. Fields of non-destructive testing and destructive therapy are the two major applications of ultrasound. In the field of non-destructive testing, ultrasonic signals can give feedback of a lot of key information invisible by the human eyes, such as surface cracks and changes in internal quality. These parameters are closely related to the unique acoustic properties of different materials. In addition, the ultrasonic wave is generated and detected by an ultrasonic transmitting device and a detector, which are mainly composed of piezoelectric materials. Fig. (**12**) is a typical diagram of an ultrasonic sensing system. There are typically two categories of ultrasonic intensity employed by ultrasonic equipment: low-power ultrasound waves with higher frequencies and high-power ultrasound waves with lower frequencies. The former has high sensitivity and has no effect on mechanical or chemical performances of material, but can lead to the vibrations of material molecules to be detected. The latter carries more sound energy, leading to physical, mechanical, and chemical variations in the material. As a physical phenomenon, cavitation brings about this variation by causing microbubbles to collapse and provide high temperature and pressure [57, 58].

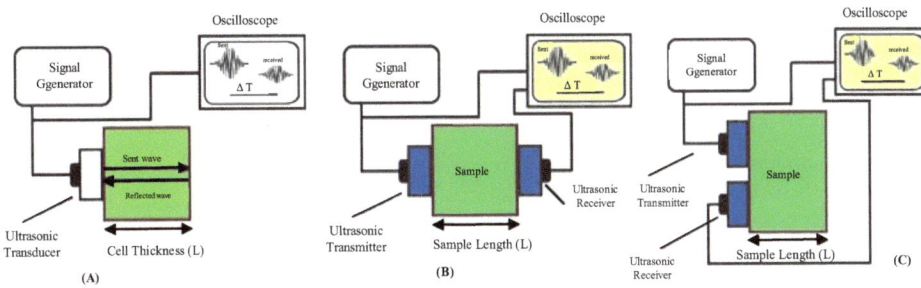

Fig. (12). A typical diagram of ultrasonic sensing systems.

Ultrasonic imaging technology is an image diagnosis technology that applies ultrasonic detection technology to the object under analysis and tests the object organization with the advantage of high frequency, wavelength, short energy concentration, good directionality, strong penetrability, safety and non-invasive. Because ultrasound can penetrate many opaque objects, the information of the acoustic characteristics of the internal structure of objects can be obtained by using ultrasound. Ultrasonic imaging technology can turn the information into the visible image of human eyes. The image directly formed by the sound wave is called the sound image. Due to the physiological limitation, the human eye cannot directly perceive the sound image. It must be transformed into the image or figure visible to the naked eye by optical or electronic or other ways. The image visible to the naked eye is called the acoustic image. The acoustic image reflects the distribution or difference of some or several sound field parameters inside the object. Conversely, for the same object, different acoustic images can be generated by using different acoustic parameters, such as acoustic impedance, sound speed or sound attenuation. An experimental configuration by ACU imaging and the resultant scan image of wooden seat is displayed in Fig. (13) [59].

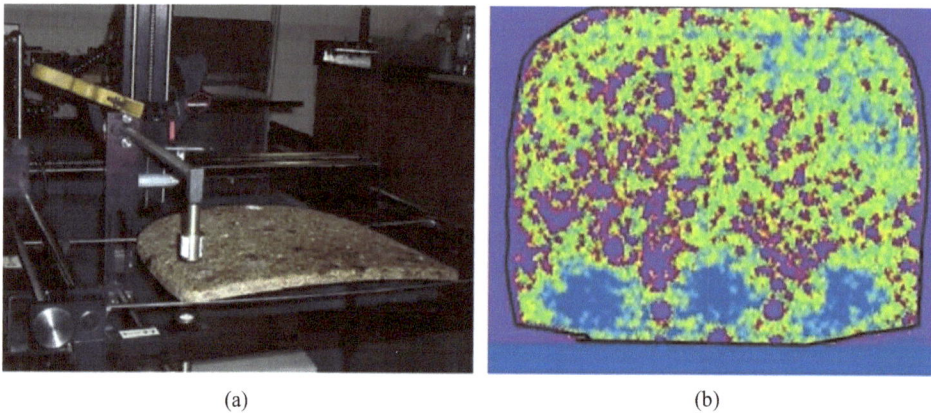

<table>
<tr><td>(a)</td><td>(b)</td></tr>
</table>

Fig. (13). (a) The experimental configuration, **(b)** images.

3.2. Research Advances and Application of Acoustic Sensing Technology

In the early 1970s, people discovered the possibility of ultrasonic technology in the detection of agricultural products [60]. Nowadays, the ultrasonic detection of agricultural products is primarily focused on the following aspects:

3.2.1. Acoustic Detection of Physical Characteristics of Agricultural Products

In the production of commercial agricultural products, the main goal is to avoid damage to the surface of agricultural products. In recent years, with the continuous increase of processing and processing procedures of agricultural products, the possibility of surface damage increases sharply. In this case, in the process of collecting, sorting and transporting agricultural products, the detection of epidermal hardness and cracks is essential to pursue the balance between skin strength and coping with load, as well as to remove defective products in time.

An online detection system for measuring eggshell strength on the basis of acoustic resonance was exploited (Fig. **14**). Si-PLS and multiple stepwise regression were used to screen the effective frequencies in the frequency domain after FFT and DCT transform, followed by the establishment of a linear model . After comparing several key parameters, the linear modeling method based on DCT domain was utilized to estimate the eggshell strength online . The results fully showed that the method proposed in their study was an effective non-invasive online testing technique for eggshell strength. Further study was concluded to be focused on the establishment of -performance analysis using multivariable models and perfect the online system [61].

Acoustic detection is a wide range of eggshell crack detection method, which can identify broken eggs by stimulating the sound or vibration signals of eggs. After signal acquisition, three processes are carried out to obtain recognition results: (1) feature extraction; (2) feature selection; (3) recognition.

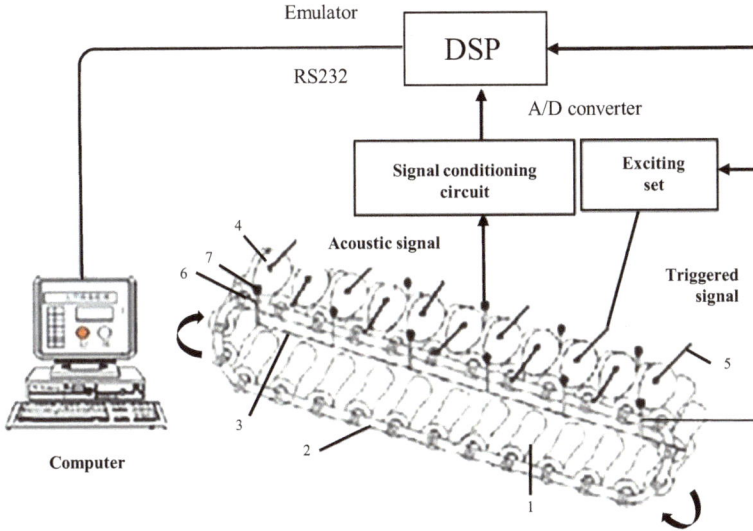

1.Support roller;2.Cycle chain;3.Work platform;4.Eggs; 5.Stick; 6.Microphone; 7.Infrared triggering device

Fig. (14). Schematic of the on-line system.

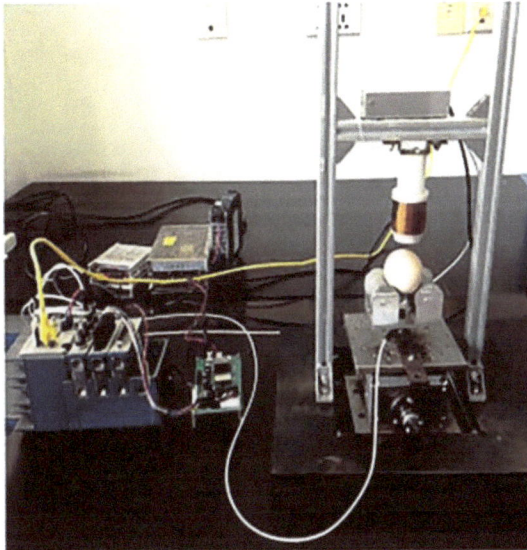

Fig. (15). Schematic diagram of the eggshell crack detection system.

The detection of egg cracks on the basis of acoustic method was studied by focusing on the extraction of varying features and the comparison between sound eggs and broken eggs (Fig. **15**). Therefore, in their study, a kind of excitation device driven by solenoid was exploited, which generated acoustic signal by striking the eggs. The time domain and frequency domain features selected and customized by predecessors were extracted. The discriminant effect of each feature on sound eggs and broken eggs was evaluated by F-Ratio, as well as the correlation between the features was studied. In highly correlated feature pairs, features with significantly low F values and features with relatively low values are ignored. The neural network is used for feature reduction, and the classification accuracy is up to 99.2%. Feature reduction helps to simplify the recognition algorithm of the online system and reduce the amount of computation [62].

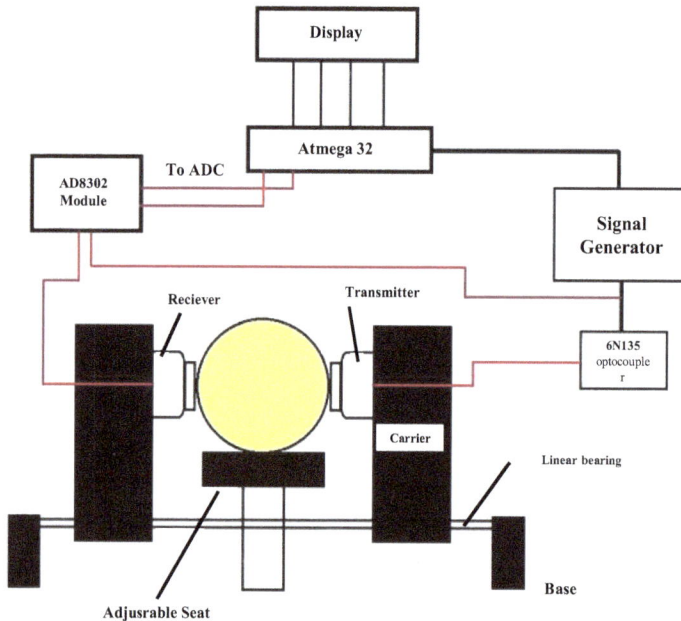

Fig. (16). The schematic diagram of the ultrasonic parameters measuring system.

Fig. (**16**) is the schematic diagram of the ultrasonic parameter measurement system. Ultrasonic method was utilized to detect the firmness and sugar content of greenhouse tomatoes (cv. 870). Through contacting the peel with the ultrasonic probe, this method was to measure the acoustic wave attenuation in the fruit tissue. The tested fruit was transferred from the greenhouse to the temperature control room, followed by a destructive osmotic hardness measurement after non-destructive ultrasonic testing. Their results displayed that the measured decay rate and hardness obviously decreased during the shelf life. Before the end of the softening process, a linear relationship between attenuation and firmness was obtained [63]. With the application of ultrasonic technology, a portable quality testing system for mechanical properties of apple was developed [64]. In their study, the relationship between ultrasonic parameters and apple hardness measured by a fruit ultrasonic transducer was evaluated, and the estimation model of apple hardness was established. The tissue of apple was treated with ultrasonic wave, and the hardness of the cell wall was measured by atomic force microscope (AFM). It was found that the hardness of the cell wall decreased with the extension of ultrasonic treatment time, and ultrasound could increase the solubilization of pectin in the cell wall [65].

3.2.2. Ultrasound for Monitoring of Agricultural Products and Extending Shelf Life

Ultrasound is one of the latest non-thermal methods used to prolong the shelf life of fresh fruits during storage. Wave frequency, treatment time and power are the three key factors for the efficacy of ultrasound. Acoustic monitoring displays great prospect in agricultural systems biodiversity monitoring because its implementation and implementation cost are relatively low in comparison with traditional methods [66, 67].

The effects of different ultrasonic power (30W, 60W, 90W) and treatment time (5min, 10min) on the quality of strawberry were studied. The oxygen concentration in strawberry packaging treated with 30W and 60W ultrasonic power were higher than that of 90-rod control (CNT) groups. During storage, the release rate of CO_2 from carbon nanotubes and 90W treatment increased sharply. The pH value, total soluble solids and color of 30W and 60W treatments were better than those of CNT and 90W treatments. The analysis of the incidence of decay confirmed that all ultrasonic treatments could effectively reduce the growth of mold. In comparison with 90W and CNT groups, 30W and 60W treatments gave better texture performances. The moisture and sugar content of strawberries were quantitatively analyzed by Fourier transform near-infrared spectroscopy (FT-NIR). The results showed that ultrasonic power of 90W had an adverse effect on the quality of strawberry, while the ultrasonic power of 30W~60W could improve the quality of strawberry and have the possibility to prolong the shelf life of strawberry [68].

The effects of frequency sweep ultrasound (SFUS), sodium hypochlorite (NAOCL) and their combination (SFUS-NAOCL) on the inhibition as well as quality preservation of natural microflora of fresh-cut Chinese cabbage during 7 days storage at 4 °C and 25 °C were studied (Fig. **17**). The results showed that the condition of 40kHz frequency sweep ultrasound and 100mg/L sodium hypochlorite could minimize and inhibit the growth of neutrophils, yeasts and molds, and reduce chlorophyll consumption. However, the color and texture properties got worse. Fourier transform near-infrared spectroscopy showed that the compound treatment inhibited the activities of polyphenol oxidase and peroxidase and showed its fresh-keeping impact. Most of the combined methods displayed synergistic effect, but the decrease of most treatments was less than 1.0 log CFU/g, especially the combined treatment significantly reduced the number of mesophilic bacteria by 2.7 log CFU/g, yeast and mold decreased by 2.0 log CFU/g respectively compared with the single method, and yeast and mold decreased by 2.0 log CFU/g and 2.0 log CFU/g respectively compared with the single method (P < 0.05). The results revealed that the combined treatment showed a synergistic effect, but the decrease of most of the combined treatment was less than 1.0 log CFU/g, especially the combined treatment significantly reduced the number of mesophilic bacteria by 2.7 log CFU/g, the yeast and mold also decreased respectively. During storage at 4 and 25 °C, the number of Chinese cabbage microorganisms produced by SFUS+NaOCl cleaning was less than that of the single treatment. However, even at the cold storage temperature of 4 °C, with increase in -the number of microorganisms during storage, post-treatment storage could not completely inhibit the survival of microorganisms. The results showed that ultrasound combined with sodium hypochlorite is promising in reducing and inhibiting microorganisms and prolonging the shelf life of Chinese cabbage while maintaining the quality characteristics of Chinese cabbage [69].

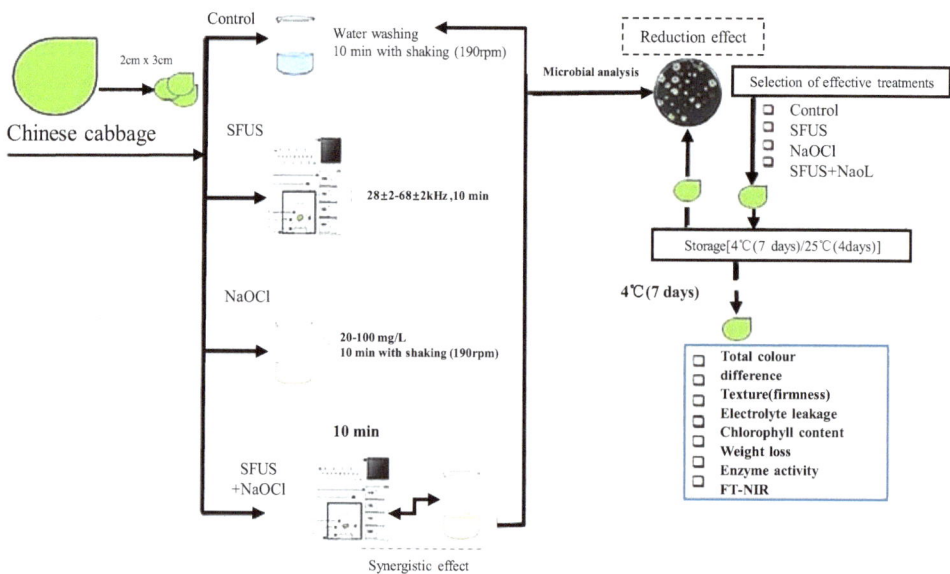

Fig. (17). Procedure flow illustration for SFUS and NAOCL treatment of fresh-cut Chinese cabbage.

3.2.3. Ultrasonic Techniques for the Recognition and Separation of Agricultural Products

An intelligent sorting system was designed and evaluated by Omid M *et al.* [70] for open and closed-shell pistachio nuts (Fig. **18**). They established a prototype system to detect closed pistachios by putting closed pistachios on the steel plate and recording the acoustic signal generated when the core hits the steel plate. The recognition adopts the principal component analysis (PCA) method which combines collision acoustics with neural network classifier. In order to generate useful features, the recorded acoustic signals were analyzed in time domain and frequency domain. After principal component analysis, the original multivariable was represented by 7 principal component (PC), which reduced the input parameters of the ANN model by more than 99%. In order to obtain the optimal classifier, offline modeling and online modeling were utilized respectively. In the offline phase, 3200 nuts were selected to train the neural network model. Various PC topologies with different numbers of ANN were designed. 300 closed-shell nuts, open-shell nuts and thin-walled nuts were tested, and the total error was 4.3%. The results showed that the CCR of the thin-walled nuts acquired by this system was higher than that of the previous system. Through further experiments, the accuracy of this system for sorting pistachios of different sizes and closed shells was verified. The results also showed that the size of pistachios doesn't affect the accuracy of the separator.

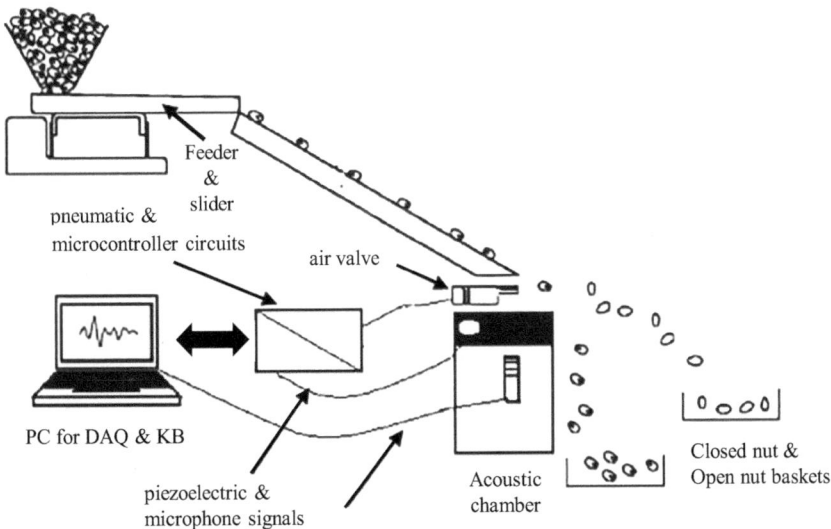

Fig. (18). Schematic of recognition and separation system.

An acoustic signal device was used to detect intact, pest-infested and moldy corn kernels (Figs. **19** and **20**). At the same time, comprehensive empirical mode decomposition method was utilized to compare different signals. These methods were used because of their great advantages in processing non-stationary signals and suppressing mode mixing. The features of time domain, frequency domain and Hilbert domain were extracted in the experiment. Then, these features were adopted as the input of support vector machine. Compared with the separate-use of features in various fields, the mixed-use of features can achieve higher classification accuracy. In their study, the classification accuracy of undamaged grains, pest-infested and moldy grains was 99.2%, 99.6% and 99.3%, respectively [71].

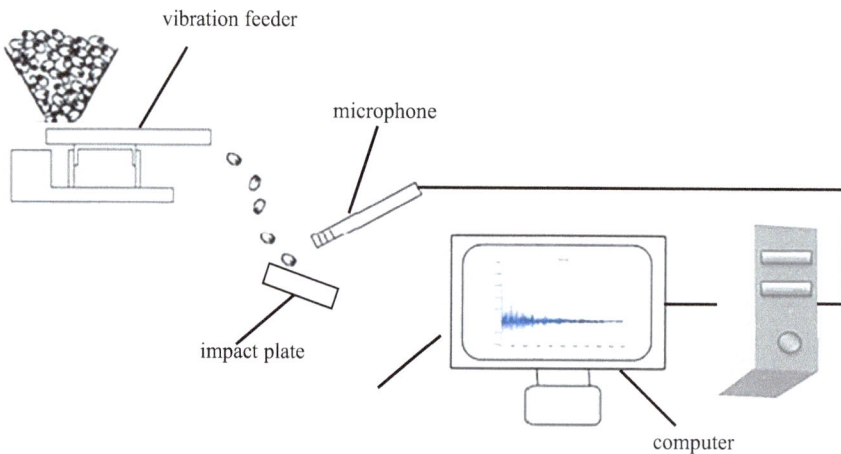

Fig. (19). Experimental apparatus with impact acoustics multi-domain patterns.

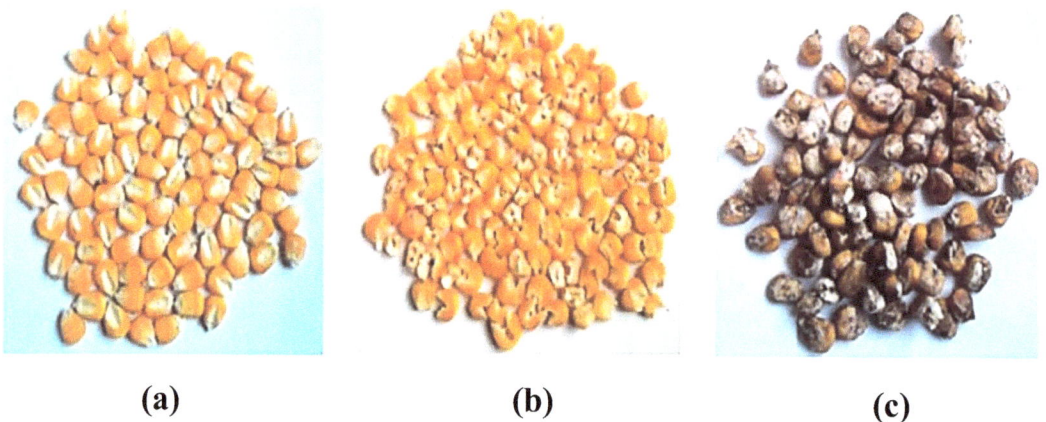

(a)   (b)   (c)

Fig. (20). Three types of corns **(a)** undamaged kernels; **(b)** pest-damaged kernels; **(c)** mildew-damaged Kernels.

The flexibility and adaptability of ultrasound have increased its possibility to be used in the dairy industry. Ultrasonic laboratory system can provide important reference data for the analysis of physical and chemical properties of dairy products, and can be used for monitoring, process control and harmless detection of dairy products. In addition, in the fermentation, extraction, pasteurization, homogenization and other processing processes, high-intensity ultrasonic system has also been applied [72].

The research results of Ouacha *et al.* [74] have proved the accuracy of ultrasonic transmission technology to select whether there are any air traces in ultra-high temperature milk packaging. They proved it by tracking the evolution of ultrasonic parameters: peak amplitude and flight time. It was also noted that flight time was the most informative parameter, while 35 °C was ideal for the characterization of ultra-high temperature milk packaging [73]. An eight-channel ultrasonic detector which can be used to detect the growth of microorganisms in carton-packed UHT milk even without opening the package was exploited. The system automatically analyzes the amplitude and delay of ultrasonic pulse in processing milk under UHT packaging, which realizes the coupling between transducer and packaging under dry conditions. Even if other physical and chemical parameters remain aseptic, variation in these parameters produced by diverse microorganisms can be perceived [74].

3.2.4. The Future Trends of Acoustic Sensing Technology

In order to replace destructive methods with non-destructive methods, ultrasonic measurement involves the rapid study of the state of food raw materials at harvest, during storage and when they reach the distribution point. Also, compared with other traditional methods, ultrasound has unique advantages such as high speed, full automation and on-line application. The convenience and ease of using - ultrasound, as well as the safety of this treatment to humans-makes it a suitable technology for analysis, control and processing of fruits, nuts, dairy products and for application in other industries.

Furthermore, the research in the field of acoustic was usually carried out under controlled laboratory conditions, so the same results cannot be expected on an industrial scale. Also, the use of acoustic in analyzing agricultural products should not result in adverse effects. Agricultural products are very sensitive, thus, ultrasonic operation on them could have adverse effect on the performance of someproducts. Therefore, further study is needed to clarify the various aspects of this technology and its impact on industrial agricultural products.

4. ELECTRICAL SENSING TECHNOLOGY IN AGRICULTURAL PRODUCTS PROCESSING

The composition and structure of agricultural products are closely related to their electrical properties. Most of the agricultural products have some kind of charge, which can form electric potential difference or electromotive force, so it is possible to conduct non-destructive and rapid detection of agricultural products through electrical characteristics. At present, the quality of agricultural products or sterilization of agricultural products are studied by using dielectric property, radio frequency, high voltage electric field and electrical resistance.

4.1. Dielectric Properties of Agricultural Products Processing

Dielectric property refers to the response of bound charge (the charge that can only move within the range of molecular linearity) in the molecule to the applied electric field. When agricultural products are stimulated by an external electric field, they produce a response, which usually uses different dielectric parameters to describe the different characteristics of the agricultural products when they are stimulated by the electric field. The method of measuring the dielectric properties of agricultural products is used to express the relationship between the moisture content, freshness and other quality attributes of agricultural products with accurate relevant parameters. At the same time, it can be used to monitor the growth of agricultural products and detect the degree of preservation of agricultural products during storage, to provide a new technical means for the deep processing and classification of materials [75]. Different agricultural products have different structures and different gene properties , so the dielectric properties of agricultural products are a comprehensive reflection of various material structures and components.

The common methods for measuring the dielectric properties [76] of agricultural products are as follows:

4.1.1. Parallel Plate Capacitor

A parallel plate capacitor is a capacitor composed of a dielectric element sandwiched between two electrodes (Fig. **21**).

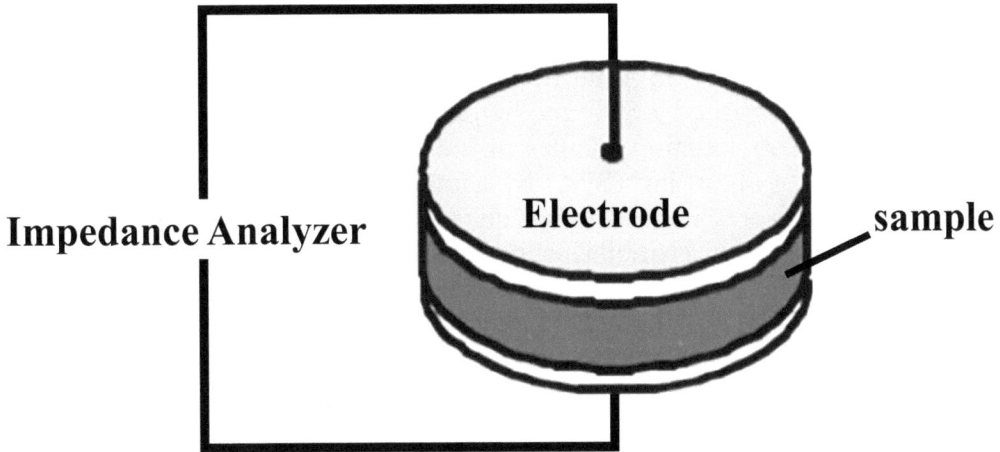

Fig. (21). Schematic design of parallel plate capacitor method.

4.1.2. Resonant Cavity Method

In the cavity method, the sample is placed at the middle-end of the waveguide (Fig. **22**). This changes the resonance frequency and quality factor of the shaft, which provides data from the interpolated sample to cover the dielectric constant.

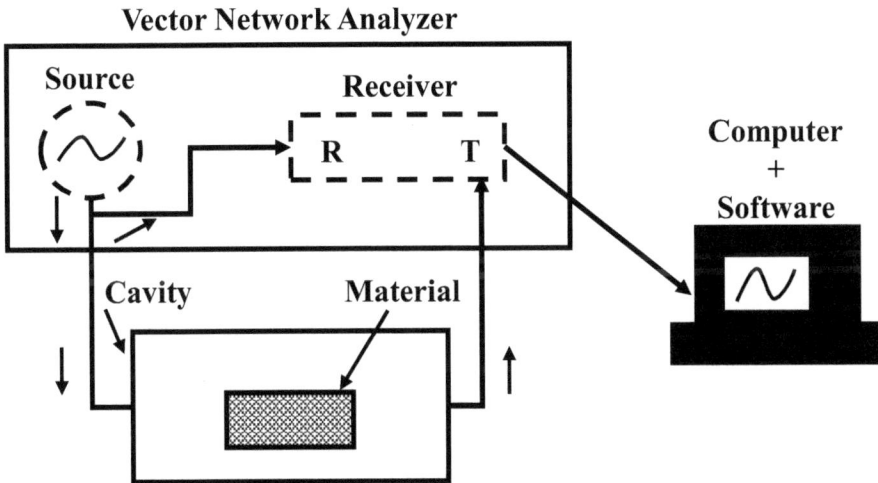

Fig. (22). Schematic design of resonant cavity method.

At present, the research on the rapid detection of the quality of agricultural products by using the dielectric characteristics mainly includes the detection of the maturity of agricultural products, the detection of water content and the detection of the internal nutrients. Guo *et al.* [77] determined the content and hardness of soluble solids in pear ripening process by means of dielectric spectroscopy and established a SSC and hardness measurement model based on the extraction of characteristic variables of the full dielectric spectroscopy and successive projection algorithm (SPA) by using multiple linear regression and partial least squares regression methods. The results showed that the combination of dielectric spectrum and artificial neural network is feasible. Sun *et al.* [78] Conducted non-destructive testing on the moisture of green tea based on dielectric properties and algorithm. Tu *et al.* [79] monitored the changes of damage volume and dielectric parameters of Fuji apple during 29 days of storage, established the equation of damage volume and relative dielectric constant with VB = -1.374 + 0.227 (R^2 = 0.848), and rapidly evaluated the damage degree of fruit by using dielectric properties. The results provide a new method for the rapid detection of agricultural products and lay a foundation for the evaluation of fruit quality by dielectric properties.

Soltani *et al.* [80] tested the freshness of poultry eggs by means of dielectric spectroscopy and machine learning technology. This paper mainly studied the detection of egg freshness, and employed several machine learning methods including artificial neural network (ANN) and Bayesian network (BNs),- to successfully predict egg freshness.

Shang *et al.* [81] obtained the dielectric performance of apple samples in the range of 10 ~ 1800 MHz, extracted feature variables by pattern recognition method, and established a model by using LVQ network to identify apple varieties. The accuracy of reached 99%, which has the potential of identifying apple varieties. Zhang *et al.* [82] used the dielectric properties of Apple to design a prototype of the automatic detection principle. The prototype used two capacitive sensors and an inductive capacitance resistance (LCR) instrument connected to the computer to measure the dielectric parameters of the apples under test and sorted the apples based on this approach.

The relationship between dielectric properties and apple quality parameters was discussed. Fig. (**23**) is a flowchart of the dielectric constant measurement. SSC and pH were measured by refractometer and pH meter respectively, and color was measured by taking two photos of the apple. Secondly, the dielectric data was preprocessed, including changing the range from 0.1 to 3 GHz to 0.1 to 1 GHz; using Savitzky-Golay second-order filters and 30-frame smoothing; dividing the range of physical and chemical parameters into 20 parts; Obtaining and saving median values of dielectric data and physicochemical parameters from each section [83].

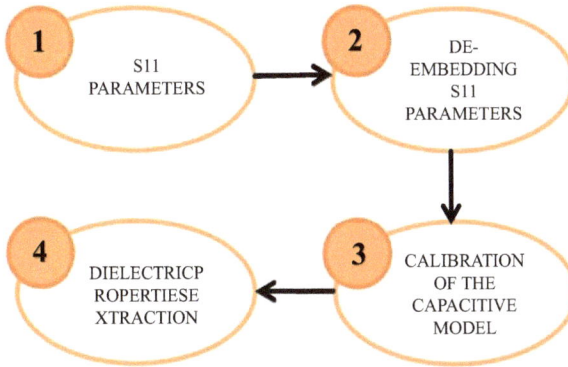

Fig. (23). Permittivity measurements flow chart.

Finally, a relationship model was established between the dielectric data and the parameters. The correlation coefficients of the physical and chemical parameters and models are shown in Table **3**. By analyzing the dielectric constant data of the apple samples of 100 MHz~1GHz, the relationship between the dielectric constant and the quality characteristics of the apples was studied. As shown in Table **3**, the PLSR model was found to have a high correlation. Therefore, the quality characteristics of apples can be predicted by using dielectric spectroscopy.

Table 3. Correlation factor for each physicochemical parameter and model.

Model	SSC	pH	L*	a*	b*
PCA-MLR	0.4647	0.7370	0.6463	0.4471	0.3780
PLSR Complete	1.000	0.7765	1.000	1.000	1.000
Optimized	0.8768	0.8402	0.8266	0.8363	0.8434

At present, researchers have paid attention to the application of dielectric properties in the field of agricultural product quality testing for the detection of moisture, freshness and internal damage of agricultural products. As the dielectric properties of each agricultural product are different under different conditions, a lot of research is needed to be explored in this field. Secondly, there is no on-line detection using dielectric property. Finally, how to combine this method with other non-destructive testing methods to make the detection methods more accurate is also an urgent problem. In general, dielectric properties are related to many internal qualities of agricultural products, so the prospects for using dielectric properties to detect the quality of agricultural products is very broad.

4.2. Radio Frequency of Agricultural Products Processing

Radio Frequency (RF) is an electromagnetic wave with a frequency between 3kHz and 300MHz [84]. When the radio frequency penetrates the interior of the object, the conversion of electric energy to thermal energy is completed by stimulating the migration of electric ions inside the object to produce the effect of heating the object. Compared with traditional heating methods such as electrified heating and microwave heating, RF technology not only uses shorter heating time but also ensures more uniform heating [85]. In the face of some large agricultural products, RF has a stronger penetration depth and better effect. At the same time, compared with the X-ray or gamma-ray heating method, the energy generated by RF does not ionize the water molecules, so it is considered as a safe heat treatment method. In recent years, with the development and optimization of related technologies, radio-frequency technology has been widely used in the field of agricultural products processing.

In order to improve the application of potato starch in the industry, Zhu [86] explored an alternative method based on RF heating, optimized the EWC parameter equation, and improved the uniformity of RF heating of potato. Jiao *et al.* [87] studied the use of RF to control the moisture in grains curb the production of insects or microorganisms in grains, selected wheat grains as sample products, and established a computer simulation model to study the impact of sample size. The simulated RF heating temperature curve was compared with experimental results to verify the computer model. The results of this study showed that RF technology could control the production of insects or microorganisms in granular agricultural products by controlling the moisture content in granular agricultural products.

Radio frequency technology is has been used to control pathogens in agricultural products. Based on radio frequency technology, an oven was developed to pasteurize walnuts in shell to maintain product quality [88]. By combining with hot air, product mixing and fixing during processing, the uniformity of RF heating was improved. The best RF processing conditions for the shelled walnuts included preheating between 16.0 cm electrode gaps and then drying for 40 minutes between 19.0 cm electrode gaps. This process was completed with a single layer of forced air-cooled walnuts. Staphylococcus aureus in walnuts with this RF treatment was significantly reduced. Therefore, RF treatment was used to control the pathogen of shelled walnut. Yang *et al.* [89] compared the inactivation effect of three heat treatment methods on eggshell. Studies have shown that using RF alone or using hot air and hot water combined with RF can inactivate *Salmonella typhimurium* in eggshells. Xu *et al.* [90] investigated the effect of RF technology combined with nano-ZnO on the germicidal efficacy and quality of carrot. The results showed that the effect of RF treatment was better, and the shelf life of carrot could be extended to 60 days.

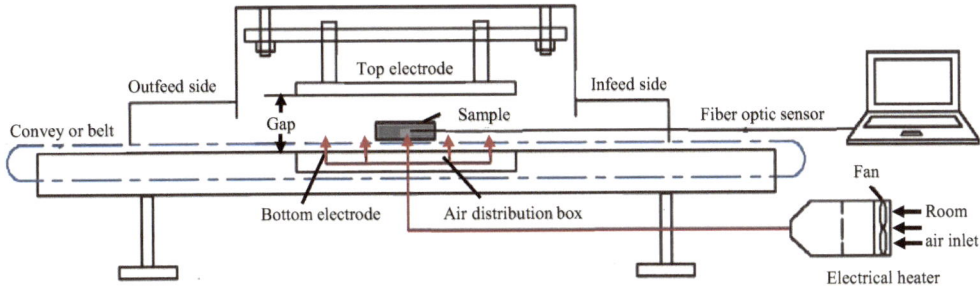

Fig. (24). Schematic diagram of RF system.

Radio frequency (RF) is has been used to eliminate pathogens in agricultural products in addition to drying them. In this study, 6 kW, 27.12 MHz RF oven (Fig. **24**) was used to pasteurize walnut (15.01% WB. Of.), and the effect of RF treatment on Staphylococcus aureus ATCC 25923 of walnut was studied. The device regulates the RF power transmission to the sample by adjusting the top electrode. The air distribution box provides forced hot air flow below the bottom electrode to maintain sample surface temperature.

Firstly, walnut with shell was selected and preserved in polyethylene bag at 4 °C. The water content (MC) of ten randomly selected walnut shells and kernels were $12.48 \pm 0.06\%$, $5.15 \pm 0.02\%$ and $8.50 \pm 0.03\%$ (WB), respectively. The water content treated sample was defined as natural water content (NMC) sample. At the same time, a series of AMC samples were prepared.

AMC shell walnuts and NMC samples were placed in two layers of polyethylene containers. The probe was inserted into six representative positions of the walnut kernel by pre-drilling (Fig. **25**). The temperature of the sample was recorded continuously until the sample temperature reached 70 °C.

The samples were tested by inoculating the samples with Staphylococcus aureus and treating the sample with radio frequency . During the storage process, the *S. aureus* in the sample was gradually reduced to below the detection limit after radio frequency treatment. Fig. (**26**) shows the survival curve of Staphylococcus aureus treated with RF and control samples during storage. When the samples were stored at 35 °C and 30% RH, there was an initial rapid decline in the number of microorganisms in both RF and control samples (Fig. **26a**). However, the microbial reduction rate was higher in the NMC control sample (Fig. **26b**). Therefore, the difference in -microbial reduction was due to the low initial MC value of NMC samples. In addition, the storage temperature also affected the survival of pathogens on the nuts, resulting in the rapid reduction of pathogens. Therefore, RF treatment can be used as an effective method to control the pathogenic bacteria of walnut in shell.

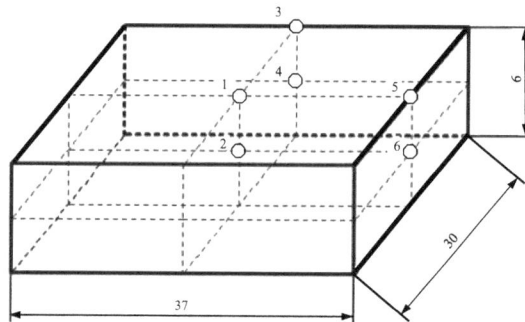

Fig. (25). The temperature of the shelled walnut is at the position (1-6) of the container. (The unit of measurement is cm).

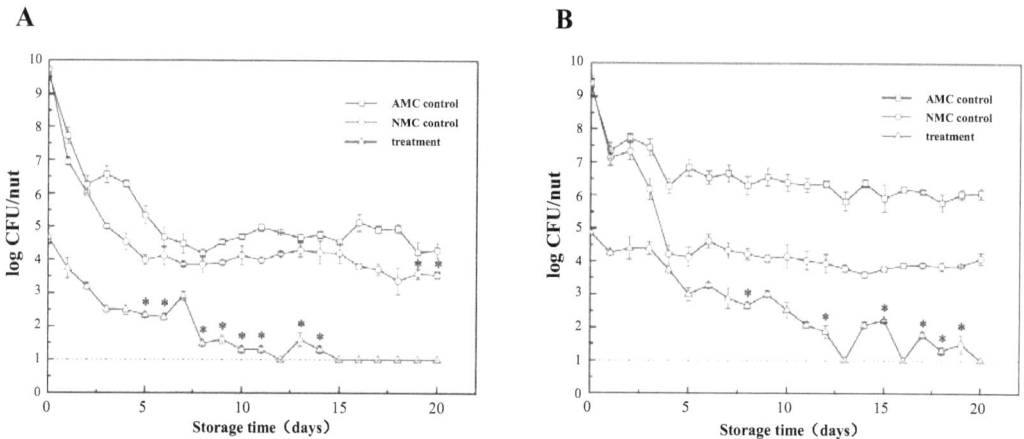

Fig. (26). The survival curve of *S. aureus* in shelled walnuts after heating with hot air assisted RF at 35°C. * indicates at least one replicate was below the LOD.

At present, RF technology has been applied to dry agricultural products, insecticide and sterilization. Especially its industrial applications to agricultural products such as rice and walnuts have been expanded. However, the operating range of RF processing has some limitations, mainly because the quality of fresh fruits is sensitive to temperature changes. Therefore, RF technology should be combined with other technologies to develop feasible and effective methods. Future research should improve the uniformity of RF heating, do a good job in the pretreatment of wetted product surface, reduce the heat resistance of pathogens, and carry out mild heat treatment at low temperature to improve the adaptability of food to heat. At the same time, explore different pretreatment steps to improve the effect of radiofrequency therapy [91]. With the application of computer technology in agricultural products and the rise of the Internet of things, future research on RF heating would turn to precision.

4.3. High Voltage Electric Field of Agricultural Products Processing

High voltage electric field technologies can be divided into high electrostatic field (HEF) and high voltage electrical discharge (HVED). As a feasible non-chemical technology, high electrostatic field has been applied in the drying and preservation of agricultural products [92]. Static electricity means that there is no current or voltage change during the test. A parallel plate electrode is usually used to generate a uniform electric field. The electric field between the two electrodes becomes equal after removing the edge of the electrode. Therefore, it can be used as an auxiliary method to improve the quality of agricultural products or to refrigerate them.

High-voltage discharge is a process in which current flows from a high-potential electrode to a neutral liquid and forms a plasma area around the electrode by ionizing the liquid. Some partial discharges in gas medium are widely used in the food industry. One of the effects of corona discharge is to produce electric field-induced flow or secondary electrodynamic flow (EHD). The EHD system is two electrodes connected to a high-voltage power supply, as shown in Fig. (**27**) [93]. Because it can enhance heat and mass transfer, non-mechanical, simple design, small energy consumption and other advantages, EHD technology has been widely used in the processing of agricultural products.

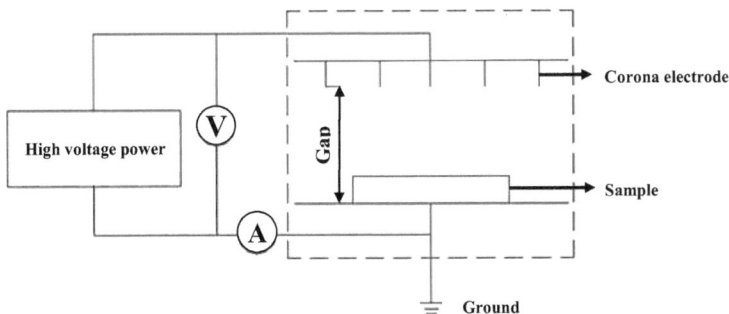

Fig. 27. Schematic diagram of the electrohydrodynamic (EHD) system.

Gao *et al*. [94]. investigated the drying rate of HVEF wheat under different drying temperatures and electric field intensities, and studied the influence of high voltage electrostatic field on the drying characteristics of wheat. It was found that the drying enhancement of HVEF was more obvious at lower drying temperatures than at higher drying temperatures, and a model describing the drying characteristics of wheat was established. Ding *et al*. [95] discussed the effect of different voltages on the drying rate of carrot slices in the EHD system. Taking carotene content and rehydration rate as quality indexes, compared with the conventional drying system, the drying curve was modeled and simulated, and the optimal formula for calculating the change of the drying curve was selected. The results showed that the drying rate of carrots in the EHD system was significantly accelerated and the quality of carrots was also improved.

Wang *et al*. [96] explored the effects of high electric field on Postharvest Physiology and quality of tomato. It was found that the negative strong electrostatic field could delay the decrease of fruit hardness and the change of fruit color. At the same time, it can affect the respiration peak and ethylene production of tomato fruits during storage, and inhibit the increase of malondialdehyde content and electrical conductivity of tomato fruits during storage. The results proved that the strong electric field can maintain the quality of tomato fruits and extend the shelf life. Nien-Yu Kao *et al*. [97] applied HVEF at 4 °C to prolong the shelf life of fresh-cut broccoli and reduce the hardness loss of fresh-cut broccoli. Studies have shown that HVEF can extend the shelf life of fresh-cut broccoli and may improve its quality. Zhao *et al*. [98] studied the effect of HVEF on the antioxidant activity of mature tomatoes during storage. When the tomatoes were exposed to HVEF, the content of O_2 and H_2O_2 in tomatoes decreased significantly. HVEF treatment can also enhance the activity of antioxidant enzymes and increase the content of non-enzymatic antioxidant components. Therefore, HVEF treatment can improve the antioxidant capacity of stored tomato fruits.

HVEF can be widely used in the drying of agricultural products. The advantages are the improvement of drying rate, the reduction of energy consumption and the reduction of cost. In the future, based on the results of existing studies, the emphasis would be placed on designing equipment suitable for specific drying processes [99]. At the same time, HVEF can affect the shelf life of fruits and vegetables. The results showed that HVEF can effectively extend the shelf life of fruits and vegetables and maintain their freshness.

These studies show that HVEF is a promising agricultural processing technology, but it still lacks industrial applicability. In the future, the operation can further be standardized to further understand the principle of the technology, so as to prove their non-destructive and sustainability in the processing of agricultural products.

4.3. Electrical Resistance of Agricultural Products Processing

Generally, agricultural products contain charged ions and mobile carriers. Agricultural products produce complex bioelectrical impedance, which is determined by the tissue composition, tissue anatomy and tissue health of agricultural products. Since the impedance of agricultural products is the same as the frequency, it is the same as any other bioelectrical impedance of other living cells or tissues. The factors affecting food resistance include salinity concentration of charge carriers, formula charge, number and mobility of charge carriers, their aggregation state, molecular weight and bonding type [100]. The current intensity is inversely proportional to the resistance of the fruit sample.

Therefore, the electrical impedance of agricultural products changes with changes in tissue structure and health conditions and also changes with changes in the frequency of the applied signal. Electrical impedance spectroscopy is an emerging technology that can characterize the physical and physiological characteristics of biological cells and tissues and has been widely used to detect the quality of agricultural products.

Chowdhury *et al.* [101] investigated the changes in electrical impedance and weight of citrus fruits under different ripening conditions. The equivalent circuit of the Nyquist curve during citrus ripening was modelled by using the nonlinear curve fitting technique. Kuson *et al.* [102] measured the impedance of the stem and skin of durian at different maturity stages by taking the days after flowering as the index. The impedance parameters of durian were selected by stepwise regression, and the electrical impedance spectrum of durian was studied by partial least square regression.

Chowdhury *et al.* [103] explored a rapid, nondestructive method for fruit ripening identification, using EIS to discuss changes in electrical impedance during banana ripening, and correlate the ripening state with the corresponding impedance. The results showed that impedance spectroscopy can be used to detect the ripening process of banana. Wu *et al.* [104] used electrical impedance spectroscopy to study the effects of drying and freeze-thaw treatments on eggplant. The results showed that due to the loss of water, the impedance of the sample increased throughout the frequency range, which confirmed that the cell membrane of the eggplant was seriously damaged during the freezing process. Chowdhury *et al.* [105] designed a portable impedance measurement system based on single-chip microcomputer to study the impedance of fruits during ripening and storage. Based on this system, the impedance characteristics of cucumber under storage conditions were studied, indicating that the developed system is also suitable for fruit EIS study.

The drying process of fruits and vegetables is the process of reducing water content. Kertész *et al.* [106] obtained the impedance spectra of carrot slices during drying and correlated the impedance parameters with the moisture content during drying. There was a good correlation between the impedance parameter and the change of moisture content in the drying process.

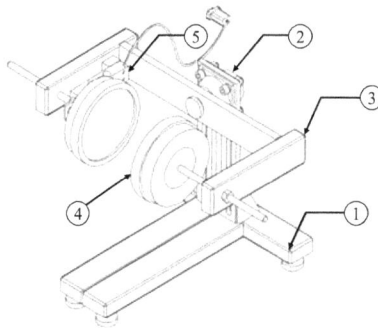

Fig. (28). The schematics of the apparatus.

When voltage is applied to agricultural products, the current flows through the agricultural products. Therefore, the resistance value of agricultural products can be measured to study the quality of agricultural products. Fig. (**28**) shows the instrument system used in the study to obtain the electrical resistance of apple fruit. In this experiment, two copper plate electrodes with thickness of 0.03 mm and diameter of 140 mm were used to measure the resistance. The utility model is made of two polyethylene screws fixed on the frame (Fig. **29**). The adjusting screw can adjust the distance between the two chambers to adapt to the size of the fruit.

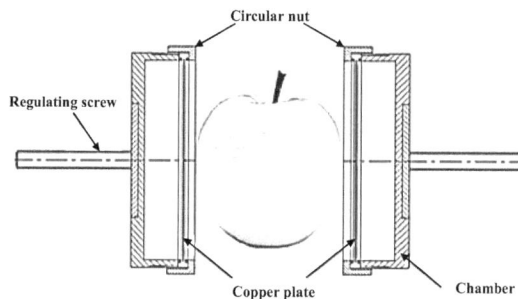

Fig. (29). The schematics of the probes.

Fig. (**30**) shows the effects of storage time on sample:(a) moisture content and (**b**) weight. The average weight of ten fresh apple fruits of 141.4 g was selected and stored at 22 ±2 °C. During storage, the electrical resistance was measured at 120 Hz and 1 kHz every 12 hours. The experiment lasted for 24 days. At 1khz and 120hz frequencies, the relationship between resistance and storage period and resistance to weight is shown in Figs. (**31** and **32**).

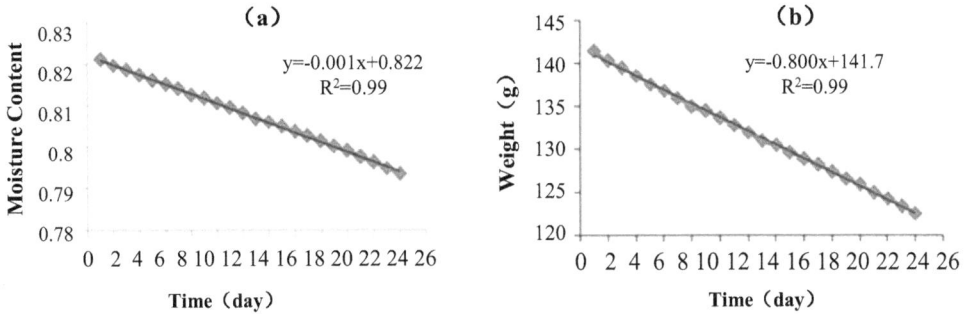

Fig. (30). Effect of storage period on samples': (**a**) Moisture content and (**b**) Weight.

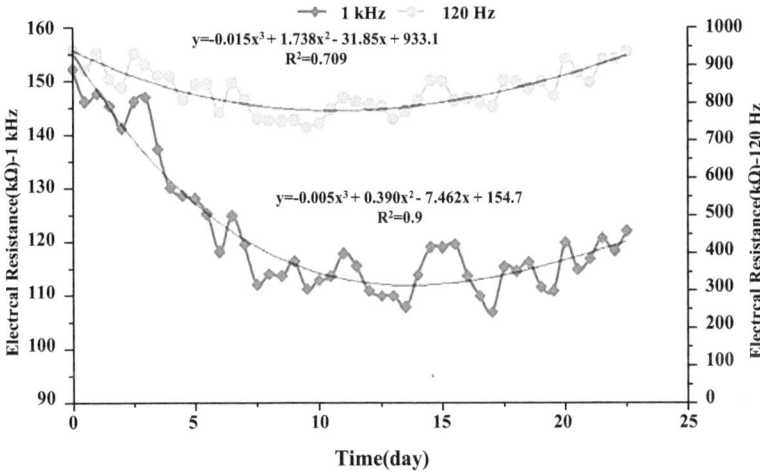

Fig. (31). The relationship between resistance and storage time at 1 kHz and 120 Hz.

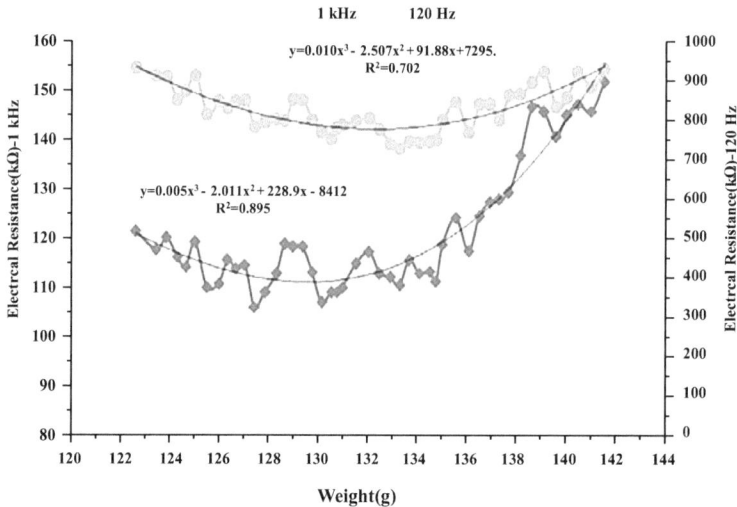

Fig. (32). Electrical resistance *versus* weight, measured at 1 kHz and 120 Hz.

The results showed that in the first 15 days of storage, the resistance decreased with the increase of storage time, but with the increase of time, the resistance also increased. The decrease in resistance at the beginning of storage may have caused the loss of water in the apple, indicating that the resistance can be used to evaluate the freshness of the apple during storage.

In summary, more and more attention has been paid to the electrical characteristics of agricultural products, and has been widely used in the processing of agricultural products. These electrical characteristics can be used to evaluate the quality of agricultural products, providing a new means of sterilization. With the advancement of agricultural modernization, more and more attention will be paid to research on the electrical characteristics of agricultural products at home and abroad. Therefore, the research on the theory and technology of the electrical characteristics of agricultural products has great potential and wide application prospect.

5. MAGNETIC SENSING TECHNOLOGY IN AGRICULTURAL PRODUCTS PROCESSING

5.1. Principle of Magnetic Sensing Technology

The magnetic sensing technology commonly used in nondestructive testing of agricultural products mainly includes nuclear magnetic resonance (NMR) and magnetic resonance imaging (MRI). NMR is produced when multiple nuclei are immersed in a fixed magnetic field and exposed to a second vibrating magnetic field [108]. On the basis of the principle of NMR and the difference between relaxation and diffusion, MRI can show the internal structure of intact food on a macro scale without interfering with the sample [109]. Over the past 30 years, MRI/NMR has become a vital technology to study the structure and composition of food materials. In particular, information about compartments, diffusion and movement can be acquired by detecting common protons caused by water in food. MRI/NMR technology provides an attractive option for noninvasive evaluation of the composition and microstructure of agricultural products [110].

A novel cross-sectional view of a hand-held single-sided MRI sensor is shown in Fig. (**33**) [111], a 3d-printed case, and a printed circuit board (PCB) containing rf coils, gradient coils, and a matching network. By installing all the coils directly to the PCB, the geometry can be optimized and then manufactured at a lower cost.

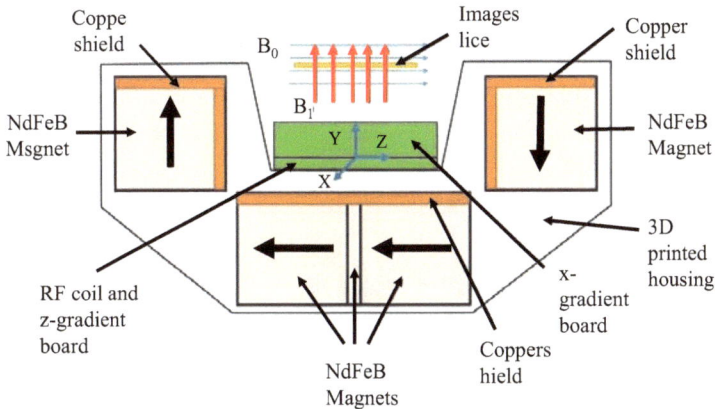

Fig. (33). A cross-sectional view of the proposed hand-held MRI sensor.

5.2. Research Advances and Application of Magnetic Sensing Technology

Magnetic sensing technology can be used for detection of contaminants or adulteration in agricultural products, component and internal defect analysis, droplet size measurement and emulsion assurance, process control and optimization because of its magnetic properties.

5.2.1. Magnetic Technology for Analysis of Internal Components and Defects of Agricultural Products

MRI may be an appropriate method for studying the development of potato, tomato, apple, kiwi and other fruits or vegetables during postharvest storage [112].

Blueberry fruit is rich in anthocyanin, flavonol, vitamins and other nutrients, which are widely welcomed by consumers (Fig. **34**). However, the deterioration of blueberry fruit during storage and transportation leads to quality decline and food safety problems. Qiao *et al.* [113] used LF-NMR to detect rotten blueberry fruits. NMR relaxation analysis suggested that three peaks exist in the relaxation spectrum of fruit. Six variables were selected for the BPNN model to distinguish the decay category of the fruit. The recognition accuracy of the training set was up to 86.7%, and the overall recognition reliability of the verification set was 90%.

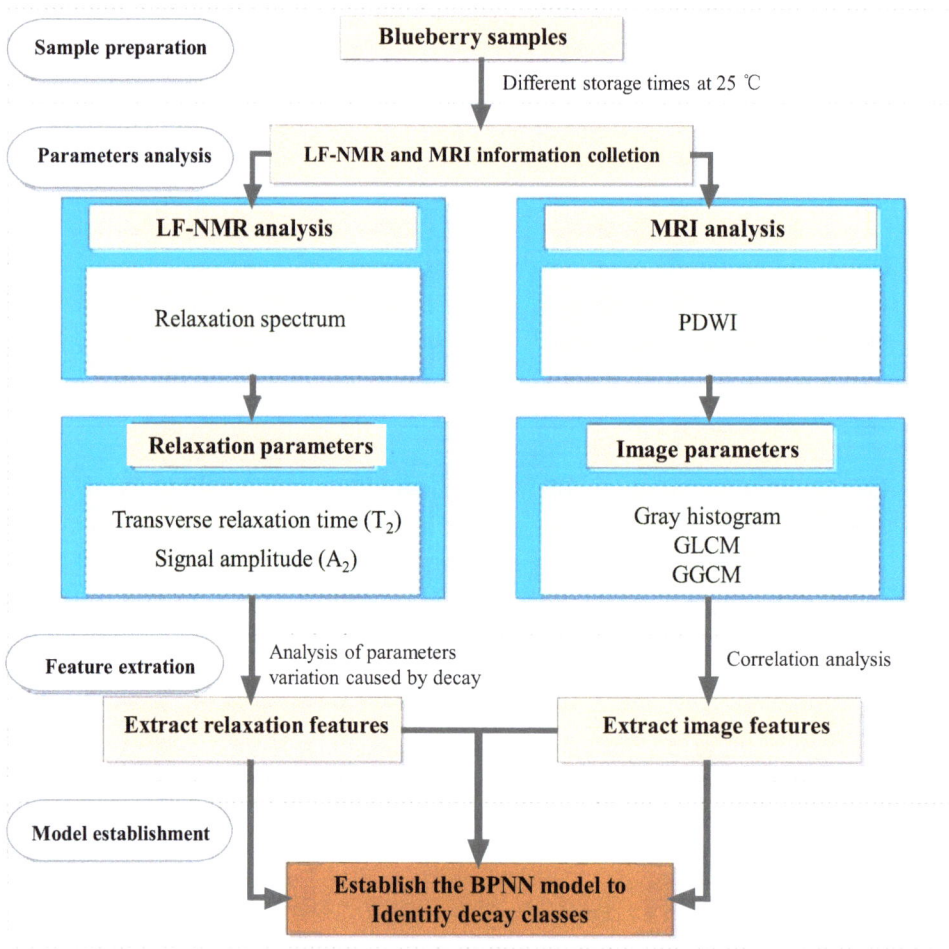

Fig. (34). Flowchart of typical steps for detecting and analyzing decayed fruit.

MRI can be utilized to identify the internal variation of harvested tomato fruit [114]. The determination of ethylene release, respiration and ion leakage showed that chilling injury (CI) occurred upon fruit storage at 0°C. MRI provides spatially resolved data. The apparent diffusion coefficient (ADC) was calculated based on magnetic resonance imaging of different parts of the fruit. The ADC value (D value) of pericarp stored at 0 °C for 1 week or 2 weeks was not significantly different from that of unrefrigerated fruit ($P > 0.05$), but the ADC value of internal tissue, that is, columnar region and chamber zone, was dramatically higher than that of unrefrigerated pericarp ($P < 0.05$). After 1 and 2 weeks of cold storage at 0 °C, the changes of D value in fruits were similar to those of respiration, ethylene release as well as ion leakage, which were higher than those of non-cold storage control ($P < 0.05$) [115].

The MRI technique was applied to the detection of the internal space of watermelon. More than 30 samples were tested by one-dimensional projection profile, and 28 samples were evaluated correctly. The measurement rate was 900 milliseconds for every sample [116]. Dellarosa *et al.* [117] processed apple tissue with pulsed electric field-and obtained the spatial distribution image of apple electroporation by MRI and computer vision system. The apple samples analyzed by MRI included PEF pretreatment of 400V/cm, VI and DIP samples. The MRI transverse relaxation time (T_2) map (Fig. **35**) and longitudinal relaxation time (T_1) weighted image (Fig. **36**) of the whole apple tissue confirmed the uneven distribution and degree of cell rupture and the release of intracellular contents to external solution.

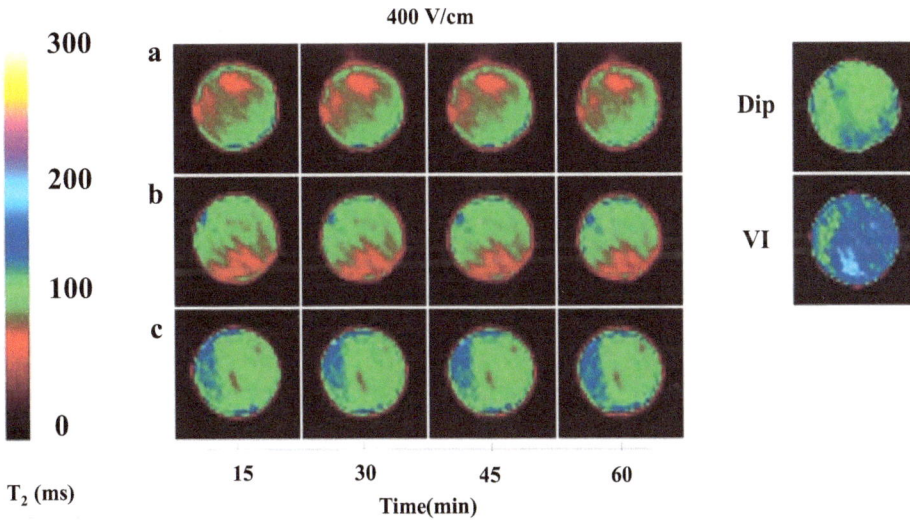

Fig. (35). MRI transverse relaxation time (T_2) map.

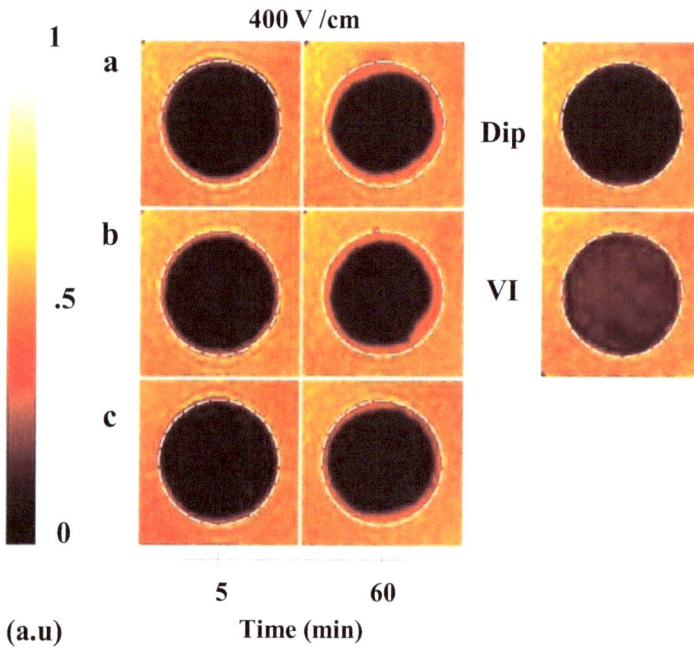

Fig. (36). Longitudinal relaxation time (T$_1$) weighted image.

Magnetic resonance techniques have been successful in helping to study the multifactor complex systems involved in plant growth and detect small changes in grape molecular composition and associated water relaxation behavior in grape berry tissues [118] (Fig. **37**).

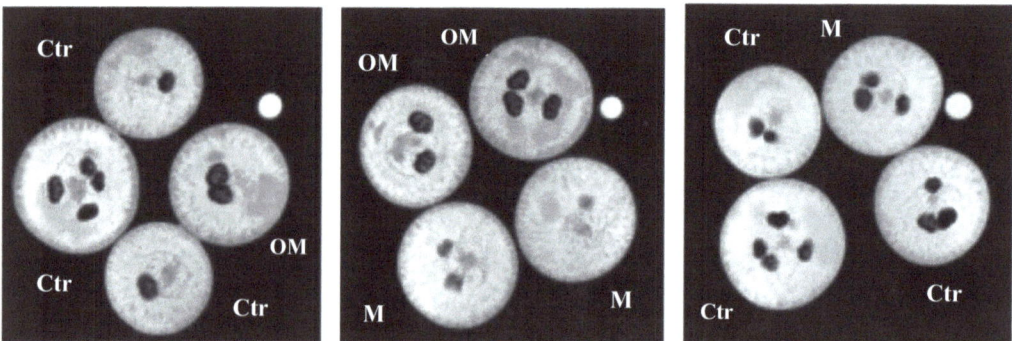

Fig. (37). (a) Images (msme_T$_2$) of 12 grape berries from diverse fertilization type (Ctr, OM, M); **(b)** morphological data: berry weight (g) and transversal diameter (mm).

The changes of kiwi tissue structure under two storage conditions (0, 90% RH for 65 days, 20 and 70 percent relative humidity for 35 days) were compared. Fig. (38) shows the MRI images of different kiwifruits. The weighted proton density (q) and transverse relaxation time (T_2) were recorded by MRI after 10,20, 35, 55 and 65 days respectively [119]. After 3 weeks of storage, the diameter of the sample decreased significantly under the condition of 20 degrees storage, and the peel showed a rough outline after 40 days. There was water accumulation in the pericarp, indicating that the fruit water moves outward, and the fruit water loss caused by evapotranspiration is slower than the migration process. For the samples stored at 0°C, except after moving to the shelf life, there were no related variations in size and structural characteristics during the whole period under consideration. Table **4** shows the T_{2f}, T_{2s} and A_f values of Kiwi at different storage times at 0°C and 20°C.

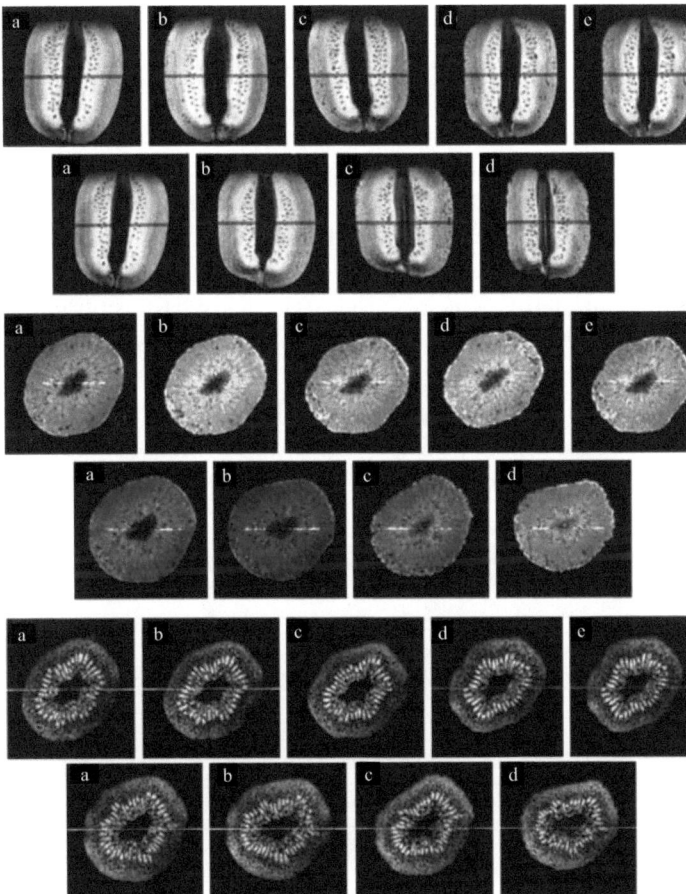

Fig. (38). GEFI (upper), GE3 D (middle) and RARE (lower) MRI images of kiwifruits.

Table 4. Values of T_{2f}, T_{2s} and A_f obtained for the kiwifruits at 0 or 20 °C at different storage times; all values are the arithmetic mean of the replicates (10).

Storage Temp. (°C)	Storage Time (days)	T_{2f} (ms)		T_{2s} (ms)		A_f (%)	
		S_1	S_2	S_1	S_2	S_1	S_2
0	0	51	53	308	309	0.48	0.47
	10	54	53	292	292	0.47	0.47
	35	52	53	296	294	0.50	0.48
	55	52	56	292	293	0.49	0.49
	65	44	47	290	288	0.49	0.50
+20	0	50	51	310	304	0.49	0.47
	10	47	49	290	299	0.49	0.49
	20	40	45	291	283	0.47	0.49
	35	41	45	287	267	0.45	0.50

The pear fruits of the "Dakazi" variety are very sensitive to bruises caused by mechanical shock and compression. Fig. (**39**) is the Magnetom Symphony 1.5T system. At different time after loading, MRI non-destructive testing technology and image processing technology were used to determine the propagation of internal damage volume of fruit. The results showed that the contusion volume expands linearly with the increase of external force and non-linearly with the increase of time. In fact, since loading, it has grown at a logarithmic rate for 26 days, and then its expansion had almost stopped. Force and time had significant effects on contusion volume respectively. At the same time. It is concluded that the best consumption time of the product is 12 days after loading and unloading or 12 days after external shock in the process of harvest and storage [120].

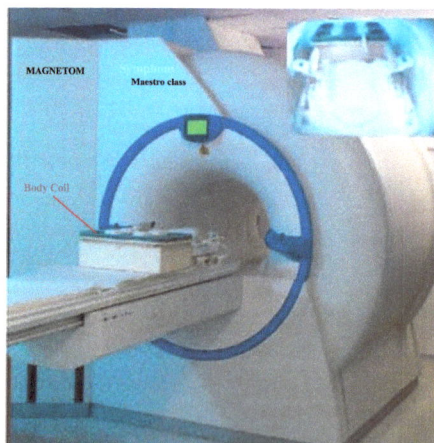

Fig. (39). Magnetom Symphony 1.5 T system at Kowsar Medical Imaging Center.

5.2.2. Magnetic Technology for Quality Control of Dairy Products

Time domain nuclear magnetic resonance (TD-NMR) has been proved to be an effective approach for detecting formaldehyde in raw milk of dairy industry. Coimbra *et al* [121] evaluated the effectiveness of TD-NMR in the determination of formaldehyde in milk samples (from 1.75 to 21% v/v) by comparing chemometrics (principal component analysis (PCA), partial least square regression (PLS), soft independent modeling of category analogies (SIMCA) with official colorimetric methods. The analysis was carried out at different cold storage time (0h and 48h) (Fig. **40**). The change in relaxation time was proportional to the concentration of formaldehyde. This is because the change of fluidity of water and fat during coagulation leads to phase separation, which results in the increase of transverse relaxation time T_2.

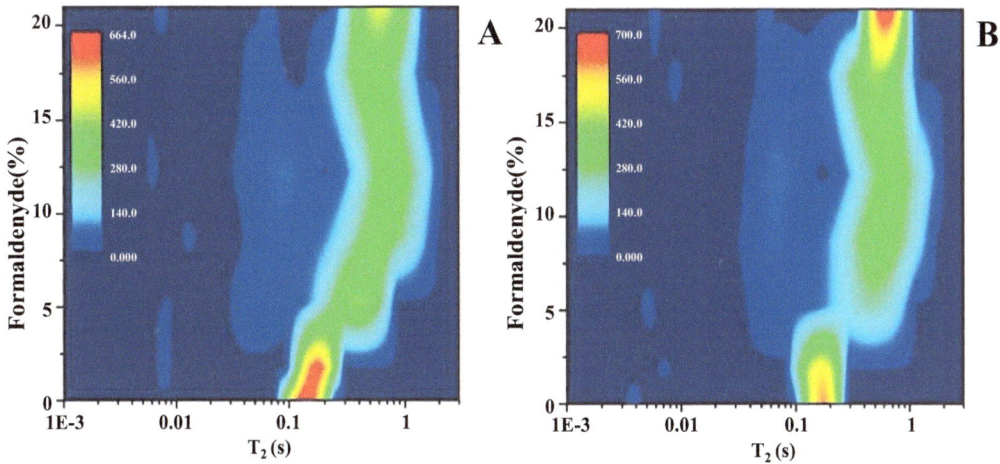

Fig. (40). Microzone shape of formaldehyde adulterated milk at 0h (**A**) and 48h (**B**).

A sensitive and rapid NMR method was established by Jin *et al.* [122] for the monitoring and control of Salmonella (Fig. **41**). In order to enhance the sensitivity of nuclear magnetic resonance sensor, the signal amplification method of streptavidin-biotin system was introduced to improve the binding capacity and binding force with the target. The membrane filtration technology was used to solve the problem of -difficulty in the separation of the target probe and the free probe, with emphasis on solving the problem of "hook effect". Under the optimum conditions, the pure culture medium and samples were prepared, filtered and detected by NMR. The total time was less than 2 hours, and the detection limit was 2.3×10^3 cfu mL^{-1}.

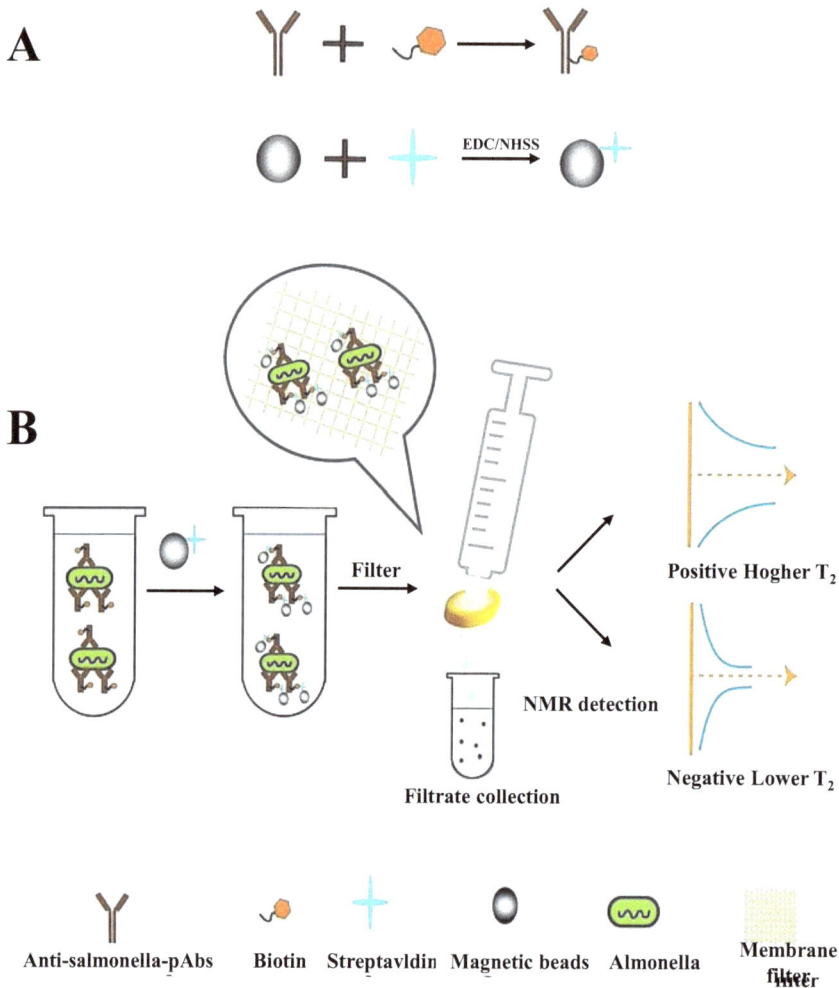

Fig. (41). NMR biosensor based on NMR for detection of Salmonella.

5.2.3. Magnetic Techniques for Detection of Lipid Deposition Patterns and Moisture in Agricultural Products

The flavor and commercial value of economic crustaceans depend on their lipid deposition patterns. In this study, MRI technique was introduced to compare the lipid deposition patterns of three types of prawns, including red swamp crayfish (procrayfish), Oriental river prawn (biogas Nippon) and white prawn (vannamei) (Fig. **42**). The results displayed that lipids were only deposited as visceral adipose tissue. Hepatopancreas is the main site of fat deposition, and fat deposition is less in Gill and muscle tissue [123].

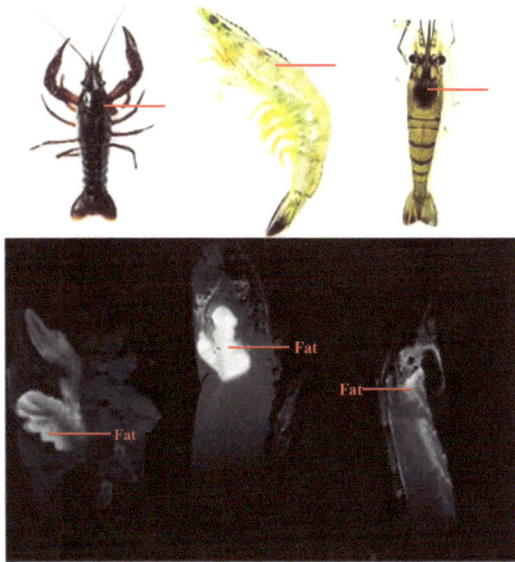

Fig. (42). The photos and MRI slices of three prawn species.

The normalized maximum signal value and fat volume measured by T_2 relaxation time and 3D LF-MRI model can be introduced as indexes to determine the fullness of four mature stages of Chinese mitten crab Eriocheir sinensis (Fig. **43**).

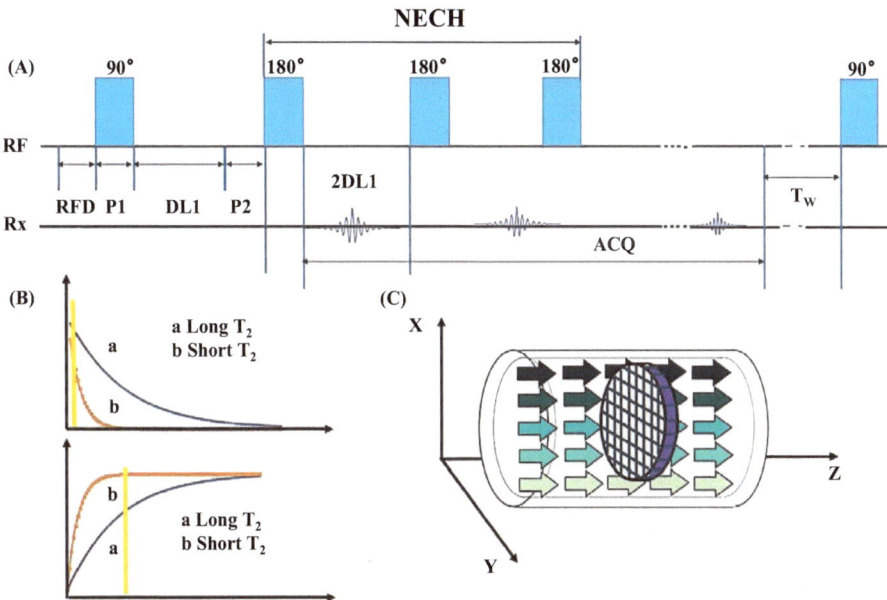

Fig. (43). Schematic diagram of quantitative determination of maturity of E. sinensis using LF-1H NMR.

Low field nuclear magnetic resonance (LF-NMR) was used by Li *et al.* [124] to detect the content of water and fat in oats. The applicability of internal standard method and external standard method for moisture determination of oat were compared. The calibration curve for the determination of oat fat was established by external standard method and compared with conventional method. Moisture and fat were expressed by the peaks of T21 (0.01–3.0 ms) and T22 (9.01–410.27 ms) in the T2 spectral curves of oat flour, respectively (Fig. **44**). Taking 3% $MnCl_2 \cdot 4H_2O$ as the standard, it was found that the internal standard method was more reliable than the external standard method.

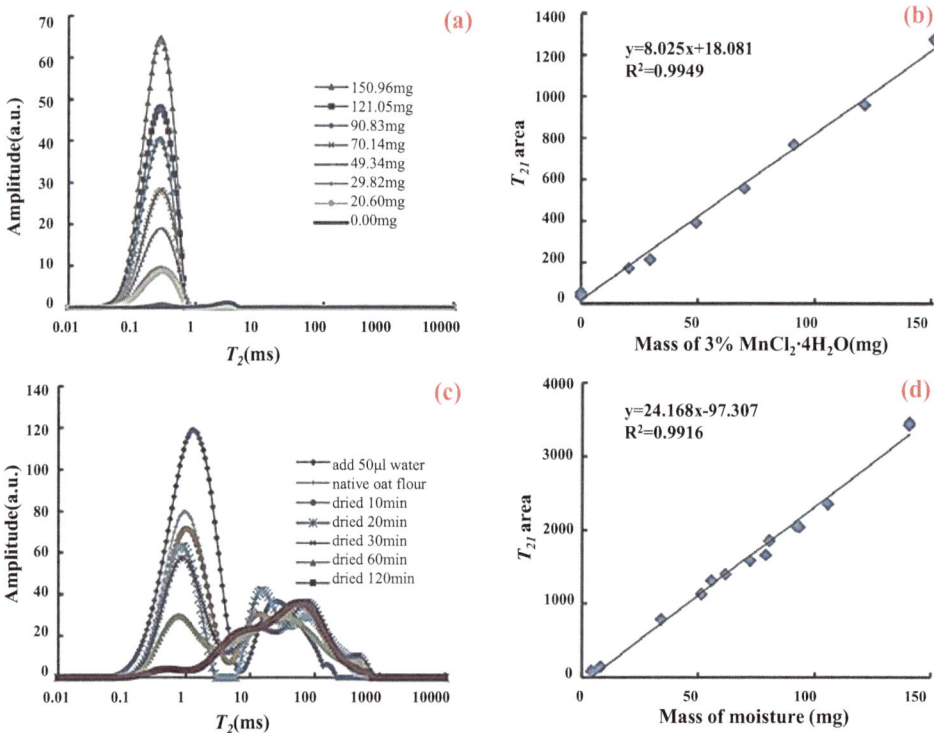

Fig. (44). T_2 distribution and calibration curves for water prepared using LF-NMR.

MRI was used to observe the water distribution of white rice grains during soaking, and the NMR signal intensity spectrum (SI-profile) was generated. Different kinds of rice had different trends in the formation of permeation morphology, and grains of the same variety in different regions and different years had the same water permeation trend. At the end of water absorption, the nuclear magnetic signal intensity in the center and periphery of the grain was higher than that in the middle region [125]. In addition, MRI has been successful in helping to study the multifactor complex systems involved in plant growth and detect small changes in grape molecular composition and the associated water relaxation behavior in grape berry tissues [126].

5.3. The Future Trends of Magnetic Sensing Technology

In terms of non-destructive testing, the cost of imaging equipment is a major factor. The price of magnetic resonance imaging equipment is too high, and the depreciation value of the equipment may even be greater than the value of agricultural products detected by the equipment. In addition, there are some engineering bugs in the integration of imaging equipment and automation pipeline. In the case of CT, it needs to collect images from different angles, and it is quite time-consuming to reconstruct an object.

Therefore, the research on equipment cost and equipment miniaturization is indispensable. In the case of LF-NMR, it has been utilized to monitor water content and distribution. Compared with the current NMR technology for bruise detection, it has the characteristics of low radiation use and low equipment cost.

In the future, with the recent exploitation of smaller, lighter and cheaper magnets, using Halbach and single side configurations, allow measurements through encapsulation. NMR technology is expected to be an important part of non-destructive testing of agricultural products. In addition, the multi-scale mathematical modeling of non-destructive testing of agricultural products using nonlinear finite element method is very promising, and it is improved by the combination of deep learning methods. Although the deep learning approach lacks explanatory power, it may be a more objective strategy to better develop magnetic detection techniques and have a positive impact on commercial applications.

6. SENSORY SENSING TECHNOLOGY IN AGRO-PRODUCTS PROCESSING

6.1. Concept and Principle of Sensory

Sensory sensing techniques are the core technologies in sensory evaluation of agricultural product quality and safety. Bionic intelligent sensory sensing detection technology is a discipline that use modern information technology and sensing technology to simulate human sensory behaviors such as vision, hearing, taste and smell, automatically acquire information reflecting the quality characteristics of the object, and simulate human's understanding and discrimination of information to process the acquired signals. Technology based on sensor development includes electronic nose (e-nose), electronic tongue and olfactory visualization sensor technology. Due to people's emphasis on the quality and safety of agricultural products, the demand for non-destructive and rapid inspection of agricultural products has increased. Compared with the traditional method, the electronic nose has the advantages of being fast and accurate, and is used by researchers in the detection of agricultural products.

6.2. Electronic Nose

The human olfactory system is an example of natural sensors. There has been a lot of work in the sensor field to simulate the function of the natural human sensors, thereby creating an artificial nose that can detect very low concentrations and chemically as diverse as the human nose. Therefore, it is necessary to understand the function of the human olfactory system before studying the sensors created by human beings. At the back of the nose is an area called epithelium, which plays a major role in odor detection. This area has specialized cells, called neurons, that allow the brain to connect with the outside world. Hair-like structures called cilia that extend from neurons extend out of each cell and come into direct contact with smell [127]. At the other end of the cell is a fiber called an axon, which enters the brain. Odor molecules usually come into contact with cilia, which have special proteins that can be sensed. The protein forms itself around odorant molecules and sends electrical signals to the olfactory bulb. The olfactory bulb transmits signals to the cerebral cortex, where it interprets odors [128].

The electronic nose detects gas by imitating the structure of the human nose. The comparison between human and electronic nose is shown in Fig. (**45**). The electronic nose is an instrument that includes a sensor and a pattern recognition system for identifying smells. The electronic nose is actually composed of interactive materials and equipment. The first is interactive materials that interact with the environment and generate responses. The second is a device that reads the response and converts it into interpretable and quantifiable terms. The interactive material is the main part of the sensor, which determines the nature, selectivity and sensitivity of the chemical sensor. As an interactive material for the interface between the environment and the transducer, when the material becomes more complex, it can respond to different molecules, which greatly improves the sensing ability. The sensing material captures vapor molecules with a certain selectivity, which cause physical changes in the material because the captured molecules chemically interact with the material [129]. The physical changes are called intermediate quantities. Based on the sensitivity, selectivity, reliability and reversibility of the relevant sensing mechanism, specific interaction materials can be selected [130].

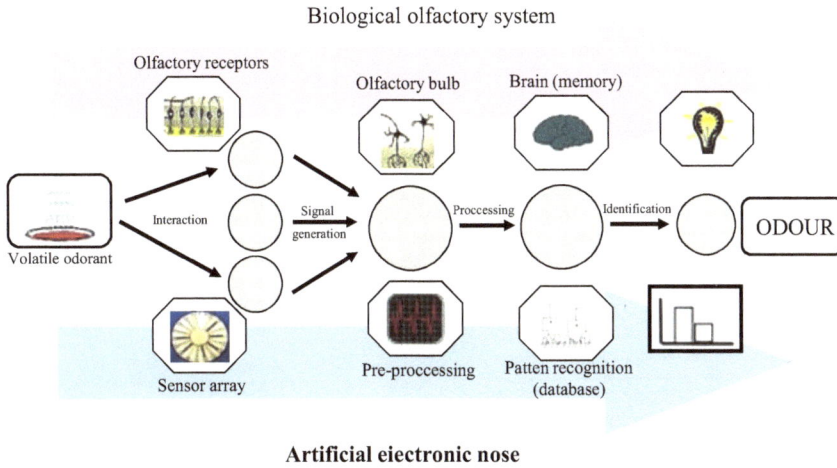

Fig. (45). Contrast between biological nose and electronic nose.

The typical electronic nose devices were based on sensor array and suitable pattern recognition techniques. For each application, this general method is implemented in a different module. Namely, fiber optic [131], semiconducting [132], conducting polymers [133], quartz crystal microbalance [134], surface acoustic wave [135]. In this section, the functions, advantages, and requirements of all the sensor technologies mentioned above will be discussed. Though many sensors such as semiconducting sensors, conducting polymer-based sensors, silicon carbide-based sensors, tin oxide-based sensors, can broadly be classified under one category, in this section all these sensors have been discussed separately as they have different applications and slightly different functioning mode which necessitates separate classification.

Electronic nose technology has been used in agricultural product safety testing, to show its application potential in safety testing. The electronic nose consists of a sample processing system, a chemical sensor array and a pattern recognition system [136]. The electronic nose can obtain the characteristic information of gas, and the gas information can be processed by using pattern recognition method to distinguish the degree and type of fruit and vegetable spoilage [137]. The electronic nose detection has the advantages of being fast, anti-interference and accurate detection [138]. The electronic nose is used to detect the freshness [139] and storage time [140] of litchi. The electronic nose technology was researched to detect the volatile components in the fermentation process of silver bacteria [141]. The freshness of pineapple [142] and fish [143] was predicted separately through the electronic nose.

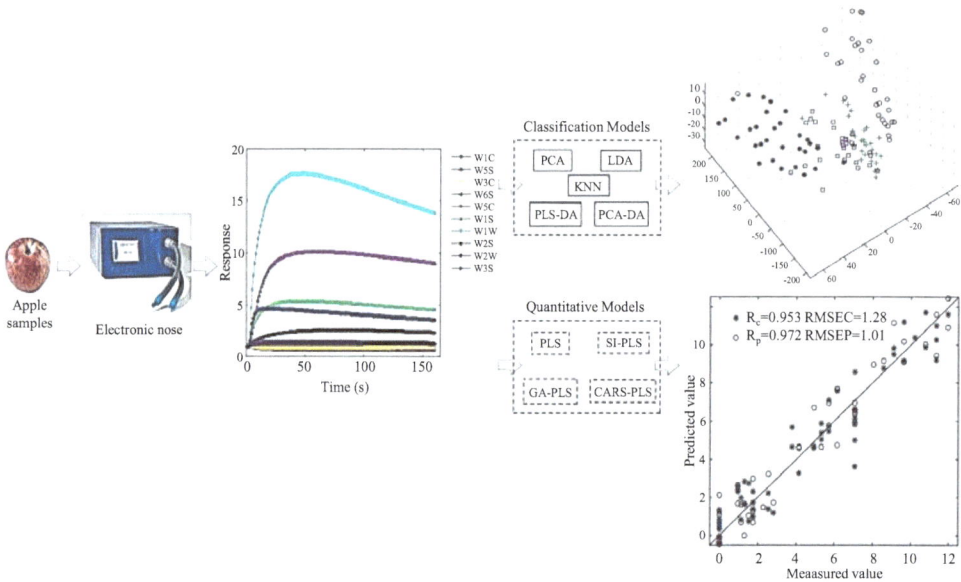

Fig. (46). The flow chart of using electronic nose technology to classify and predict *Penicillium expansum* defects in apples.

Electronic nose has also been developed to detect the pathogenic fungal of fruit and vegetable varieties. In the past, the aroma data of four kinds of fruits including strawberry, lemon, cherry and melon has been collected using electronic nose [144] and three kinds of fish was also successfully classified based on the aroma data [145]. Electronic nose combined with GC-MS was used to detect and identify pathogenic fungal diseases in strawberry fruits after harvest, and the detection rate reached 96.6% [146], In addition, electronic nose was applied to identify bacterial foodborne pathogens [147] combined with the pattern recognition tools [148]. Bacterial foodborne pathogens such as the potato ring rot and brown rot have been successfully discriminated using electronic nose, and the discrimination accuracy of the linear discriminant analysis (LDA) obtained 81.3% [149]. Apples with multiple deterioration levels have also been detected and identified effectively by electronic nose [150], Fig. (**46**) shows the schematic procedures, form the results, it can be seen that the electronic nose has great potential to distinguish rotten apples and quantitatively detect spoilage area of apples.

With the development of scientific research and industry, researchers are aware of the function of electronic nose, therefore innovative electronic nose application is of great development potential. In the future, the development of the electronic nose for specific purposes or narrow application range would obtain more and more attention. In addition, e-noses are of greater potential to be applied in portability field due to its advantages of low-cost, simple operation and labor-saving. What's more, the electronic nose is easy to operate and can be used by unskilled personnel , therefore, it has greater potential for practical application in residential and public places. There are also some disadvantages associated with e-nose sensing, such as poor reproducibility and recovery, humidity and temperature can cause negative effects on the sensor responses, and it is difficult to discriminate individual chemical component based on the sample gases. Therefore, electronic nose would never completely replace the complex analytical equipment or odor panel for all applications, and would be more likely to be applied to the accurate, rapid and repeated determination of samples to achieve the purpose of rapid real-time detection and identification.

6.3. Colorimetric Sensors

Traditional sensors have some defects, such as being easily interfered by substances with similar structures or chemical characteristics. In addition, it is impractical to synthesize highly selective receptors for each of the complex analytes [151]. Although the previous electronic nose system has potential applications in the field of fruit and vegetable quality detection, the limited range of interaction between the sensor array and measured substances limits the wide application of electronic nose, especially in the environment of low concentration relative to vapor pressure, the sensitivity, anti-interference and selectivity of traditional sensors to detect the differences between substances are greatly limited [152]. In order to solve the above problems, one solution is to develop a disposable sensor that is not integrated into the readout device, which can relieve the opposite demand [153], and consider only the cost and portability of the sensor.

At present, colorimetric sensor array technology has been used to detect odor components of various agricultural products. Colorimetric sensor technology simulates the mammalian olfactory system by producing a special response to the odor of each tested substance. The intermolecular interaction between the tested substance and the active center is usually a chemical reaction [154], that results in a colorimetric change. Therefore, the colorimetric sensor array can generate special patterns through chemically responsive colorants and odorants or their mixtures to realize odor discrimination. The colorimetric sensor array uses cross-response and chemically responsive dyes to produce an olfactory response, while using digital imaging to quantify the concentration of odors (Fig. **47**). Colorimetric sensor arrays have been used in many fields such as food science [155]. For example, the introduction of inorganic carrier materials, including pH indicator and Lewis acid and other optical dyes, has realized the monitoring of meat spoilage gas products [156]. Similar ideas have been used for monitoring the freshness of pork [157], fish [158] and chicken [159] under the package atmosphere.

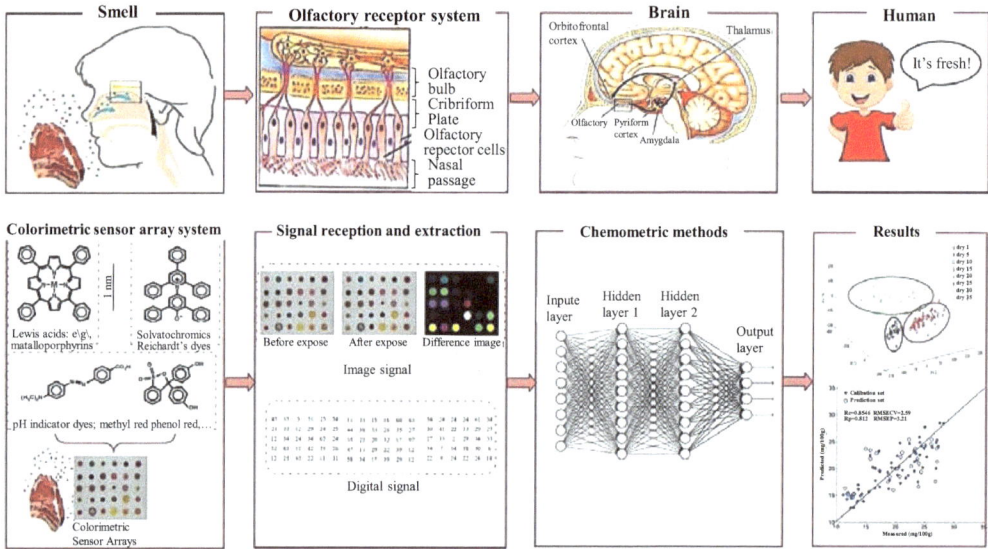

Fig. (47). Comparison of biological and colorimetric sensor array systems.

Compared with the traditional method, colorimetry has many advantages. The colorimetric method does not need complicated instruments, and can realize on-site visual inspection based on color change. At the same time, on the basis of digital imaging technology, it provides a simple and efficient method for the rapid detection and identification of substances. In addition, in terms of portability, the colorimetric sensor array is easy to be miniaturized, which can be used for multiple analysis with a single control instrument in the detection position, and can also detect flammable, explosive and toxic compounds. In addition, it has the advantages of good selectivity, high sensitivity, and fast response speed. Therefore, the colorimetric sensor array technology has a broad application prospect in food "odor visualization". There are a large number of researchers interested in this field.

All in all, this technology is a new detection method in the field of agricultural product detection and has a large space for development. There has been a lot of literature showing the usability of this technology and researchers are considering to study various types of colorimetric sensor arrays on a global scale. However, due to the diversity of food samples and the complexity of volatile components, the establishment of models is difficult. The colorimetric sensor array technology poses a great challenge to the accurate quantification of target analytes. In order to make the colorimetric sensor array technology more useful in analysis, some aspects should be addressed, including reproducibility, sensitivity, and portability.

CONCLUSION

Due to the nutrition and safety considerations in the food supply chain, the demand for rapid and accurate detection of agricultural products is gradually increasing. At the same time, due to the low cost, good efficiency and repeatability of sensing operation, the development trend of intelligent sensors in the field of agricultural product quality detection would continue and expand. The purpose of this chapter is to summarize the non-destructive testing technology of food and agricultural products. Compared with other traditional detection methods such as instrumental analysis and chemical analysis, the intelligent sensing technology of agricultural products processing has unique advantages and broad application prospects and development potential.

The new research findings of intelligent sensing operation would promote the expansion of optical, acoustic, electrical, magnetic and analog sensing technologies, and solve the existing problems in food and agriculture. Therefore, the application of intelligent sensor technology in the field of food and agricultural products quality detection was studied. In order to replace the traditional analysis method, it needs to be further improved and improved. Therefore, the use of intelligent sensing technology discussed in this chapter will reduce the disadvantages of traditional analytical instruments. As the related technical challenges range from physics to computer science to food and agricultural sciences, it is essential for researchers from all fields to participate in research. Through such joint efforts, it is possible to develop a system suitable for food and agricultural products detection. Therefore, in the future, advanced intelligent sensing technology can be used to enhance the ability of sensing tools, which can quickly and easily detect the quality of food and agricultural products.

CONSENT FOR PUBLICATION

Not applicable.

CONFLICT OF INTEREST

TThe authors confirm that this chapter contents have no conflict of interest.

ACKNOWLEDGEMENT

The author acknowledges the financial support provided by the National Key

R&D Program of China (2017YFC1600802, 2018YFC1604401), National Natural Science Foundation of China (31972151, 31501216), Key R&D Project of Jiangsu Province (BE2019359, BE2018307), the Priority Academic Program Development of Jiangsu Higher Education Institutions. Meanwhile, the authors also thank the colleagues for their assistance.

REFERENCES

[1]   H. Tian, T. Wang, and Y. Liu, "Computer vision technology in agricultural automation —A review", *Inf. Process. Agric.,* vol. 7, no. 1, pp. 1-19, 2020.
[http://dx.doi.org/10.1016/j.inpa.2019.09.006]

[2]   M. Kundu, P. Krishnan, and R.K. Kotnala, "Recent developments in biosensors to combat agricultural challenges and their future prospects", *Trends Food Sci. Technol.,* vol. 88, pp. 157-178, 2019.
[http://dx.doi.org/10.1016/j.tifs.2019.03.024]

[3]   B. Jia, W. Wang, and X. Ni, "Essential processing methods of hyperspectral images of agricultural and food products", *Chemometr Intell Lab,* vol. 198, p. 103936, 2020.
[http://dx.doi.org/10.1016/j.chemolab.2020.103936]

[4]   R.J.B. Peters, H. Bouwmeester, and S. Gottardo, "Nanomaterials for products and application in agriculture, feed and food", *Trends Food Sci. Technol.,* vol. 54, pp. 155-164, 2016.
[http://dx.doi.org/10.1016/j.tifs.2016.06.008]

[5]   M. Rezvanyvardom, T.G. Nejad, and E. Farshidi, "A 5-bit time to digital converter using time to voltage conversion and integrating techniques for agricultural products analysis by Raman spectroscopy", *Inf. Process. Agric.,* vol. 1, no. 2, pp. 124-130, 2014.
[http://dx.doi.org/10.1016/j.inpa.2014.11.003]

[6]   N. Hussain, D-W. Sun, and H. Pu, "Classical and emerging non-destructive technologies for safety and quality evaluation of cereals: A review of recent applications", *Trends Food Sci. Technol.,* vol. 91, pp. 598-608, 2019.
[http://dx.doi.org/10.1016/j.tifs.2019.07.018]

[7]   M. Mohd Ali, N. Hashim, and S.K. Bejo, "Rapid and nondestructive techniques for internal and external quality evaluation of watermelons: A review", *Sci. Hortic. (Amsterdam),* vol. 225, pp. 689-699, 2017.
[http://dx.doi.org/10.1016/j.scienta.2017.08.012]

[8]   B. Zhang, B. Gu, and G. Tian, "Challenges and solutions of optical-based nondestructive quality inspection for robotic fruit and vegetable grading systems: A technical review", *Trends Food Sci. Technol.,* vol. 81, pp. 213-231, 2018.
[http://dx.doi.org/10.1016/j.tifs.2018.09.018]

[9]   A. Hussain, H. Pu, and D-W. Sun, "Innovative nondestructive imaging techniques for ripening and maturity of fruits – A review of recent applications", *Trends Food Sci. Technol.,* vol. 72, pp. 144-152, 2018.
[http://dx.doi.org/10.1016/j.tifs.2017.12.010]

[10]   A.M. Rady, and D.E. Guyer, "Rapid and/or nondestructive quality evaluation methods for potatoes: A review", *Comput. Electron. Agric.,* vol. 117, pp. 31-48, 2015.
[http://dx.doi.org/10.1016/j.compag.2015.07.002]

[11]   S. Gunasekaran, M.R. Paulsen, and G.C. Shove, "Optical methods for nondestructive quality evaluation of agricultural and biological materials", *J. Agric. Eng. Res.,* vol. 32, no. 3, pp. 209-241, 1985.
[http://dx.doi.org/10.1016/0021-8634(85)90081-2]

[12]   K. Wang, D-W. Sun, and H. Pu, "Emerging non-destructive terahertz spectroscopic imaging

technique: Principle and applications in the agri-food industry", *Trends Food Sci. Technol.*, vol. 67, pp. 93-105, 2017.
[http://dx.doi.org/10.1016/j.tifs.2017.06.001]

[13]   M-T. Sánchez, J-A. Entrenas, and I. Torres, "Monitoring texture and other quality parameters in spinach plants using NIR spectroscopy", *Comput. Electron. Agric.*, vol. 155, pp. 446-452, 2018.
[http://dx.doi.org/10.1016/j.compag.2018.11.004]

[14]   S. Verdú, J.M. Barat, and R. Grau, "Non destructive monitoring of the yoghurt fermentation phase by an image analysis of laser-diffraction patterns: Characterization of cow's, goat's and sheep's milk", *Food Chem.*, vol. 274, pp. 46-54, 2019.
[http://dx.doi.org/10.1016/j.foodchem.2018.08.091] [PMID: 30372965]

[15]   M. Zhang, C. Li, and F. Yang, "Optical properties of blueberry flesh and skin and Monte Carlo multi-layered simulation of light interaction with fruit tissues", *Postharvest Biol. Technol.*, vol. 150, pp. 28-41, 2019.
[http://dx.doi.org/10.1016/j.postharvbio.2018.12.006]

[16]   G. Lv, C. Du, F. Ma, Y. Shen, and J. Zhou, "Rapid and nondestructive detection of pesticide residues by depth-profiling fourier transform infrared photoacoustic spectroscopy", *ACS Omega*, vol. 3, no. 3, pp. 3548-3553, 2018.
[http://dx.doi.org/10.1021/acsomega.8b00339] [PMID: 31458606]

[17]   P. Ma, L. Wang, and L. Xu, "Rapid quantitative determination of chlorpyrifos pesticide residues in tomatoes by surface-enhanced Raman spectroscopy", *Eur. Food Res. Technol.*, vol. 246, no. 1, pp. 239-251, 2019.
[http://dx.doi.org/10.1007/s00217-019-03408-8]

[18]   F.Y.H. Kutsanedzie, A.A. Agyekum, V. Annavaram, and Q. Chen, "Signal-enhanced SERS-sensors of CAR-PLS and GA-PLS coupled AgNPs for ochratoxin A and aflatoxin B1 detection", *Food Chem.*, vol. 315, p. 126231, 2020.
[http://dx.doi.org/10.1016/j.foodchem.2020.126231] [PMID: 31991258]

[19]   S. Zhang, X. Wu, and S. Zhang, "An effective method to inspect and classify the bruising degree of apples based on the optical properties", *Postharvest Biol. Technol.*, vol. 127, pp. 44-52, 2017.
[http://dx.doi.org/10.1016/j.postharvbio.2016.12.008]

[20]   X. He, X. Fu, and X. Rao, "Assessing firmness and SSC of pears based on absorption and scattering properties using an automatic integrating sphere system from 400 to 1150 nm", *Postharvest Biol. Technol.*, vol. 121, pp. 62-70, 2016.
[http://dx.doi.org/10.1016/j.postharvbio.2016.07.013]

[21]   J. Lim, G. Kim, C. Mo, K. Oh, H. Yoo, H. Ham, and M.S. Kim, "Classification of fusarium-infected korean hulled barley using near-infrared reflectance spectroscopy and partial least squares discriminant analysis", *Sensors (Basel)*, vol. 17, no. 10, p. E2258, 2017.
[http://dx.doi.org/10.3390/s17102258] [PMID: 28974012]

[22]   M. Nishino, S. Kuroki, and Y. Deguchi, "Dual-beam spectral measurement improves accuracy of nondestructive identification of internal rot in onion bulbs", *Postharvest Biol. Technol.*, p. 156, 2019.
[http://dx.doi.org/10.1016/j.postharvbio.2019.110935]

[23]   S. Tian, J. Zhang, and Z. Zhang, "Effective modification through transmission Vis/NIR spectra affected by fruit size to improve the prediction of moldy apple core", *Infrared Phys. Technol.*, vol. 100, pp. 117-124, 2019.
[http://dx.doi.org/10.1016/j.infrared.2019.05.015]

[24]   P. Theanjumpol, K. Wongzeewasakun, and N. Muenmanee, "Non-destructive identification and estimation of granulation in 'Sai Num Pung' tangerine fruit using near infrared spectroscopy and chemometrics", *Postharvest Biol. Technol.*, vol. 153, pp. 13-20, 2019.
[http://dx.doi.org/10.1016/j.postharvbio.2019.03.009]

[25]   A. Wang, and L. Xie, "Technology using near infrared spectroscopic and multivariate analysis to

determine the soluble solids content of citrus fruit", *J. Food Eng.,* vol. 143, pp. 17-24, 2014.
[http://dx.doi.org/10.1016/j.jfoodeng.2014.06.023]

[26]   R. Sheng, W. Cheng, and H. Li, "Model development for soluble solids and lycopene contents of cherry tomato at different temperatures using near-infrared spectroscopy", *Postharvest Biol. Technol.,* p. 156, 2019.
[http://dx.doi.org/10.1016/j.postharvbio.2019.110952]

[27]   Y. Huang, R. Lu, and K. Chen, "Assessment of tomato soluble solids content and pH by spatially-resolved and conventional Vis/NIR spectroscopy", *J. Food Eng.,* vol. 236, pp. 19-28, 2018.
[http://dx.doi.org/10.1016/j.jfoodeng.2018.05.008]

[28]   J-H. Cheng, and D-W. Sun, "Partial Least Squares Regression (PLSR) Applied to NIR and HSI Spectral Data Modeling to Predict Chemical Properties of Fish Muscle", *Food Eng. Rev.,* vol. 9, no. 1, pp. 36-49, 2016.
[http://dx.doi.org/10.1007/s12393-016-9147-1]

[29]   J. Sun, X. Ge, and X. Wu, "Identification of pesticide residues in lettuce leaves based on near infrared transmission spectroscopy", *J. Food Process Eng.,* vol. 41, no. 6, 2018.
[http://dx.doi.org/10.1111/jfpe.12816]

[30]   Z. Xin, S. Jun, and L. Bing, "Study on pesticide residues classification of lettuce leaves based on polarization spectroscopy", *J. Food Process Eng.,* vol. 41, no. 8, 2018.
[http://dx.doi.org/10.1111/jfpe.12903]

[31]   J. Jiang, Z. Li, Y. Wang, X. Zhang, K. Yu, H. Zhang, J. Zhang, J. Gao, X. Liu, H. Zhang, W. Wu, and N. Li, "Rapid determination of cadmium in rice by portable dielectric barrier discharge-atomic emission spectrometer", *Food Chem.,* vol. 310, p. 125824, 2020.
[http://dx.doi.org/10.1016/j.foodchem.2019.125824] [PMID: 31732245]

[32]   P. Chapanya, P. Ritthiruangdej, and R. Mueangmontri, "Temperature compensation on sugar content prediction of molasses by near-infrared spectroscopy (NIR)", *Sugar Tech,* vol. 21, no. 1, pp. 162-169, 2018.
[http://dx.doi.org/10.1007/s12355-018-0635-x]

[33]   Q. Chen, M. Chen, Y. Liu, J. Wu, X. Wang, Q. Ouyang, and X. Chen, "Application of FT-NIR spectroscopy for simultaneous estimation of taste quality and taste-related compounds content of black tea", *J. Food Sci. Technol.,* vol. 55, no. 10, pp. 4363-4368, 2018.
[http://dx.doi.org/10.1007/s13197-018-3353-1] [PMID: 30228436]

[34]   R. Hu, L. Zhang, and Z. Yu, "Optimization of soluble solids content prediction models in 'Hami' melons by means of Vis-NIR spectroscopy and chemometric tools", *Infrared Phys. Technol.,* p. 102, 2019.
[http://dx.doi.org/10.1016/j.infrared.2019.102999]

[35]   J.C. Keresztes, E. Diels, and M. Goodarzi, "Glare based apple sorting and iterative algorithm for bruise region detection using shortwave infrared hyperspectral imaging", *Postharvest Biol. Technol.,* vol. 130, pp. 103-115, 2017.
[http://dx.doi.org/10.1016/j.postharvbio.2017.04.005]

[36]   L. Feng, S. Zhu, F. Lin, Z. Su, K. Yuan, Y. Zhao, Y. He, and C. Zhang, "Detection of oil chestnuts infected by blue mold using near-infrared hyperspectral imaging combined with artificial neural networks", *Sensors (Basel),* vol. 18, no. 6, p. E1944, 2018.
[http://dx.doi.org/10.3390/s18061944] [PMID: 29914074]

[37]   A. Folch-Fortuny, J.M. Prats-Montalbán, and S. Cubero, "VIS/NIR hyperspectral imaging and N-way PLS-DA models for detection of decay lesions in citrus fruits", *Chemom. Intell. Lab. Syst.,* vol. 156, pp. 241-248, 2016.
[http://dx.doi.org/10.1016/j.chemolab.2016.05.005]

[38]   Y. Xiao, X. Gu, and Y. Niu, "Hyperspectral classification for identifying decayed oranges infected by fungi", *Emir. J. Food Agric.,* 2017.

[39]   X. Yu, H. Lu, and D. Wu, "Development of deep learning method for predicting firmness and soluble solid content of postharvest Korla fragrant pear using Vis/NIR hyperspectral reflectance imaging", *Postharvest Biol. Technol.,* vol. 141, pp. 39-49, 2018.
[http://dx.doi.org/10.1016/j.postharvbio.2018.02.013]

[40]   Y. Sun, K. Wei, Q. Liu, L. Pan, and K. Tu, "Classification and discrimination of different fungal diseases of three infection levels on peaches using hyperspectral reflectance imaging analysis", *Sensors (Basel),* vol. 18, no. 4, p. E1295, 2018.
[http://dx.doi.org/10.3390/s18041295] [PMID: 29690625]

[41]   Y. Sun, H. Xiao, and S. Tu, "Detecting decayed peach using a rotating hyperspectral imaging testbed", *Lwt,* vol. 87, pp. 326-332, 2018.
[http://dx.doi.org/10.1016/j.lwt.2017.08.086]

[42]   W. Liu, P. Zhao, C. Wu, C. Liu, J. Yang, and L. Zheng, "Rapid determination of aflatoxin B_1 concentration in soybean oil using terahertz spectroscopy with chemometric methods", *Food Chem.,* vol. 293, pp. 213-219, 2019.
[http://dx.doi.org/10.1016/j.foodchem.2019.04.081] [PMID: 31151603]

[43]   S. Munera, J. Blasco, and J.M. Amigo, "Use of hyperspectral transmittance imaging to evaluate the internal quality of nectarines", *Biosyst. Eng.,* vol. 182, pp. 54-64, 2019.
[http://dx.doi.org/10.1016/j.biosystemseng.2019.04.001]

[44]   X. Ye, K. Iino, and S. Zhang, "Monitoring of bacterial contamination on chicken meat surface using a novel narrowband spectral index derived from hyperspectral imagery data", *Meat Sci.,* vol. 122, pp. 25-31, 2016.
[http://dx.doi.org/10.1016/j.meatsci.2016.07.015] [PMID: 27471794]

[45]   A. López-Maestresalas, J.C. Keresztes, and M. Goodarzi, "Non-destructive detection of blackspot in potatoes by Vis-NIR and SWIR hyperspectral imaging", *Food Control,* vol. 70, pp. 229-241, 2016.
[http://dx.doi.org/10.1016/j.foodcont.2016.06.001]

[46]   H. Zhongzhi, and D. Limiao, "Aflatoxin contaminated degree detection by hyperspectral data using band index", *Food Chem. Toxicol.,* vol. 137, p. 111159, 2020.
[http://dx.doi.org/10.1016/j.fct.2020.111159] [PMID: 31991198]

[47]   Y. Liu, Q. Wang, and X. Gao, "Total phenolic content prediction in Flos Lonicerae using hyperspectral imaging combined with wavelengths selection methods", *J. Food Process Eng.,* vol. 42, no. 6, 2019.
[http://dx.doi.org/10.1111/jfpe.13224]

[48]   N. Arens, A. Backhaus, S. Döll, S. Fischer, U. Seiffert, and H.P. Mock, "Non-invasive presymptomatic detection of *Cercospora beticola* infection and identification of early metabolic responses in sugar beet", *Front. Plant Sci.,* vol. 7, p. 1377, 2016.
[http://dx.doi.org/10.3389/fpls.2016.01377] [PMID: 27713750]

[49]   F. Nekvapil, I. Brezestean, D. Barchewitz, B. Glamuzina, V. Chiş, and S. Cîntă Pinzaru, "Citrus fruits freshness assessment using Raman spectroscopy", *Food Chem.,* vol. 242, pp. 560-567, 2018.
[http://dx.doi.org/10.1016/j.foodchem.2017.09.105] [PMID: 29037730]

[50]   R. Hara, M. Ishigaki, Y. Kitahama, Y. Ozaki, and T. Genkawa, "Excitation wavelength selection for quantitative analysis of carotenoids in tomatoes using Raman spectroscopy", *Food Chem.,* vol. 258, pp. 308-313, 2018.
[http://dx.doi.org/10.1016/j.foodchem.2018.03.089] [PMID: 29655738]

[51]   V. Egging, J. Nguyen, and D. Kurouski, "Detection and identification of fungal infections in intact wheat and sorghum grain using a hand-held raman spectrometer", *Anal. Chem.,* vol. 90, no. 14, pp. 8616-8621, 2018.
[http://dx.doi.org/10.1021/acs.analchem.8b01863] [PMID: 29898358]

[52]   C. Sunli, S. Jun, M. Hanping, W. Xiaohong, W. Pei, and Z. Xiaodong, "Non-destructive detection for

mold colonies in rice based on hyperspectra and GWO-SVR", *J. Sci. Food Agric.,* vol. 98, no. 4, pp. 1453-1459, 2018.
[http://dx.doi.org/10.1002/jsfa.8613] [PMID: 28786119]

[53]   S. Mittal, M.K. Dutta, and A. Issac, "Non-destructive image processing based system for assessment of rice quality and defects for classification according to inferred commercial value", *Measurement,* p. 148, 2019.
[http://dx.doi.org/10.1016/j.measurement.2019.106969]

[54]   X. Lu, Y. Liu, and Q. Zhang, "Study on tea harvested in different seasons based on laser-induced breakdown spectroscopy", *Laser Phys. Lett.,* vol. 17, no. 1, 2020.
[http://dx.doi.org/10.1088/1612-202X/ab5c23]

[55]   W. Zhang, Q. Zhu, and M. Huang, "Detection and classification of potato defects using multispectral imaging system based on single shot method", *Food Anal. Methods,* vol. 12, no. 12, pp. 2920-2929, 2019.
[http://dx.doi.org/10.1007/s12161-019-01654-w]

[56]   W. Hu, J. Li, and X. Zhu, "Nondestructive detection of underlying moldy lesions of apple using frequency domain diffuse optical tomography", *Postharvest Biol. Technol.,* vol. 153, pp. 31-42, 2019.
[http://dx.doi.org/10.1016/j.postharvbio.2019.03.014]

[57]   X. Fu, T. Belwal, G. Cravotto, and Z. Luo, "Sono-physical and sono-chemical effects of ultrasound: Primary applications in extraction and freezing operations and influence on food components", *Ultrason. Sonochem.,* vol. 60, p. 104726, 2020.
[http://dx.doi.org/10.1016/j.ultsonch.2019.104726] [PMID: 31541966]

[58]   M. Soltani Firouz, A. Farahmandi, and S. Hosseinpour, "Recent advances in ultrasound application as a novel technique in analysis, processing and quality control of fruits, juices and dairy products industries: A review", *Ultrason. Sonochem.,* vol. 57, pp. 73-88, 2019.
[http://dx.doi.org/10.1016/j.ultsonch.2019.05.014] [PMID: 31208621]

[59]   Y. Fang, L. Lin, and H. Feng, "Review of the use of air-coupled ultrasonic technologies for nondestructive testing of wood and wood products", *Comput. Electron. Agric.,* vol. 137, pp. 79-87, 2017.
[http://dx.doi.org/10.1016/j.compag.2017.03.015]

[60]   M.J.W. Povey, "Ultrasonics in food engineering Part II: Applications", *J. Food Eng.,* vol. 9, no. 1, pp. 1-20, 1989.
[http://dx.doi.org/10.1016/0260-8774(89)90047-2]

[61]   L. Sun, J-r. Cai, and H. Lin, "On-line estimation of eggshell strength based on acoustic impulse response analysis", *Innov. Food Sci. Emerg. Technol.,* vol. 18, pp. 220-225, 2013.
[http://dx.doi.org/10.1016/j.ifset.2013.01.010]

[62]   H. Wang, J. Mao, and J. Zhang, "Acoustic feature extraction and optimization of crack detection for eggshell", *J. Food Eng.,* vol. 171, pp. 240-247, 2016.
[http://dx.doi.org/10.1016/j.jfoodeng.2015.10.030]

[63]   A. Mizrach, "Nondestructive ultrasonic monitoring of tomato quality during shelf-life storage", *Postharvest Biol. Technol.,* vol. 46, no. 3, pp. 271-274, 2007.
[http://dx.doi.org/10.1016/j.postharvbio.2007.05.012]

[64]   H. Vasighi-Shojae, M. Gholami-Parashkouhi, and D. Mohammadzamani, "Ultrasonic based determination of apple quality as a nondestructive technology", *Sens. Biosensing Res.,* vol. 21, pp. 22-26, 2018.
[http://dx.doi.org/10.1016/j.sbsr.2018.09.002]

[65]   P.M. Pieczywek, A. Kozioł, and D. Konopacka, "Changes in cell wall stiffness and microstructure in ultrasonically treated apple", *J. Food Eng.,* vol. 197, pp. 1-8, 2017.
[http://dx.doi.org/10.1016/j.jfoodeng.2016.10.028]

[66]   B. Doohan, S. Fuller, and S. Parsons, "The sound of management: Acoustic monitoring for agricultural industries", *Ecol. Indic.,* vol. 96, pp. 739-746, 2019.
[http://dx.doi.org/10.1016/j.ecolind.2018.09.029]

[67]   B. Khadhraoui, A-S. Fabiano-Tixier, and P. Robinet, "Ultrasound technology for food processing, preservation, and extraction", In: *Green Food Processing Techniques.,* F. Chemat, E. Vorobiev, Eds., Academic Press, 2019, pp. 23-56.
[http://dx.doi.org/10.1016/B978-0-12-815353-6.00002-1]

[68]   M.S. Aday, R. Temizkan, and M.B. Büyükcan, "An innovative technique for extending shelf life of strawberry: Ultrasound", *Lebensm. Wiss. Technol.,* vol. 52, no. 2, pp. 93-101, 2013.
[http://dx.doi.org/10.1016/j.lwt.2012.09.013]

[69]   E.A. Alenyorege, H. Ma, I. Ayim, F. Lu, and C. Zhou, "Efficacy of sweep ultrasound on natural microbiota reduction and quality preservation of Chinese cabbage during storage", *Ultrason. Sonochem.,* vol. 59, p. 104712, 2019.
[http://dx.doi.org/10.1016/j.ultsonch.2019.104712] [PMID: 31421620]

[70]   M. Omid, A. Mahmoudi, and M.H. Omid, "Development of pistachio sorting system using principal component analysis (PCA) assisted artificial neural network (ANN) of impact acoustics", *Expert Syst. Appl.,* vol. 37, no. 10, pp. 7205-7212, 2010.
[http://dx.doi.org/10.1016/j.eswa.2010.04.008]

[71]   X. Sun, M. Guo, and M. Ma, "Identification and classification of damaged corn kernels with impact acoustics multi-domain patterns", *Comput. Electron. Agric.,* vol. 150, pp. 152-161, 2018.
[http://dx.doi.org/10.1016/j.compag.2018.04.008]

[72]   V. Mohammadi, M. Ghasemi-Varnamkhasti, and R. Ebrahimi, "Ultrasonic techniques for the milk production industry", *Measurement,* vol. 58, pp. 93-102, 2014.
[http://dx.doi.org/10.1016/j.measurement.2014.08.022]

[73]   E. Ouacha, B. Faiz, and A. Moudden, "Non-destructive detection of air traces in the uht milk packet by using ultrasonic waves", *Phys. Procedia,* vol. 70, pp. 406-410, 2015.
[http://dx.doi.org/10.1016/j.phpro.2015.08.120]

[74]   L. Elvira, L. Sampedro, and J. Matesanz, "Non-invasive and non-destructive ultrasonic technique for the detection of microbial contamination in packed UHT milk", *Food Res. Int.,* vol. 38, no. 6, pp. 631-638, 2005.
[http://dx.doi.org/10.1016/j.foodres.2004.12.001]

[75]   W. Routray, and V. Orsat, "Recent advances in dielectric properties–measurements and importance", *Curr. Opin. Food Sci.,* vol. 23, pp. 120-126, 2018.
[http://dx.doi.org/10.1016/j.cofs.2018.10.001]

[76]   A. Al Faruq, M. Zhang, and B. Bhandari, "New understandings of how dielectric properties of fruits and vegetables are affected by heat-induced dehydration: A review", *Dry. Technol.,* vol. 37, no. 14, pp. 1780-1792, 2018.
[http://dx.doi.org/10.1080/07373937.2018.1538157]

[77]   W. Guo, L. Fang, and D. Liu, "Determination of soluble solids content and firmness of pears during ripening by using dielectric spectroscopy", *Comput. Electron. Agric.,* vol. 117, pp. 226-233, 2015.
[http://dx.doi.org/10.1016/j.compag.2015.08.012]

[78]   J Sun, Y Tian, and X Wu, "Nondestructive detection for moisture content in green tea based on dielectric properties and VISSA-GWO-SVR algorithm", *J Food Processing and Preservation,* p. e14421, .
[http://dx.doi.org/10.1111/jfpp.14421]

[79]   TU P, "BIAN H-x. Prediction of damage volume on apple basing the dielectric property", *Sci. Tech. Food Industry,* vol. 2018, no. 16, p. 40, 2018.

[80]   M. Soltani, and M. Omid, "Detection of poultry egg freshness by dielectric spectroscopy and machine

learning techniques", *Lebensm. Wiss. Technol.,* vol. 62, no. 2, pp. 1034-1042, 2015.
[http://dx.doi.org/10.1016/j.lwt.2015.02.019]

[81] L. Shang, W. Guo, and S.O. Nelson, "Apple variety identification based on dielectric spectra and chemometric methods", *Food Anal. Methods,* vol. 8, no. 4, pp. 1042-1052, 2015.
[http://dx.doi.org/10.1007/s12161-014-9985-5]

[82] Z. Libin, X. Fang, and J. Shiming, "Principle and implementation of automatically nondestructive inspection system for apple internal quality based on dielectric property", *Nongye Gongcheng Xuebao (Beijing),* p. 1, 2001.

[83] A. Reyes, M. Yarlequé, and W. Castro, "Determination of dielectric properties of the red delicious apple and its correlation with quality parameters", In: *2017 Progress in Electromagnetics Research Symposium-Fall (PIERS-FALL).* IEEE, 2017, pp. 2067-2072.
[http://dx.doi.org/10.1109/PIERS-FALL.2017.8293478]

[84] J Peng, X Yin, and S Jiao, "Air jet impingement and hot air-assisted radio frequency hybrid drying of apple slices", *Lwt,* p. 116, 2019.
[http://dx.doi.org/10.1016/j.lwt.2019.108517]

[85] F. Marra, L. Zhang, and J.G. Lyng, "Radio frequency treatment of foods: Review of recent advances", *J. Food Eng.,* vol. 91, no. 4, pp. 497-508, 2009.
[http://dx.doi.org/10.1016/j.jfoodeng.2008.10.015]

[86] H. Zhu, D. Li, and S. Li, "A novel method to improve heating uniformity in mid-high moisture potato starch with radio frequency assisted treatment", *J. Food Eng.,* vol. 206, pp. 23-36, 2017.
[http://dx.doi.org/10.1016/j.jfoodeng.2017.03.001]

[87] S. Jiao, Y. Deng, and Y. Zhong, "Investigation of radio frequency heating uniformity of wheat kernels by using the developed computer simulation model", *Food Res. Int.,* vol. 71, pp. 41-49, 2015.
[http://dx.doi.org/10.1016/j.foodres.2015.02.010]

[88] L. Zhang, J.G. Lyng, and R. Xu, "Influence of radio frequency treatment on in-shell walnut quality and Staphylococcus aureus ATCC 25923 survival", *Food Control,* vol. 102, pp. 197-205, 2019.
[http://dx.doi.org/10.1016/j.foodcont.2019.03.030]

[89] Y. Yang, and D.J. Geveke, "Shell egg pasteurization using radio frequency in combination with hot air or hot water", *Food Microbiol.,* vol. 85, p. 103281, 2020.
[http://dx.doi.org/10.1016/j.fm.2019.103281] [PMID: 31500700]

[90] J. Xu, M. Zhang, and B. Bhandari, "ZnO nanoparticles combined radio frequency heating: A novel method to control microorganism and improve product quality of prepared carrots", *Innov. Food Sci. Emerg. Technol.,* vol. 44, pp. 46-53, 2017.
[http://dx.doi.org/10.1016/j.ifset.2017.07.025]

[91] L. Hou, J.A. Johnson, and S. Wang, "Radio frequency heating for postharvest control of pests in agricultural products: A review", *Postharvest Biol. Technol.,* vol. 113, pp. 106-118, 2016.
[http://dx.doi.org/10.1016/j.postharvbio.2015.11.011]

[92] M. Dalvi-Isfahan, N. Hamdami, A. Le-Bail, and E. Xanthakis, "The principles of high voltage electric field and its application in food processing: A review", *Food Res. Int.,* vol. 89, no. Pt 1, pp. 48-62, 2016.
[http://dx.doi.org/10.1016/j.foodres.2016.09.002] [PMID: 28460942]

[93] X He, G Jia, and E Tatsumi, "Effect of corona wind, current, electric field and energy consumption on the reduction of the thawing time during the high-voltage electrostatic-field (HVEF) treatment process", *Innovative Food Science & Emerging Technologies..* S1466856416000102.

[94] W. Cao, Y. Nishiyama, and S. Koide, "Electrohydrodynamic drying characteristics of wheat using high voltage electrostatic field", *J. Food Eng.,* vol. 62, no. 3, pp. 209-213, 2004.
[http://dx.doi.org/10.1016/S0260-8774(03)00232-2]

[95] C. Ding, J. Lu, and Z. Song, "Electrohydrodynamic drying of carrot slices", *PLoS One,* vol. 10, no. 4,

p. e0124077, 2015.
[http://dx.doi.org/10.1371/journal.pone.0124077] [PMID: 25874695]

[96] Y. Wang, B. Wang, and L. Li, "Keeping quality of tomato fruit by high electrostatic field pretreatment during storage", *J. Sci. Food Agric.,* vol. 88, no. 3, pp. 464-470, 2008.
[http://dx.doi.org/10.1002/jsfa.3108]

[97] N-Y Kao, Y-F Tu, and K Sridhar, "Effect of a high voltage electrostatic field (HVEF) on the shelf-life of fresh-cut broccoli (Brassica oleracea var. italica)", *Lwt,* vol. 116, 2019.

[98] R. Zhao, J. Hao, J. Xue, H. Liu, and L. Li, "Effect of high-voltage electrostatic field pretreatment on the antioxidant system in stored green mature tomatoes", *J. Sci. Food Agric.,* vol. 91, no. 9, pp. 1680-1686, 2011.
[http://dx.doi.org/10.1002/jsfa.4369] [PMID: 21480264]

[99] Q. Wang, Y. Li, D.W. Sun, and Z. Zhu, "Enhancing food processing by pulsed and high voltage electric fields: principles and applications", *Crit. Rev. Food Sci. Nutr.,* vol. 58, no. 13, pp. 2285-2298, 2018.
[http://dx.doi.org/10.1080/10408398.2018.1434609] [PMID: 29393667]

[100] S.N. Jha, K. Narsaiah, A.L. Basediya, R. Sharma, P. Jaiswal, R. Kumar, and R. Bhardwaj, "Measurement techniques and application of electrical properties for nondestructive quality evaluation of foods-a review", *J. Food Sci. Technol.,* vol. 48, no. 4, pp. 387-411, 2011.
[http://dx.doi.org/10.1007/s13197-011-0263-x] [PMID: 23572764]

[101] A. Chowdhury, P. Singh, and T.K. Bera, "Electrical impedance spectroscopic study of mandarin orange during ripening", *J. Food Meas. Charact.,* vol. 11, no. 4, pp. 1654-1664, 2017.
[http://dx.doi.org/10.1007/s11694-017-9545-y]

[102] P. Kuson, and A. Terdwongworakul, "Minimally-destructive evaluation of durian maturity based on electrical impedance measurement", *J. Food Eng.,* vol. 116, no. 1, pp. 50-56, 2013.
[http://dx.doi.org/10.1016/j.jfoodeng.2012.11.021]

[103] A. Chowdhury, T. Kanti Bera, and D. Ghoshal, "Electrical impedance variations in banana ripening: an analytical study with electrical impedance spectroscopy", *J. Food Process Eng.,* vol. 40, no. 2, 2017.
[http://dx.doi.org/10.1111/jfpe.12387]

[104] L. Wu, Y. Ogawa, and A. Tagawa, "Electrical impedance spectroscopy analysis of eggplant pulp and effects of drying and freezing–thawing treatments on its impedance characteristics", *J. Food Eng.,* vol. 87, no. 2, pp. 274-280, 2008.
[http://dx.doi.org/10.1016/j.jfoodeng.2007.12.003]

[105] A. Chowdhury, S. Datta, and T.K. Bera, "Design and development of microcontroller based instrumentation for studying complex bioelectrical impedance of fruits using electrical impedance spectroscopy", *J. Food Process Eng.,* vol. 41, no. 1, pp. 1-13, 2017.
[http://dx.doi.org/10.1111/jfpe.12640]

[106] Á. Kertész, Z. Hlaváčová, and E. Vozáry, "Relationship between moisture content and electrical impedance of carrot slices during drying", *Int. Agrophys.,* vol. 29, no. 1, pp. 61-66, 2015.
[http://dx.doi.org/10.1515/intag-2015-0013]

[107] J. Massah, F. Hajiheydari, and M.H. Derafshi, "Application of electrical resistance in nondestructive postharvest quality evaluation of apple fruit", *J. Agric. Sci. Technol.,* vol. 19, no. 5, pp. 1031-1039, 2017.

[108] E. Kirtil, S. Cikrikci, and M.J. McCarthy, "Recent advances in time domain NMR & MRI sensors and their food applications", *Curr. Opin. Food Sci.,* vol. 17, pp. 9-15, 2017.
[http://dx.doi.org/10.1016/j.cofs.2017.07.005]

[109] C.J. Clark, P.D. Hockings, and D.C. Joyce, "Application of magnetic resonance imaging to pre- and post-harvest studies of fruits and vegetables", *Postharvest Biol. Technol.,* vol. 11, no. 1, pp. 1-21,

1997.
[http://dx.doi.org/10.1016/S0925-5214(97)01413-0]

[110] M.F. Marcone, S. Wang, and W. Albabish, "Diverse food-based applications of nuclear magnetic resonance (NMR) technology", *Food Res. Int.,* vol. 51, no. 2, pp. 729-747, 2013.
[http://dx.doi.org/10.1016/j.foodres.2012.12.046]

[111] M. Greer, C. Chen, and S. Mandal, "An easily reproducible, hand-held, single-sided, MRI sensor", *J. Magn. Reson.,* vol. 308, p. 106591, 2019.
[http://dx.doi.org/10.1016/j.jmr.2019.106591] [PMID: 31546179]

[112] A.K. Thybo, S.N. Jespersen, P.E. Laerke, and H.J. Stødkilde-Jørgensen, "Nondestructive detection of internal bruise and spraing disease symptoms in potatoes using magnetic resonance imaging", *Magn. Reson. Imaging,* vol. 22, no. 9, pp. 1311-1317, 2004.
[http://dx.doi.org/10.1016/j.mri.2004.08.022] [PMID: 15607104]

[113] S. Qiao, Y. Tian, and P. Song, "Analysis and detection of decayed blueberry by low field nuclear magnetic resonance and imaging", *Postharvest Biol. Technol.,* vol. 156, p. 110951, 2019.
[http://dx.doi.org/10.1016/j.postharvbio.2019.110951]

[114] Z. Du, X. Zeng, and X. Li, "Recent advances in imaging techniques for bruise detection in fruits and vegetables", *Trends Food Sci. Technol.,* vol. 99, pp. 133-141, 2020.
[http://dx.doi.org/10.1016/j.tifs.2020.02.024]

[115] F. Tao, L. Zhang, and M.J. McCarthy, "Magnetic resonance imaging provides spatial resolution of Chilling Injury in Micro-Tom tomato (*Solanum lycopersicum* L.) fruit", *Postharvest Biol. Technol.,* vol. 97, pp. 62-67, 2014.
[http://dx.doi.org/10.1016/j.postharvbio.2014.06.005]

[116] K. Saito, T. Miki, and S. Hayashi, "Application of magnetic resonance imaging to non-destructive void detection in watermelon", *Cryogenics,* vol. 36, no. 12, pp. 1027-1031, 1996.
[http://dx.doi.org/10.1016/S0011-2275(96)00087-2]

[117] N. Dellarosa, L. Laghi, and L. Ragni, "Pulsed electric fields processing of apple tissue: Spatial distribution of electroporation by means of magnetic resonance imaging and computer vision system", *Innov. Food Sci. Emerg. Technol.,* vol. 47, pp. 120-126, 2018.
[http://dx.doi.org/10.1016/j.ifset.2018.02.010]

[118] A. Ciampa, M.T. Dell'Abate, A. Florio, L. Tarricone, D. Di Gennaro, G. Picone, A. Trimigno, F. Capozzi, and A. Benedetti, "Combined magnetic resonance imaging and high resolution spectroscopy approaches to study the fertilization effects on metabolome, morphology and yeast community of wine grape berries, cultivar Nero di Troia", *Food Chem.,* vol. 274, pp. 831-839, 2019.
[http://dx.doi.org/10.1016/j.foodchem.2018.09.056] [PMID: 30373017]

[119] A. Taglienti, R. Massantini, and R. Botondi, "Postharvest structural changes of Hayward kiwifruit by means of magnetic resonance imaging spectroscopy", *Food Chem.,* vol. 114, no. 4, pp. 1583-1589, 2009.
[http://dx.doi.org/10.1016/j.foodchem.2008.11.066]

[120] M.S. Razavi, A. Asghari, and M. Azadbakh, "Analyzing the pear bruised volume after static loading by Magnetic Resonance Imaging (MRI)", *Sci. Hortic. (Amsterdam),* vol. 229, pp. 33-39, 2018.
[http://dx.doi.org/10.1016/j.scienta.2017.10.011]

[121] P.T. Coimbra, C.F. Bathazar, and J.T. Guimarães, "Detection of formaldehyde in raw milk by time domain nuclear magnetic resonance and chemometrics", *Food Control,* vol. 110, p. 107006, 2020.
[http://dx.doi.org/10.1016/j.foodcont.2019.107006]

[122] L. Jin, T. Li, and B. Wu, "Rapid detection of Salmonella in milk by nuclear magnetic resonance based on membrane filtration superparamagnetic nanobiosensor", *Food Control,* vol. 110, p. 107011, 2020.
[http://dx.doi.org/10.1016/j.foodcont.2019.107011]

[123] S. Sun, Y. Wang, and H. Fu, "Application of magnetic resonance imaging to study lipid deposition

patterns in three commercial prawn species", *Aquaculture,* vol. 519, p. 734714, 2020.
[http://dx.doi.org/10.1016/j.aquaculture.2019.734714]

[124] L. Song, H. Zhang, and S. Chen, "A novel non-destructive manner for quantitative determination of plumpness of live Eriocheir sinensis using low-field nuclear magnetic resonance", *Food Res. Int.,* vol. 105, pp. 298-304, 2018.
[http://dx.doi.org/10.1016/j.foodres.2017.11.043] [PMID: 29433219]

[125] Y. Li, M. Obadi, and J. Shi, "Determination of moisture, total lipid, and bound lipid contents in oats using low-field nuclear magnetic resonance", *J. Food Compos. Anal.,* vol. 87, p. 103401, 2020.
[http://dx.doi.org/10.1016/j.jfca.2019.103401]

[126] A.K. Horigane, K. Suzuki, and M. Yoshida, "Moisture distribution of soaked rice grains observed by magnetic resonance imaging and physicochemical properties of cooked rice grains", *J. Cereal Sci.,* vol. 57, no. 1, pp. 47-55, 2013.
[http://dx.doi.org/10.1016/j.jcs.2012.09.009]

[127] J.W. Gardner, H.V. Shurmer, and T.T. Tan, "Application of an electronic nose to the discrimination of coffees", *Sens. Actuators B Chem.,* vol. B6, pp. 71-75, 1992.
[http://dx.doi.org/10.1016/0925-4005(92)80033-T]

[128] P. Mielle, "'Electronic noses': Towards the objective instrumental characterization of food aroma", *Trends Food Sci. Technol.,* vol. 7, p. 432, 1996.
[http://dx.doi.org/10.1016/S0924-2244(96)10045-5]

[129] S. Capone, P. Siciliano, and F. Quaranta, "Analysis of vapours and foods by means of an electronic nose based on a sol-gel metal oxide sensors array", *Sens. Actuators B Chem.,* vol. 69, pp. 230-235, 2000.
[http://dx.doi.org/10.1016/S0925-4005(00)00496-2]

[130] C. Di Natale, A. Macagnano, and R. Paolesse, "Electronic nose and sensorial analysis: comparison of performances in selected cases", *Sens. Actuators B Chem.,* vol. B50, pp. 246-252, 1998.
[http://dx.doi.org/10.1016/S0925-4005(98)00242-1]

[131] D. James, S.M. Scott, and Z. Ali, "Chemical sensors for electronic nose systems", *Mikrochim. Acta,* vol. 149, pp. 1-17, 2005.
[http://dx.doi.org/10.1007/s00604-004-0291-6]

[132] H.M. Zhang, J. Wang, and X.J. Tian, "Optimization of sensor array and detection of stored duration of wheat by electronic nose", *J. Food Eng.,* vol. 82, pp. 403-408, 2007.
[http://dx.doi.org/10.1016/j.jfoodeng.2007.02.005]

[133] M.E. Hossain, G.M.A. Rahman, M.S. Freund, D.S. Jayas, N.D. White, C. Shafai, and D.J. Thomson, "Fabrication and optimization of a conducting polymer sensor array using stored grain model volatiles", *J. Agric. Food Chem.,* vol. 60, no. 11, pp. 2863-2873, 2012.
[http://dx.doi.org/10.1021/jf204631q] [PMID: 22332842]

[134] J.H. Sung, H.J. Ko, and T.H. Park, "Piezoelectric biosensor using olfactory receptor protein expressed in *Escherichia coli*", *Biosens. Bioelectron.,* vol. 21, no. 10, pp. 1981-1986, 2006.
[http://dx.doi.org/10.1016/j.bios.2005.10.002] [PMID: 16297612]

[135] C. Wu, L. Du, D. Wang, L. Wang, L. Zhao, and P. Wang, "A novel surface acoustic wave-based biosensor for highly sensitive functional assays of olfactory receptors", *Biochem. Biophys. Res. Commun.,* vol. 407, no. 1, pp. 18-22, 2011.
[http://dx.doi.org/10.1016/j.bbrc.2011.02.073] [PMID: 21333624]

[136] A. Berna, "Metal oxide sensors for electronic noses and their application to food analysis", *Sensors (Basel),* vol. 10, no. 4, pp. 3882-3910, 2010.
[http://dx.doi.org/10.3390/s100403882] [PMID: 22319332]

[137] H. Shi, M. Zhang, and B. Adhikari, "Advances of electronic nose and its application in fresh foods: A review", *Crit. Rev. Food Sci. Nutr.,* vol. 58, no. 16, pp. 2700-2710, 2018.

[http://dx.doi.org/10.1080/10408398.2017.1327419] [PMID: 28665685]

[138] C. Dai, X. Huang, D. Huang, R. Lv, J. Sun, Z. Zhang, M. Ma, and J.H. Aheto, "Detection of submerged fermentation of *Tremella aurantialba* using data fusion of electronic nose and tongue", *J. Food Process Eng.,* vol. 42, p. e13002, 2019.
[http://dx.doi.org/10.1111/jfpe.13002]

[139] X. Ying, W. Liu, and G. Hui, "Litchi freshness rapid non-destructive evaluating method using electronic nose and non-linear dynamics stochastic resonance model", *Bioengineered,* vol. 6, no. 4, pp. 218-221, 2015.
[http://dx.doi.org/10.1080/21655979.2015.1011032] [PMID: 25920547]

[140] S. Xu, E. Lü, H. Lu, Z. Zhou, Y. Wang, J. Yang, and Y. Wang, "Quality detection of litchi stored in different environments using an electronic nose", *Sensors (Basel),* vol. 16, no. 6, p. 852, 2016.
[http://dx.doi.org/10.3390/s16060852] [PMID: 27338391]

[141] C. Dai, X. Huang, R. Lv, Z. Zhang, J. Sun, and J.H. Aheto, "Analysis of volatile compounds of *Tremella aurantialba* fermentation *via* electronic nose and HS-SPME-GC-MS", *J. Food Saf.,* vol. 38, p. e12555, 2018.
[http://dx.doi.org/10.1111/jfs.12555]

[142] L. Torri, N. Sinelli, and S. Limbo, "Shelf life evaluation of fresh-cut pineapple by using an electronic nose", *Postharvest Biol. Technol.,* vol. 56, pp. 239-245, 2010.
[http://dx.doi.org/10.1016/j.postharvbio.2010.01.012]

[143] J. Sun, R. Zhang, Y. Zhang, Q. Liang, G. Li, N. Yang, P. Xu, and J. Guo, "Classifying fish freshness according to the relationship between EIS parameters and spoilage stages", *J. Food Eng.,* vol. 219, pp. 101-110, 2018.
[http://dx.doi.org/10.1016/j.jfoodeng.2017.09.011]

[144] M.F. Adak, and N. Yumusak, "Classification of E-nose aroma data of four fruit types by ABC-based neural network", *Sensors (Basel),* vol. 16, no. 3, p. 304, 2016.
[http://dx.doi.org/10.3390/s16030304] [PMID: 26927124]

[145] S. Güney, and A. Atasoy, "Study of fish species discrimination *via* electronic nose", *Comput. Electron. Agric.,* vol. 119, pp. 83-91, 2015.
[http://dx.doi.org/10.1016/j.compag.2015.10.005]

[146] L. Pan, W. Zhang, N. Zhu, S. Mao, and K. Tu, "Early detection and classification of pathogenic fungal disease in post-harvest strawberry fruit by electronic nose and gas chromatography–mass spectrometry", *Food Res. Int.,* vol. 62, pp. 162-168, 2014.
[http://dx.doi.org/10.1016/j.foodres.2014.02.020]

[147] E. Bonah, X. Huang, R. Yi, J.H. Aheto, R. Osae, and M. Golly, "Electronic nose classification and differentiation of bacterial foodborne pathogens based on support vector machine optimized with particle swarm optimization algorithm", *J. Food Process Eng.,* vol. 42, p. e13236, 2019.
[http://dx.doi.org/10.1111/jfpe.13236]

[148] E. Bonah, X. Huang, J.H. Aheto, and R. Osae, "Application of electronic nose as a non-invasive technique for odor fingerprinting and detection of bacterial foodborne pathogens: a review", *J. Food Sci. Technol.,* pp. 1-14, 2019.
[PMID: 32431324]

[149] E. Biondi, S. Blasioli, A. Galeone, F. Spinelli, A. Cellini, C. Lucchese, and I. Braschi, "Detection of potato brown rot and ring rot by electronic nose: from laboratory to real scale", *Talanta,* vol. 129, pp. 422-430, 2014.
[http://dx.doi.org/10.1016/j.talanta.2014.04.057] [PMID: 25127615]

[150] Z. Guo, C. Guo, Q. Chen, Q. Ouyang, J. Shi, H.R. El-Seedi, and X. Zou, "Classification for *Penicillium expansum* spoilage and defect in apples by electronic nose combined with chemometrics", *Sensors (Basel),* vol. 20, no. 7, p. 2130, 2020.
[http://dx.doi.org/10.3390/s20072130] [PMID: 32283830]

[151] B.A. Suslick, L. Feng, and K.S. Suslick, "Discrimination of complex mixtures by a colorimetric sensor array: coffee aromas", *Anal. Chem.,* vol. 82, no. 5, pp. 2067-2073, 2010.
[http://dx.doi.org/10.1021/ac902823w] [PMID: 20143838]

[152] X. Huang, X. Zou, J. Shi, Z. Li, and J. Zhao, "Colorimetric sensor arrays based on chemo-responsive dyes for food odor visualization", *Trends Food Sci. Technol.,* vol. 81, pp. 90-107, 2018.
[http://dx.doi.org/10.1016/j.tifs.2018.09.001]

[153] J. Sun, Y. Lu, L. He, J. Pang, F. Yang, and Y. Liu, "Colorimetric sensor array based on gold nanoparticles: design principles and recent advances", *Trends Analyt. Chem.,* vol. 122, p. 115754, 2020.
[http://dx.doi.org/10.1016/j.trac.2019.115754]

[154] M.C. Janzen, J.B. Ponder, D.P. Bailey, C.K. Ingison, and K.S. Suslick, "Colorimetric sensor arrays for volatile organic compounds", *Anal. Chem.,* vol. 78, no. 11, pp. 3591-3600, 2006.
[http://dx.doi.org/10.1021/ac052111s] [PMID: 16737212]

[155] Q. Ouyang, J. Zhao, Q. Chen, and H. Lin, "Classification of rice wine according to different marked ages using a novel artificial olfactory technique based on colorimetric sensor array", *Food Chem.,* vol. 138, no. 2-3, pp. 1320-1324, 2013.
[http://dx.doi.org/10.1016/j.foodchem.2012.11.124] [PMID: 23411249]

[156] Z. Li, and K. Suslick, "Portable optoelectronic nose for monitoring meat freshness", *ACS Sens.,* vol. 1, no. 11, pp. 1330-1335, 2016.
[http://dx.doi.org/10.1021/acssensors.6b00492]

[157] X.W. Huang, X.B. Zou, J.Y. Shi, Y. Guo, J.W. Zhao, J. Zhang, and L. Hao, "Determination of pork spoilage by colorimetric gas sensor array based on natural pigments", *Food Chem.,* vol. 145, pp. 549-554, 2014.
[http://dx.doi.org/10.1016/j.foodchem.2013.08.101] [PMID: 24128513]

[158] M. Morsy, N. Kostesha, T. Alstrøm, A. Heiskanen, and H. El-Tanahi, "Development and validation of a colorimetric sensor array for fish spoilage monitoring", *Food Control,* vol. 60, pp. 346-352, 2016.
[http://dx.doi.org/10.1016/j.foodcont.2015.07.038]

[159] U. Khulal, J. Zhao, W. Hu, and Q. Chen, "Comparison of different chemometric methods in quantifying total volatile basic-nitrogen (TVB-N) content in chicken meat using a fabricated colorimetric sensor array", *RSC Advances,* vol. 6, no. 6, pp. 4663-4672, 2016.
[http://dx.doi.org/10.1039/C5RA25375F]

CHAPTER 8

Automation on Fruit and Vegetable Grading System and Traceability

Devrim Ünay[*]

Electrical-Electronics Engineering, Faculty of Engineering, İzmir Demokrasi University, İzmir, Turkey

Abstract: Automated sorting and quality grading of agricultural produce are crucial for providing commodities with consistent quality to the consumers and markets. Machine vision has been playing a key role in this quest by presenting technological solutions that provide robust, consistent, and accurate decisions with minimal human intervention. An end-to-end quality inspection system should recognize the type of agricultural product and then perform quality grading. Accordingly, in this proof-o--concept study, a deep learning-based end-to-end solution for quality inspection of agricultural produce is presented, where an initial system automatically sorts fruits-vegetables, while a second system grades apples by skin quality. Experimental evaluations show that the presented end-to-end solution achieves accurate and promising results, and thus holds high-potential for offering high-impact, traceable and generalizable answers for the industry.

Keywords: Computer vision, Deep learning, Grading, Fruit and vegetable, Machine vision, Quality inspection.

1. INTRODUCTION

Recent advances in the fields of mechanics, optics, electronics, computers, and software have led to the birth of machine vision, an engineering technology proposing high-throughput, integrated mechanical-optical-electronic-software solutions for examining, monitoring and controlling applications [1]. Automated quality inspection of food and agricultural products is one such application where accurate, fast, and objective determination of product quality is required due to high standards of safety and quality expected by the industry [2].

In machine vision-based quality inspection of food and agricultural produce, systems are typically composed of a light source, a device to capture images, and

[*] **Corresponding author Devrim Ünay:** Electrical-Electronics Engineering, Faculty of Engineering, İzmir Demokrasi University, İzmir, Turkey; Tel: +90 2322601001 (ext: 502); Fax: +90 2322601004; E-mail: devrim.unay@idu.edu.tr

Jiangbo Li & Zhao Zhang (Eds.)

an image processing/computer vision-based software to process the images [3 - 6]. Such machine vision systems can be categorized by the technological differences they contain or by the agricultural or food product they are put together for inspection, such as vegetables, grains, fruits, meat, and fish or industrialized products [3].

In the first part of this study, we focused on the machine vision systems dedicated to the inspection of fruit and vegetables. To this end, several solutions have been proposed in the literature [7]. Most of these solutions extract color, texture, and/or shape features from the images [5 - 8] and realize inspection by using a machine learning algorithm such as random forest [8], support vector machines [9], neural networks [10] and the recently popular deep learning [11].

Then, in the following part, we focus on the quality grading of a single type of agricultural product, namely the apple fruit. Quality grading of apple fruits using machine vision is challenging due to numerous apple cultivars existing, various defect types present in the fruit, and the natural variability in its skin color [12]. Many of the machine vision-based apple grading solutions proposed in the literature benefit from different sensing techniques or dedicated lighting/equipment (s) [13 - 16]. Other studies employ ordinary machine vision to automatically grade apples using approaches like thresholding [17], Naive Bayes classifier [18], decision trees [19], support vector machines [20], and neural networks [21].

The recently popular deep learning techniques, which eliminate feature engineering and learn representative features from the data, have dramatically improved the state-of-the-art in several domains [22] including the food industry [23]. However, applications of deep learning in the domains of fruit and vegetable sorting as well as apple grading are still limited. Accordingly, here we propose a deep learning-based automated, end-to-end solution for quality inspection agricultural products, and present a proof-of-concept study where an initial system automatically sorts fruits-vegetables while a second system grades apples by skin quality.

2. METHODS

We propose a deep learning-based automated, end-to-end solution for quality inspection of agricultural produce. The proof-of-concept is presented as a cascaded solution where an initial system automatically sorts fruit and vegetables while a second, subsequent system realizes quality grading of the sorted produce, *i.e.* apple fruits. Details of these two systems will be explained below, and the experimental results obtained will be reported in the following section.

2.1. Automated Fruit-Vegetable Sorting

Initially, we will be addressing the problem of automatically recognizing the type of fruit and vegetable from images by using a deep learning-based system. Below, the image dataset will be introduced first, and then the details of the proposed solution will be explained.

2.1.1. The Supermarket Produce Dataset

In order to evaluate the sorting performance of our proposed deep learning system, we decided to use the Supermarket Produce dataset [24]. The dataset comprises a total of 2633 RGB images from 15 different fruit and vegetable categories - Plum (264), Agata Potato (201), Asterix Potato (182), Cashew (210), Onion (75), Orange (103), Taiti Lime (106), Kiwi (171), Fuji Apple (212), Granny-Smith Apple (155), Watermelon (192), Honeydew Melon (145), Nectarine (247), Williams Pear (159), and Diamond Peach (211) - captured on a clear background at a resolution of 1024x768 pixels. Some example images from the dataset can be seen in Fig. (**1**).

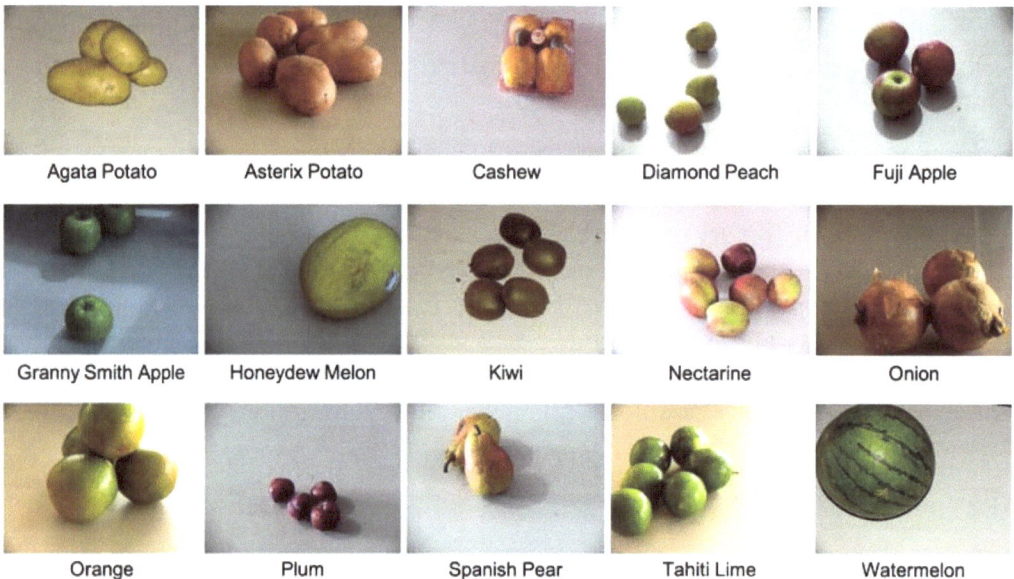

Agata Potato	Asterix Potato	Cashew	Diamond Peach	Fuji Apple
Granny Smith Apple	Honeydew Melon	Kiwi	Nectarine	Onion
Orange	Plum	Spanish Pear	Tahiti Lime	Watermelon

Fig. (1). Example images from the Supermarket Produce dataset.

Previous studies that employed this dataset proposed to use dedicated feature extraction techniques together with machine learning solutions. For example, in a study [24] several color and texture-based features extracted from the images were fed to various conventional classifiers (linear discriminant analysis, support vector machines, k-nearest neighbors, and decision trees to be more precise). A late fusion strategy was then utilized to obtain a decision from the outputs of these multiple classifiers, and around 95% classification accuracy was achieved. In a later study, automated classification of this dataset was realized by a dedicated solution, consisting of a k-means based segmentation in the HSV colorspace, extraction of dedicated texture features, and the use of multiclass support vector machines, and above 98% classification accuracy was achieved [25].

Here, we propose to use deep learning - for the first time in the literature - to automatically sort/classify the Supermarket Produce dataset.

2.1.2. Proposed Deep Learning-based Sorting System

In order to automatically classify the images in the Supermarket Produce dataset into the respective 15 fruit/vegetable categories, we propose to use the GoogLeNet (a.k.a. Inception v1) deep learning architecture (Fig. **2**), which is a 22 layer deep convolutional neural network that qualified as the winner of the 2014 ImageNet Large Scale Visual Recognition Challenge with performance very close to human experts.

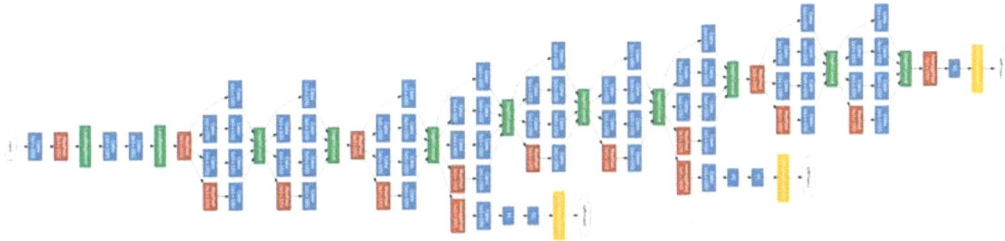

Fig. (2). The GoogLeNet deep learning architecture proposed for fruit-vegetable sorting. Convolution, pooling and softmax layers marked with blue, green and orange, respectively.

We first downsized the images in the dataset to 224x244 by bilinear interpolation to avoid memory issues. We then split the dataset into training and validation sets by the ratio of 70/30, where the training data is further augmented by translation with ±30 pixels in vertical and horizontal directions, and by scaling in the 0.9-1.1 range. We freezed all the networks except for the final layer, which is re-trained by this augmented data. The hyperparameters used in the network are stochastic gradient descent with momentum as an optimizer, minibatch size of 100, a learning rate of 0.0003, and the number of epochs as 6 with data shuffled in every epoch. Implementation of the network and its performance evaluation was realized in Matlab R2019b.

2.2. Automated Quality Inspection

The output of the initial fruit-vegetable sorting system can be connected to a second succeeding deep learning-based system that can further perform grading of the recognized agricultural produce. Details of this second system will be elaborated, after the introduction of the dataset used below.

2.2.1. The CAPA Dataset

The CAPA dataset [26] consists of multispectral images of 526 bi-colored apples of 'Jonagold' variety acquired in a diffusely illuminated environment using a high-resolution camera and four bandpass filters. Each filter image has a resolution of 430x560 pixels with 8 bits-per-pixel. A detailed explanation regarding image acquisition of the dataset can be found [27, 28]. 280 of the fruits in the dataset were healthy, while the rest contained several skin defects such as bruise, russet and rot (Fig. **3**). Furthermore, defective apples were manually categorized by an expert into four quality grades by considering the severity of the defects and the marketing standard of the European Commission [29] as fruits containing slight defects (D4, 76 fruits), serious defects (D3, 55 fruits), bruise defects (D2, 55 fruits) and defects leading to rejection (D1, 60 fruits). As a result, this dataset can be used to automatically grade apples to five quality categories as healthy, and these four defect categories.

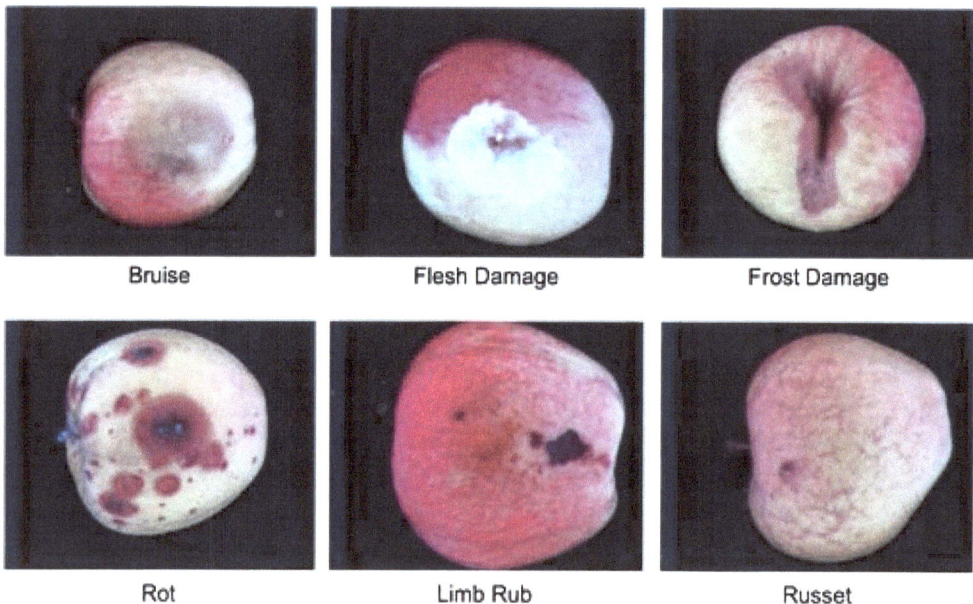

Bruise   Flesh Damage   Frost Damage

Rot   Limb Rub   Russet

Fig. (3). Example images from the CAPA dataset corresponding to apples with various defect types.

In this study, images of three filters (centered at 750, 500, and 450nm corresponding to RGB, respectively) are used only to allow for a more realistic, end-to-end machine vision solution.

2.2.2. Proposed Deep Learning-based Grading System

In order to classify the apple fruits into five quality grades, we propose to use a convolutional neural network with 3 stages of 2D convolutional and max-pooling layers followed by flattening and dense layers as shown in Fig. (4). The convolutional layers consist of 32, 64, and 128 filters with a filter size of 3x3 successively, while the dense layers have sizes of 256, 16, and 5 at the output. The rectified linear unit [30] is used as an activation function throughout the network, except for the output where sigmoid is employed. Training of the network is realized using categorical cross-entropy as loss function, adaptive moment estimation with learning, and two decay rates set as 10^{-4}, 0.9, and 0.999, respectively, for 500 epochs with early stopping patience of 50, and minibatch size of 16.

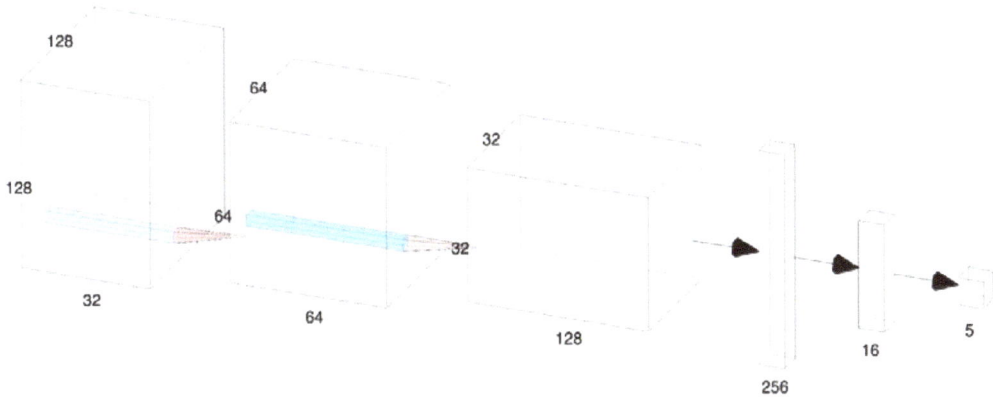

Fig. (4). Convolutional neural network proposed for apple grading. The network consists of three 2D convolutional and pooling layers followed by fully connected flattening and dense layers.

The network is implemented in Python 3.7 and its performance evaluation is realized using Keras with Tensorflow over the Google Colaboratory platform.

2.3. Experimental Evaluation

Classification performance of both deep learning networks introduced in this study are evaluated using a 5-fold cross-validation scheme, and the samples are randomly ordered before being introduced to the networks in order to generate an unbiased classifier.

Regarding the second network used for multi-category apple grading, the dataset employed is highly imbalanced with 280 fruits in the healthy category while the other categories contain around 60 fruits each. Accordingly, this second network is trained by applying class-balanced weighting to obtain an unbiased model.

3. RESULTS

In this study, a machine vision-based end-to-end solution is proposed for quality inspection of agricultural products, and its proof-of-concept is realized *via* an automated fruit and vegetable sorting system followed by an automated apple fruit grading system. In order to effectuate the dedicated inspection, both systems benefit from deep learning that takes RGB images as input and output multi-category classification results. The first system consists of the pre-trained GoogLeNet architecture, where the last layer is trained by the Supermarket Produce dataset to classify images into one of the 15 fruit and vegetable categories. Fig. (**5**) displays the plots of accuracy and loss on the training and validation sets for the first system, where it is observed that the network quickly converges (after the second epoch) and does not indicate any overfitting.

A similar observation is valid for the second network, which consists of a specific convolutional neural network architecture, and therefore corresponding plots are not presented here.

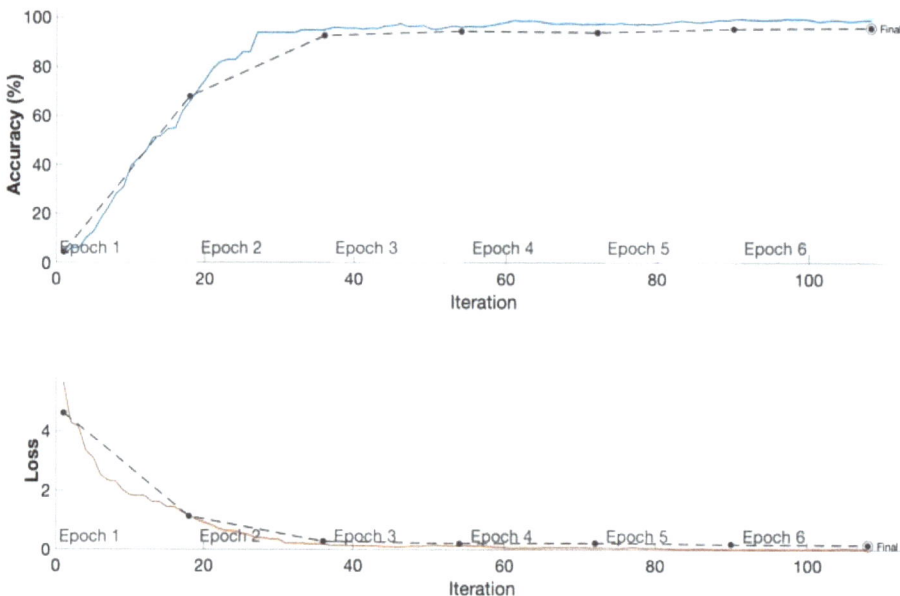

Fig. (5). Plot of accuracy (top) and loss of the training and validation sets for the proposed deep learning-based sorting system.

The performance of the fruit-vegetable sorting (first) network is evaluated on the validation data and reported as a confusion matrix in Table **1**. We notice that over 97% accuracy is achieved by the network for all categories, except the Granny Smith Apple category where the confusions are with the Taiti Lime category most probably due to the similar appearance of the two types (both having greenish, homogeneous skin color). A similar observation is also noticeable for the confusion between Nectarine and Plum categories. The overall performance of the network reaches a highly accurate score of almost 98%. Additionally, Fig. (**6**) displays some exemplary outputs of the sorting network.

Table 1. Performance of the proposed sorting system represented as a confusion matrix. Columns show the true categories, while rows correspond to the categories assigned by the network. Class-specific and overall accuracy scores of the system are also reported.

Graded in	True Categories														
	Agata Potato	Asterix Potato	Cashew	Diamond Peach	Fuji Apple	Granny Smith Apple	Honeydew Melon	Kiwi	Nectarine	Onion	Orange	Plum	Spanish Pear	Tahiti Lime	Watermelon
Agata Potato	60	0	0	0	0	0	0	0	0	0	0	0	0	0	0
Asterix Potato	0	55	0	0	0	0	0	0	0	0	0	0	0	0	0
Cashew	0	0	63	0	0	0	0	0	0	0	0	0	0	0	0
Diamond Peach	0	0	0	63	0	0	0	0	0	0	0	0	0	0	0
Fuji Apple	0	0	0	0	64	0	0	0	0	0	0	0	0	0	0
Granny Smith Apple	0	0	0	0	0	32	0	0	0	0	0	0	0	0	0
Honeydew Melon	0	0	0	0	0	0	43	0	0	0	0	0	0	0	0
Kiwi	0	0	0	0	0	0	0	51	0	0	0	1	0	0	0
Nectarine	0	0	0	0	0	0	0	0	72	0	0	0	0	0	0
Onion	0	0	0	0	0	0	0	0	0	22	0	0	0	0	0
Orange	0	0	0	0	0	1	0	0	0	0	31	0	0	0	0
Plum	0	0	0	0	0	0	0	0	2	0	0	78	0	0	0
Spanish Pear	0	0	0	0	0	0	0	0	0	0	0	0	48	0	0
Tahiti Lime	0	0	0	0	0	13	0	0	0	0	0	0	0	32	0
Watermelon	0	0	0	0	0	0	0	0	0	0	0	0	0	0	58
Accuracy (%)	100.0	100.0	100.0	100.0	100.0	69.6	100.0	100.0	97.3	100.0	100.0	98.7	100.0	100.0	100.0
	97.8														

Fig. (**7**) displays exemplary outputs of the apple grading (second) network, while Table **2** presents the performance of the network as a confusion matrix. The network achieves good overall accuracy (over 85%) with the highest performance observed in the healthy category that has the most training samples. The lowest accuracies on the other hand are seen in the seriously (D3) and slightly (D4) defected categories. Regarding the confusions, the network generally misclassified the fruits with the closest (*i.e.* most similar) category. For example, healthy and slightly defected fruits are confused reciprocally.

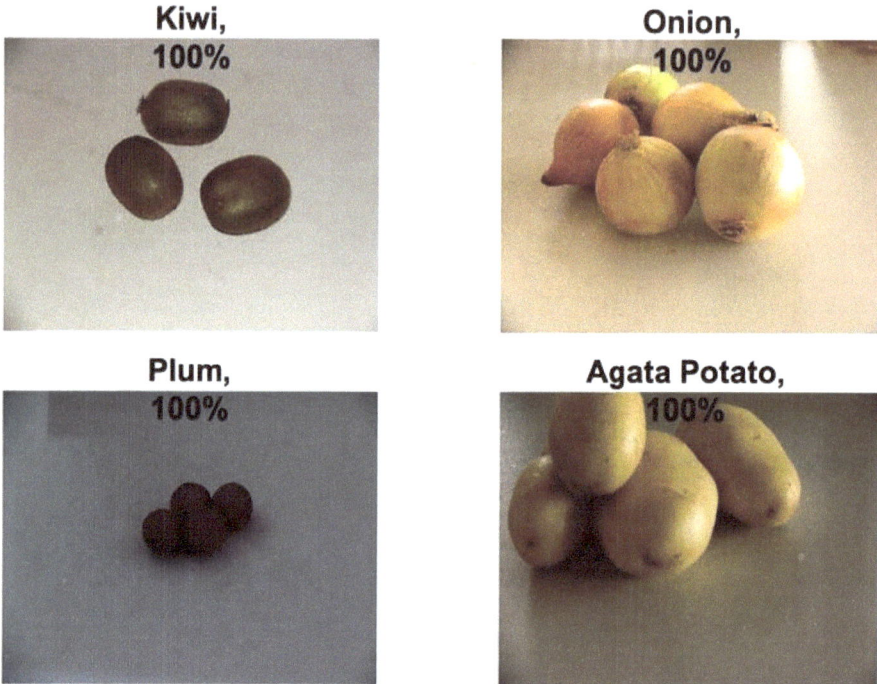

Fig. (6). Exemplary results of the sorting system with corresponding estimates of category probability.

Table 2. Performance of the proposed grading system represented as a confusion matrix. Columns show the true categories, while rows correspond to the categories assigned by the network. Class-specific and overall accuracy scores of the system are also reported.

Graded in	True Categories				
	D1	D2	D3	D4	Healthy
D1	48	3	2	0	1
D2	0	43	0	0	0
D3	12	1	40	12	4
D4	0	3	10	55	12
Healthy	0	5	3	9	263
Accuracy (%)	80.0	78.2	72.7	72.4	93.9
	85.4				

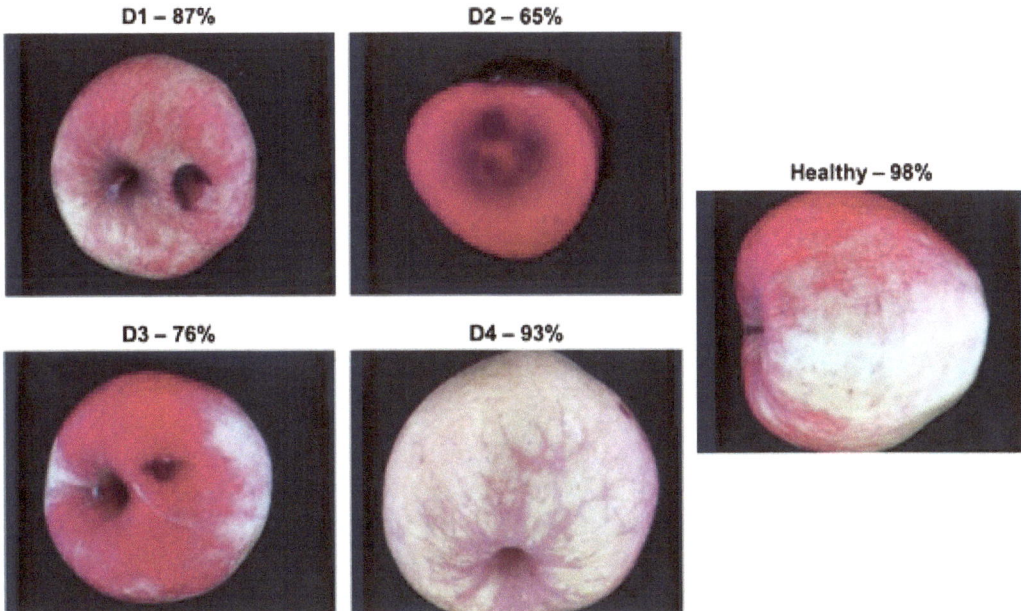

Fig. (7). Exemplary results of the grading system with corresponding estimates of category probability.

The machine vision-based solution proposed for quality inspection of agricultural produce in this study consists of two cascaded deep learning architectures: an initial fruit-vegetable sorting system followed by an apple grading system, where the former has a deeper architecture achieving almost 98% accuracy on the 15-category Supermarket Produce dataset while the latter is relatively shallow and attains 85% accuracy on the 5-category CAPA dataset.

This proof-of-concept study demonstrates that given an RGB image of agricultural produce (apple fruit in this case) as an input to our cascaded solution, the first network will accurately recognize which fruit-vegetable it is and the second network will further realize its quality inspection and assign it to one of the many quality categories.

CONCLUDING REMARKS

The solution proposed in this study is a proof-of-concept towards machine vision-based quality inspection for agricultural produce. Its effectiveness is validated for the case of apple fruits, where the first network recognizes apples (*i.e.* separates them from other fruits and vegetables) and the second one classifies them by quality (*i.e.* assigns them to appropriate quality categories). Our experimental evaluations demonstrate that both networks achieve high-performance scores, and therefore show the promising nature of the proposed solution.

Practical Use of the Proposed Solution

For our proposed framework to be used as a general quality inspection solution for agricultural products, some improvements need to be effectuated. For example, the initial network can be re-trained by larger datasets with more categories to cover and recognize/sort more fruits and vegetables.

Additionally, in order to increase the generalization capability of our proposed framework, the second stage of our solution should include one network for each product to be graded or in other words, it should be composed of as many networks as the products to be graded.

Traceability of the Proposed Solution

In the context of quality inspection of agricultural produce, traceability can be interpreted as the ability to trace and verify the history and location of an item as it is being processed. Accordingly, tracing a fruit/vegetable within the cascaded solution we proposed here a means to keep track of the product at every stage, which can be done either by capturing pictures of the fruit/vegetable first and then attaching an RFID-like device for tracking or by using mini transparent devices that will minimally impair the image quality.

The aforementioned tracing solution can be implemented at the single item level (*i.e.* one tracer per fruit/vegetable) or at the level of a group of items (*i.e.* one tracer per batch). In either case, the candidate tracer devices should be as affordable as possible to have a cost-efficient solution so that it can be adopted by the industry.

CONSENT FOR PUBLICATION

Not applicable

CONFLICT OF INTEREST

The author confirm that this chapter contents have no conflict of interest.

ACKNOWLEDGEMENTS

This work is partially funded by the General Directorate of Technology, Research, and Energy of the Walloon Region of Belgium with Convention No 9813783.

REFERENCES

[1]   K.K. Patel, A. Kar, S.N. Jha, and M.A. Khan, "Machine vision system: a tool for quality inspection of food and agricultural products", *J. Food Sci. Technol.,* vol. 49, no. 2, pp. 123-141, 2012. [http://dx.doi.org/10.1007/s13197-011-0321-4] [PMID: 23572836]

[2] S. Cubero, N. Aleixos, E. Moltó, J. Gómez-Sanchis, and J. Blasco, "Advances in machine vision applications for automatic inspection and quality evaluation of fruits and vegetables", *Food Bioprocess Technol.,* pp. 1-18, 2010.

[3] J.F.S. Gomes, and F.R. Leta, *Applications of computer vision techniques in the agriculture and food industry: A review.* vol. 235. Springer, 2012.

[4] A. Vibhute, and S.K. Bodhe, "Application of image processing in agriculture: a survey", *Int. J. Comput. Appl.,* vol. 52, no. 2, pp. 34-40, 2012.
[http://dx.doi.org/10.5120/8176-1495]

[5] B. Zhang, *Principles, developments and applications of computer vision for external quality inspection of fruits and vegetables: A review.* vol. 62. Elsevier Ltd, 2014.

[6] T.U. Rehman, M.S. Mahmud, Y.K. Chang, J. Jin, and J. Shin, *Current and future applications of statistical machine learning algorithms for agricultural machine vision systems.* vol. 156. Elsevier B.V., 2019.

[7] K. Hameed, D. Chai, and A. Rassau, "A comprehensive review of fruit and vegetable classification techniques", *Image Vis. Comput.,* vol. 80, pp. 24-44, 2018.
[http://dx.doi.org/10.1016/j.imavis.2018.09.016]

[8] H.M. Zawbaa, M. Hazman, M. Abbass, and A.E. Hassanien, "Automatic fruit classification using random forest algorithm", *2014 14th International Conference on Hybrid Intelligent Systems HIS.,* pp. 164-168, 2014.
[http://dx.doi.org/10.1109/HIS.2014.7086191]

[9] Y. Zhang, and L. Wu, "Classification of fruits using computer vision and a multiclass support vector machine", *Sensors (Basel),* vol. 12, no. 9, pp. 12489-12505, 2012.
[http://dx.doi.org/10.3390/s120912489] [PMID: 23112727]

[10] Y. Zhang, S. Wang, G. Ji, and P. Phillips, "Fruit classification using computer vision and feedforward neural network", *J. Food Eng.,* vol. 143, pp. 167-177, 2014.
[http://dx.doi.org/10.1016/j.jfoodeng.2014.07.001]

[11] M. Shamim Hossain, M. Al-Hammadi, and G. Muhammad, "Automatic fruit classification using deep learning for industrial applications", *IEEE Trans. Industr. Inform.,* vol. 15, no. 2, pp. 1027-1034, 2019.
[http://dx.doi.org/10.1109/TII.2018.2875149]

[12] Elzebroek A., and Ton G., *Guide to cultivated plants.* CABI, 2008.

[13] R.G. Diener, J.P. Mitchell, and M.L. Rhoten, "Using an X-ray image scan to sort bruised apples", *Agric. Eng.,* vol. 51, pp. 356-361, 1970.

[14] A. Davenel, C. Guizard, T. Labarre, and F. Sevila, "Automatic detection of surface defects on fruit by using a vision system", *J. Agric. Eng. Res.,* vol. 41, pp. 1-9, 1988.
[http://dx.doi.org/10.1016/0021-8634(88)90198-9]

[15] T.G. Crowe, and M.J. Delwiche, "Real-time defect detection in fruit - Part I: Design concepts and development of prototype hardware", *Trans. ASAE,* vol. 39, pp. 2299-2308, 1996.
[http://dx.doi.org/10.13031/2013.27740]

[16] G. ElMasry, N. Wang, and C. Vigneault, "Detecting chilling injury in Red Delicious apple using hyperspectral imaging and neural networks", *Postharvest Biol. Technol.,* vol. 52, no. 1, pp. 1-8, 2009.
[http://dx.doi.org/10.1016/j.postharvbio.2008.11.008]

[17] Z. Xiao-bo, Z. Jie-wen, L. Yanxiao, and M. Holmes, "In-line detection of apple defects using three color cameras system", *Comput. Electron. Agric.,* vol. 70, no. 1, pp. 129-134, 2010.
[http://dx.doi.org/10.1016/j.compag.2009.09.014]

[18] I. Kavdir, and D.E. Guyer, "Comparison of artificial neural networks and statistical classifiers in apple sorting using textural features", *Biosyst. Eng.,* vol. 89, pp. 331-344, 2004.
[http://dx.doi.org/10.1016/j.biosystemseng.2004.08.008]

[19]   M.M. Sofu, O. Er, M. Kayacan, and B. Cetişli, "Design of an automatic apple sorting system using machine vision", *Comput. Electron. Agric.,* vol. 127, pp. 395-405, 2016.
[http://dx.doi.org/10.1016/j.compag.2016.06.030]

[20]   D. Unay, B. Gosselin, O. Kleynen, V. Leemans, M-F. Destain, and O. Debeir, "Automatic grading of Bi-colored apples by multispectral machine vision", *Comput. Electron. Agric.,* vol. 75, no. 1, pp. 204-212, 2011.
[http://dx.doi.org/10.1016/j.compag.2010.11.006]

[21]   V. Leemans, H. Magein, and M.F. Destain, "On-line fruit grading according to their external quality using machine vision", *Biosyst. Eng.,* vol. 83, pp. 397-404, 2002.
[http://dx.doi.org/10.1006/bioe.2002.0131]

[22]   Y. LeCun, Y. Bengio, and G. Hinton, "Deep learning", *Nature,* vol. 521, no. 7553, p. 436, 2015.
[http://dx.doi.org/10.1038/nature14539]

[23]   L. Zhou, C. Zhang, F. Liu, Z. Qiu, and Y. He, *Application of Deep Learning in Food: A Review.* vol. 18. Blackwell Publishing Inc., 2019.

[24]   A. Rocha, D.C. Hauagge, J. Wainer, and S. Goldenstein, "Automatic fruit and vegetable classification from images", *Comput. Electron. Agric.,* vol. 70, no. 1, pp. 96-104, 2010.
[http://dx.doi.org/10.1016/j.compag.2009.09.002]

[25]   S.R. Dubey, and A.S. Jalal, "Species and variety detection of fruits and vegetables from images", *Int. J. Appl. Pattern Recognit.,* vol. 1, no. 1, pp. 108-126, 2013.
[http://dx.doi.org/10.1504/IJAPR.2013.052343]

[26]   D. Unay, M-F. Destain, B. Gosselin, O. Kleynen, and V. Leemans, *The CAPA Apple Quality Grading Multi-Spectral Image Database.* Zenodo, 2018.

[27]   O. Kleynen, V. Leemans, and M.F. Destain, "Selection of the most efficient wavelength bands for 'Jonagold' apple sorting", *Postharvest Biol. Technol.,* vol. 30, pp. 221-232, 2003.
[http://dx.doi.org/10.1016/S0925-5214(03)00112-1]

[28]   O. Kleynen, V. Leemans, and M.F. Destain, "Development of a multi-spectral vision system for the detection of defects on apples", *J. Food Eng.,* vol. 69, pp. 41-49, 2005.
[http://dx.doi.org/10.1016/j.jfoodeng.2004.07.008]

[29]   Anonymous, ""Commission Regulation (EC) No 85/2004 of 15 January 2004 on marketing standards for apples," Off", *J. Eur. Union,* vol. L, no. 13, pp. 3-18, 2004.

[30]   V. Nair, and G.E. Hinton, "Rectified linear units improve restricted boltzmann machines", In: *ICML,* 2010, pp. 807-814.

CHAPTER 9

Robotic Harvesting of Orchard Fruits

Fangfang Gao[1] and **Longsheng Fu**[1,2,3,4,*]

[1] *College of Mechanical and Electronic Engineering, Northwest A&F University, Yangling 712100, China*

[2] *Key Laboratory of Agricultural Internet of Things, Ministry of Agriculture and Rural Affairs, Yangling 712100, China*

[3] *Shaanxi Key Laboratory of Agricultural Information Perception and Intelligent Service, Yangling 712100, China*

[4] *Centre for Precision and Automated Agricultural Systems, Washington State University, Prosser, WA 99350, USA*

Abstract: Harvesting is one of the most challenging tasks in fruit production. Robotic fruit harvesting technologies are being studied because of labor-intensive and costly handpicking. Due to the unstructured and dynamic characteristics of both the target fruit and its surrounding environment, current harvesting robots have limited performance. Therefore, the commercial applications of most fruit harvesting robots are unrealized. The application and research progress of fruit harvesting robots in apple and kiwifruit harvesting have been reported in this chapter. The applications and development of fruit detection and end-effector design for complex orchard are focused. The main methods used in fruit detection are reviewed, including single feature detection methods, multi-features fusion detection methods, deep learning methods, and 3D reconstruction methods. The technology of end-effector design for selective harvesting with apple and kiwifruit, and shake-and-catch mechanism for bulk harvesting with apple are also reviewed. Existing research problems of fruit harvesting robots in robotic harvesting applications are mentioned, and future development directions of agriculture robots are described.

Keywords: Apple, End-effector, Fruit detection, Kiwifruit, Selective harvesting.

1. INTRODUCTION

The most labor-intensive and time-sensitive task in tree fruit crop production is harvesting. Local growers report that harvesting labor takes about one-third of their annual variable costs, equivalent to the sum of pruning and thinning [1].

---
* **Corresponding author Longsheng Fu:** College of Mechanical and Electronic Engineering, Northwest A&F University, Yangling 712100, China; Tel: +86 15502909963; Fax: +86298 7092391; E-mails: fulsh@nwafu.edu.cn & longsheng.fu@outlook.com

Jiangbo Li & Zhao Zhang (Eds.)
All rights reserved-© 2021 Bentham Science Publishers

Harvesting is a time-sensitive work in which variable weather patterns can cause uncertainty during employment planning [2]. For example, the threat of an early frost may lead to the demand for fruit picking to surge in the short term. Fruit picking will expose workers to fall hazards and ergonomic injuries by heavy lifting and repeated operations [3]. Therefore, except for the risks associated with labor supply and rising costs, the focus on worker safety has also stimulated interest and research in harvesting mechanization for fruit production. The lack of mechanical harvesting is a serious problem that threatens the long-term sustainability of the fruit tree industry.

To reduce dependence on seasonal labor and harvesting costs, researchers began to devote themselves to the research and development of fruit harvesting robots in the 1980s. Two methods, selective harvesting and bulk harvesting, are used to fully mechanize fruit harvesting [3]. The typical approach for selective harvesting is merging a machine vision system with an end-effector and manipulator to pick a single mature fruit selectively. The typical approach for bulk harvesting is applying vibration to the trunk or branch of the tree with shake-and-catch systems to separate the fruit. Although many attempts have been made in the past thirty years to directly incorporate industrial robot technology in this field-based bio-driven environment, the robotic harvesting system for special crop agriculture has not been commercialized.

The current state of the art robotic harvester is developed for orchard fruits, apples, and kiwifruits that have been trained in modern planning orchards to improve uniformity in size, color, and maturity of fruits on individual trees and across a single varietal block of trees. In modern planning orchards, fruits with simple, narrow, accessible, and productive (SNAP) systems are relatively more convenient to pick compared to apples in traditional orchards [4]. Kiwifruit has been planted on strong supporting structures such as T-bars and pergolas, which is more structured than other fruit trees, and thus easier to perform mechanical operations [5]. Both of them are promising to be harvested robotically in the orchard where significant researches have been conducted.

2. ROBOTIC HARVESTING OF APPLE

To resolve the problem of lack of mechanical harvesting for apples, researchers have proposed two different methods for fully mechanized harvesting of fruit trees [3]. The first method is selective fruit harvesting with robotics technology. Selective fruit harvesting technology aims to use robotic arms in conjunction with sensors to locate apples individually. The second method is bulk harvesting, where vibration is applied to the trunk or branch of the tree to detach the fruits. In all, various technologies such as machine vision, image processing, robot kinematics, sensors, and controls are required to be integrated for robotic harvest systems.

2.1. Fruit Detection for Apple

Apple harvesting robots are required to be able to detect and locate apples in the canopy. However, detecting apples under natural conditions poses complex challenges, including fruit overlap, occlusion, shadows, and bright areas. Numerous researches report that various detection algorithms have been used in apple robot harvesting systems.

2.1.1. Single Feature Detection Methods

Color, shape, and texture are the basic characteristics of fruit detection. The color is one of the most distinctive features used to distinguish between complex natural backgrounds and ripe fruits [6]. In the study of color-based fruit detection and segmentation, image pixels are divided into two categories according to the threshold value, which determines whether the pixel belongs to the background or the fruit object. For alleviating the influence of varying illumination, several color spaces (such as L*a*b*, HIS) are adopted to extract color features [7]. Besides the color feature, studies have also employed shape-based and texture-based detection methods [8], [9]. Fruit detection methods based on extracting geometric features are usually used to detect apple-like spherical fruits. Due to its independent color features, the methods of shape-based analysis are not affected by changes in illuminations. Moreover, images taken under natural orchards have a certain texture difference, which can be used to promote detachment of fruits from the background. Therefore, texture features play a significant role in fruit detection, especially when the fruit is occluded. However, robustness is reduced when detection is based on a single feature, *i.e.*, the method is sensitive to changes in the field environment. Bulanon *et al.* [10] applied a red color difference between the objects to detect apples in different lighting conditions, and obtained a detection rate of 88.0%. Kelman and Linker [11] proposed a localization method based on the convexity of ripe apples in trees, and correctly detected 94% of visible apples. Lv *et al.* [12] used Otsu dynamic threshold segmentation method with color characteristic to segment apple image and detected 86% badly occluded apples. The variable lighting condition in the orchard will affect the intensity of reflected light, while the partial occlusion of fruit by fruits, branches, and leaves will affect the geometric features of fruit in the image. Although the detection method based on a single feature can detect apples in the natural orchard, it cannot completely distinguish between target features. In such a scenario, detection based on a single feature may not be the best approach.

Therefore, multi-feature methods are commonly applied to improve efficiency and robustness.

2.1.2. Multi-features Fusion Detection Methods

Feature fusion methods combine different features to distinguish different targets. The texture difference combined with image color space, geometric features, and other algorithms makes it easier to segment the target from the background [13]. Histogram is used to distinguish the color and texture of the occluded fruit. The shape information can be used to implement a feature of circular Gabor texture, thus improving the detection rate of similar background features, uneven illumination conditions, and partially occluded surfaces. Sun *et al.* [14] corrected segmented images using the K-means clustering algorithm according to shape, color, and texture features, and achieved an F1 score of 94.12%, a detection error of 7.37%. Liu *et al.* [15] used a histogram of an oriented gradient to describe the shape of apples, which is adopted to detect apples in candidate regions determined according to the color information and locate the position of apples further. Moreover, the study reached average values of F1, recall, and precision 92.38%, 89.80%, and 95.12%, respectively. Experimental results showed that Multi-features fusion detection methods could accurately detect apples under natural light conditions.

These features are widely used in fruit detection in conjunction with segmentation algorithms. Artificial Neural Networks (ANN) and Support Vector Machines (SVM) are the most popular choices in the literature for robotic fruit harvesting. The ANN is a supervised learning algorithm that can improve its performance by the iterative training process [54]. The ANN transfers information between networks through connections between neurons to calculate the result of a given input. Multiple layers of neurons are connected to form the network, and the number of layers depends on the complexity of the system. The more complex the system, the more input or output neuron layers require [16]. Bhargava and Barisal [17] used the ANN classifier to classify apples by extracting color, statistical, textural, and geometrical features of the fruits with a detection rate of 91.03%, while k-nearest neighbor classifier and sparse representative classifier have detection rates of 80.00% and 85.51%, respectively. However, the detection results were also compared with the SVM classifier (98.48%). The SVM, also a supervised statistical learning algorithm, has been applied in pattern classification, linear and non-linear regression analysis. For linearly separable categories, different classes are separated by the SVM through a hyper-linear plane at the largest interval. Similarly, non-linear separable classification first maps feature vector to a new feature space that can be linearly separable and then performs the classification according to linear SVM separation [18]. Ji *et al.* [19] used color and shape features based on the SVM to detect the square of apple fruit in real-time and finally acquired an average detection rate of 89%, with an average processing time of 352 ms per frame. Tao and Zhou [20] applied the SVM to detect apples, branches, and leaves in an image, and acquired 94.64%, 47.05%, and 75.00% detection rates, respectively.

2.1.3. Deep Learning Methods

A deep learning network is an artificial neural network with multiple layers of perceptron and multiple hidden layers, which can form more abstract high-level attribute categories or functions. A high-level function is combined with a low-level function to express a distributed functional representation of data. The technologies have been used in various fields, especially in fruit detection. Compared with traditional machine-learning methods, the deep-learning method is highly adaptable to differences in working scenarios [21], making it a promising method for many visual tasks. Kang and Chen [22] developed a deep neural network named DaSNet-v2 based on resnet-101, which achieved a value of 0.87 on recall and a value of 0.88 of detection accuracy for apple fruits. Kang and Chen [23] also developed a deep-learning-based detector 'LedNet', which achieved 0.82 on recall and 0.85 on detection rate for apple in orchards.

Deep learning based object detection can be classified into two classes [24]: one-stage detector and two-stage detector. The representative of the two-stage detector is Regional Convolutional Neural Network (RCNN), including RCNN [25], Fast/Faster RCNN [26], and Mask RCNN [27]. The RCNN model has two network branches: a classification branch and a Regional Propose Network (RPN). The classification branch classifies all Regions of Interest (ROI) proposed by the RPN and estimates their bounding boxes, which makes the ROI searching a trainable task compared with traditional ROI searching strategies, thereby improving the model performance and computational capacity [28]. Gené-Mola *et al.* [29] used the Faster RCNN based on the VGG16 network model to detect apples with an AP value of 94.8%. Zhang *et al.* [30] developed transfer learning and fine-tuning for the VGG19 network and activated the feature of different layers to realize the detection of apples, branches, and trunks, which achieved the highest mean AP of 82.4%. In addition, R-FCN is also a commonly used two-stage detector. Wang and He [31] improved the R-FCN (ResNet-44) for apple detection based on the structure and detection results of R-FCN (ResNet-50) and R-FCN (ResNet-101), and obtained a recall rate of 85.7%. The one-stage detector was developed more recently than the two-stage detector [24]. The RPN branch and the classification branch are combined into one network, which makes the architecture more concise and computationally efficient. You Only Look Once (YOLO) network is one of the most representative aspects of the one-stage detector, which achieves the latest performance of target detection with high calculation speed. Zhao *et al*. [32] used YOLO v3 deep convolutional neural network to detect apple fruits with a detection rate of 97%, and a recall rate of 90%.

2.1.4. 3D Reconstruction Methods

3D reconstruction, an established mathematical model, is the basis for analyzing, manipulating, and processing object properties in a computational environment. In the machine vision of robotic technologies, the 3D reconstruction refers to the process of reversing the target 3D information through a set of vision sensors [33]. The purpose of the 3D reconstruction scheme found in fruit-picking robots is multi-faceted: first, to acquire spatial coordinates of fruit; second, to guide the picking robot to the corresponding coordinates of the fruit; third, to obtain the fruit posture and shape information; fourth, to provide information that can be used to help robot end-effector establish behavioral decisions. Over the years, researchers have conducted extensive research on orchard, visual recognition, and the 3D positioning of picking objects. Mai *et al*. [34] obtained the 3D space position information and fruit radius of apple fruit by point cloud segmentation based on the color threshold, thus acquiring every fruit centroid. The study obtained a calculated error for fruit average depth positioning of 8.1 mm, and a calculation error for fruit average radius of 4.5 mm. Tao and Zhou [20] proposed a method to detect apple fruits based on point cloud data, which obtained point cloud data fused color features for the apples *via* an RGB-D (Red Green Blue–Depth) camera. A detection rate of 92.30% was acquired according to 3D geometric features extracted from the data. Gené-Mola *et al*. [35] acquired 2D images using the structure-from-motion photogrammetry method from the apple tree at multiple angles to reconstruct 3D information, improving the fruit detection rate from 0.76 (2D fruit detection) to 0.86 (3D fruit detection).

In the process of harvesting crops, surrounding backgrounds related to positioning include branches and leaves in addition to environmental noise. The backgrounds will cause interference and may collide, resulting in inaccurate positioning in the location of fruits. Therefore, these branches and leaves are regarded as obstacles. An obstacle map can be built based on the detection results to describe 3D spatial information. Decision-making behavior is used for picking robots making them avoid the obstacles or select locations for shaking. Majeed *et al*. [33] segmented apple tree branches and trunks using a convolutional neural network, SegNet, in RGB-D image and obtained branch and trunk segmentation accuracy of 0.92 and

0.93, respectively. Besides, Majeed *et al.* [36] used both Simple- and Foreground-RGB images for training the SegNet to segment trellis wire, branch, and trunk to obtain a favorable result. The segmentation accuracy of the simple RGB and foreground RGB images was 0.92 and 0.97, respectively. The use of semantic segmentation based on deep learning in branch detection and skeleton estimation has brought great potential for developing effective shake-and-catch apple harvester in formally trained apple orchards.

By comparing and analyzing the 3D reconstruction literature of the harvesting robot, a binocular stereo vision system [37, 38] or an RGB-D based vision system [39 - 41] are required in the current 3D reconstruction. The 3D reconstruction based on binocular stereo vision sensor is mainly dedicated to fruit feature point extraction and visual stereo matching in a complex orchard. Besides, the algorithm based on RGB-D visual 3D reconstruction mainly focuses on point cloud data processing and target extraction.

2.2. Fruit Harvesting for Apple

Another requirement for apple harvesting robots is to detach the fruits from the tree. The selective fruit harvesting technology is intended to detach apples individually using the robotic arm, in conjunction with sensors, which is the core responsibility of the end-effector. Compared with robotic picking of individual fruit, bulk harvesting with the shake-and-catch technique can achieve higher harvest efficiency but it will also damage the fruits more seriously.

2.2.1. End-effector Design for Selective Harvesting

The end-effector, a key part of apple harvest robots, detaches apples from limbs. The use of the end-effector to directly contact the fruit to achieve low-destructive fruit picking is one of the key technologies of picking robots [42, 43]. Fruits with complex shape are usually fragile and soft, leading to picking clamping force difficult to control. The fruit may be damaged when the clamping force is too large. But the fruit may not be grasped if the clamping force is insufficient. Therefore, in order to achieve low-destructive fruit picking, the design of the end-effector must be reasonably optimized to achieve a smooth grasp on the fruit.

Traditional robot manipulators with rigid mechanical structures can achieve precise target grasping. However, the fruit may be damaged by picking because of the strong structural rigidity. Zhao *et al.* [44] researched a pneumatic, spoon-shaped end-effector that detached apples by cutting the fruit stem. The end-effector contained two main parts: a cutting device and a gripper. The gripper was used to grasp fruits, and the cutting device was used to cut the stem. Silwal *et al.* [2] developed an underactuated, tendon-driven end-effector with a three-fingered grasp to pick apples like manual picking. The underactuated robotic hands have a high degree of freedom so that the fingers can adapt to the fruit shape when clamping but still cannot overcome the flexibility of the rigid mechanism. Therefore, in order to safely grasp fragile and deformable targets with proper output force, high-precision sensors and complex control algorithms are employed in end-effector systems. The harvesting equipment is greatly restricted due to its complex and robust system.

Recently, soft robotic grippers have emerged in the research field of picking robots such as pneumatic soft robotic grippers, EAP (electro active polymer) actuators and SMA (shape memory alloy) actuators. Hohimer *et al.* [1] developed a pneumatic 3D-printed soft-robotic end-effector to promote the detachment of apples. The end-effector consists of three highly compliant, pneumatic actuators mounted on a soft, flexible palm designed to support the fruit on the actuators. Kang *et al.* [45] also designed a soft-finger based end-effector to grasp the apples. The soft robotic grippers with high environmental adaptability, high flexibility, and simple mechanical structure have the potential to become a solution for compliant fruit gripping. However, there is a problem that the force and posture for grasping cannot be accurately controlled. The poor structural rigidity leads to a small load-bearing capacity for the soft manipulator. Therefore, these actuators may not reach the fruit when encountering obstacles such as tree branches. To solve the problem, Miao and Zheng [46] proposed the concept of a constant force compliance mechanism to design an apple picking actuator. The compliant mechanism is a mechanism that relies on the deformation of elastic members to perform all movements, thereby realizing the transmission of force or energy [47]. The study designed the constant-force compliant mechanism by a distributed method. Higher structural rigidity, better controllability, and better environment adaptability are the characteristics of the compliant mechanisms compared to the soft robotic grippers. Besides, the compliant mechanisms can also meet the constant clamping demands of different types of fruits by appropriately adjusting the parameters of the compliant beam. This design has only been tested in the laboratory, but the size of the gripper is too large, which will limit its practical application in the orchard.

2.2.2. Shake-and-catch for Bulk Harvesting

The bulk harvesting with a shaking or vibration mechanism is one of the widely studied approaches for tree fruit mechanical harvesting. The basic principle of harvesting using the shake-and-catch mechanism is to transfer kinetic energy to the fruit branch, which is adapted to generate a detachment force on the fruit interface to detach the fruit from the tree [3]. During the shaking process, trees have different responses to different amplitudes and excitation frequencies. Therefore, the fruits can be removed through one or a combination of twisting motion, tilting motion, beam-column motion, and pendulum motion [48]. Meanwhile, stem fatigue during repeated bending motion was also found to play a vital role in preventing fruit detachment [4]. Input vibration energy (vibration frequency and amplitude) and the biophysical characteristics of the fruit/tree system (including fruit type, branch dimension, tree structure, *etc.*) will be the main factors affecting the efficiency of fruit removal.

The position of the fruit on the branch will affect the separation of the fruit. Point of fruit attachment in the different fruit/branch structures, *e.g.*, hangers, free branch, and being directly attached, also affects the effectiveness of the fruit detachment techniques. He *et al.* [4] found that most of the unharvested fruits were far away from the shaking position in the experiment of shake-and-catch apple fruits. A technique based on this discovery can be developed to determine the points of shake-and-catch by detecting trunks, branches, and apples in orchards. However, the current machines used for bulk harvesting with shake-an--catch are not fully automated. The operator is required to engage the shaker to the target branch. Therefore, the realization of automated harvesting is the focus of current research. Zhang *et al.* [49] proposed a branch detection method based on the depth features of apple fruits and the R-CNN to estimate the tree skeleton. The results provide a foundation for selecting shaking locations automatically. Zhang *et al.* [30] developed an automated estimation algorithm to choose shaking positions based on the detection results of the Faster R-CNN automatically, which provided a baseline technology for the development of a fully automatic shake-and-catch apple harvesting system.

The orchard under the unified and modern system structure (SNAP) has good fruit quality and high harvesting efficiency [50]. However, the shake-and-catch method usually results in unacceptable damage to the fresh market fruit. Millier *et al.* [51] developed a Cornell fresh market apple harvester, which included a capture system that decreased apple free-fall distance, a roll-pad feeder that transferred fruits from an under-tree catcher conveyor to the main conveyor, and a bulk bin filler. However, the results of field experiments indicated that the method of shake and multi-level capture was not suitable for fruit picking due to its high incidence of bruises. Although fruit bruises caused by the bending, pulling, and twisting movements of the gripper in contact with the fruit can be reduced due to adding elastic materials to the vibrating screen [52], [53], the bruises cannot be avoided. Minimizing bruises caused by fruit contact with branch/fruit and bruises caused by fruit contact with the capture surface are subjects of active researches.

3. ROBOTIC HARVESTING OF KIWIFRUIT

Different from the planting model of apple, kiwifruit is planted in a relatively standardized growth environment through scaffolding, vines, or bushes. This cultivation method is thus convenient for mechanical harvesting.

3.1. Fruit Detection for Kiwifruit

Researches on kiwifruit detection are being carried out in two counties, China and New Zealand, which mainly produce kiwifruit. Fruit detection is an important prerequisite for achieving higher-level agricultural tasks, and it is also a major obstacle for developing commercial harvesting systems [54]. Fruit detection is the basis as once the fruit is located, the end-effector is guided regarding the position of the fruit to be harvested.

Traditional image processing methods such as Sobel edge extraction, Hough transform, color network recognition classifier, fruit calyx feature recognition, and K-means multi-target recognition have been used for kiwifruit detection in the past. Cui *et al.* [55] employed the Otsu threshold in 0.9R-G component to segment the fruit in the side-viewed color image of kiwifruit, and detected fruit with a detection rate of 89.1%. Zhan *et al.* [56] proposed an algorithm based on Adaboost to segment kiwifruit image but needed to improve recognition speed. Mu *et al.* [57] applied the canny operator and elliptical Hough transformation to detect kiwifruit in side-viewed images, and obtained a detection rate of 88.5%. These studies solved the scenario of picking in daylight by placing a camera under sunlight and adjusting it so that the central axis is parallel to the canopy and the images of kiwifruit are obtained for detection. However, the method of acquiring images will cause the background of the acquired images to contain distant non-target fruits and pendulous leaves, which will add excessive noise to the image, thereby affecting the accuracy of image segmentation. Besides, there may be overlaps between kiwifruits that grow in clusters, which will also increase the difficulty of fruit detecting, thereby reducing detection accuracy.

Kiwifruit is grown commercially on a solid support structure. Therefore, the kiwifruit image can also be obtained by placing the central axis of the camera perpendicular to the canopy under the fruit. Most kiwifruits and their calyxes will appear in the camera field of the bottom-up view, and the kiwifruits in a cluster will appear adjacent to each other instead of overlapping with each other. Fu *et al.* [58] employed the Otsu threshold in the 1.1R-G color component to segment kiwifruit obtained by the camera field of bottom-up view and achieved a detection rate of 88.3% using elliptical Hough transform and minimal bounding rectangle.

Fu *et al.* [59] developed a kiwifruit detection system based on detecting the fruit calyx at night using artificial lighting, which reached a success rate of 94.3% and took 0.5 s on average to detect a fruit. However, the above studies are limited for use in nighttime conditions when the light cannot be controlled. Fu *et al.* [5] scanned each detected fruit cluster to seek the contact points between the adjacent fruits and drew a dividing line between the two closest contact points to separate linear clusters of kiwifruits. The method divided the fruits correctly and counted 92.0% of fruits. However, these studies mainly detect kiwifruit based on the color and shape of the calyx and fruit. Besides, the detection effect is only good for images with few fruits, and the detection effect is poor when there are multiple clusters of fruits on the image. To overcome these limitations, a general method that remains invariant and robust to highly discriminative feature representations, different viewpoints, and brightness is required.

As mentioned before, CNN has a good ability to classify and characterize fruit images, which shows superior performance in object detection applications. Fu *et al.* [7] adopted the LeNet to detect kiwifruits in the orchard, which achieved a detection rate of 89.29% and took 0.27 s on average, indicating that the CNN has good application prospects in the field of kiwifruit detection. Fu *et al.* [60] used ZFNet to detect kiwifruit image, which achieved a 92.3% detection rate and took 0.005 s on average to detect a kiwifruit. The model proposed in this work can process an image in a short time, and the model has relatively good robustness to leaves occlusion and light changes, being a strong support for the study on robotic picking kiwifruit with multi-arm operations and crop-load estimation. Mu *et al.* [61] designed a target detection algorithm based on the Im-AlexNet network to solve the problem of low detection rate caused by occlusion of branches and leaves or overlapping occlusion of some fruits in a wide-area complex environment. The network achieved a detection rate of 96.00%, which was 5.74 percentage points higher than the average detection accuracy of LeNet, AlexNet, and VGG16 networks. These studies showed a good prospect for using CNN to detect fruits in RGB images.

The single sensor can hardly provide the needed information on various changes of lighting, occlusion to complete the detection under uncontrolled field conditions. Therefore, a multi-mode sensor that can provide supplementary information about all aspects of the fruit will obtain better detection results. Liu *et al.* [62] presented a method to detect fruits by fusing NIR and RGB images as input to the Faster R-CNN. The study obtained the highest AP value of 90.7% in integration with color and NIR images, which was higher than that of the color image (88.4%) and NIR image (89.2%). The results showed that the proposed kiwifruit detection method of fusing multi-information has better fruit detection potential.

3.2. End-effector for Kiwifruit

The complex environment in which kiwifruits grow and the ease with which their peel can be damaged make it hard for conducting robotic harvesting automatically. Because kiwifruit grows on a vine, inertial shakers, such as the one suggested by Polat *et al.* [63], are not feasible. Simply grasping a kiwifruit and picking it from the vine will exert excessive force on the vine, thereby causing unnecessary shaking to the vine. Shaking also can cause other fruits to sway and lead to accidental detachment of nearby fruit. The movement in the vine will also cause the position of fruit to change, making previous detections meaningless. Furthermore, the operation will increase the possibility of tearing the stem from the plant instead of the fruit, thereby increasing the possibility of bruises during storage and transportation. Commercial practitioners estimated that at least 80% of the kiwifruit in the canopy needs to be picked at an average rate of 4 fruits per second [64]. However, many challenges need to be overcome before harvesting robots are commercially viable. There are three main aspects of harvesting: approaching a target while avoiding obstacles [65], non-destructive separation of the fruit from the tree [66], and establishing a mechanical structure suitable for farm environment and fruit characteristics [67].

The industrial requirement is to remove the stem when the fruit is harvested, which reduces the possibility of bruises to the fruit in the storage box. Besides, the connection between the kiwifruit and the stem is most likely to break. Therefore, currently, the kiwifruit and the stem are mainly separated mechanically by cutting, twisting, and bending the stem. Some studies chose to use end-effectors to grasp the kiwifruit from both sides for operation. Chen *et al.* [68] designed an end-effector of harvesting to grasp kiwifruit and separate the fruit and stem by repeated twisting. The success rate of kiwifruit harvesting by this end-effector was 90%, and it took 9 s to pick the fruit. The design of the kiwifruit harvester is a combination of a linear positioner and a four-degree-of-freedom robot with two rotary joints and two translation joints [69]. Graham *et al.* [70] developed an end-effector for a kiwifruit harvester based on integrating the physical characteristics of the fruit into the design. The end-effector only took 0.1 s to pick the fruit. However, those studies were only tested in the laboratory and did not investigate the picking of clustered kiwifruit under natural growth conditions. Moreover, challenges with providing sufficient space near the vines were foreseen, as the bulk of the robot was mounted near the vines. Supporting the weight of the suspended robot in the orchard was another challenge for those studies.

Another method to harvest kiwifruit is to use the end-effector to grab the fruit from the bottom. Fu *et al.* [66] designed an end-effector to separate fruits based on a bending method. The end-effector approached the fruit from the bottom, separated clustered kiwifruit, and automatically grabbed individual fruits. The model had been proved by field experiments that the separation of fruit and stem could be achieved with a success rate of 96.0% and an average time per fruit of 22 s. Mu *et al.* [71] designed an end-effector with two bionic fingers to separate the fruit from below using the bending method. The model took on average 4–5 s to pick the fruit, with a successful picking rate of 94.2% in an orchard test containing 240 samples. To improve the harvesting speed of the robot, Williams *et al.* [64] studied a four-armed kiwifruit harvesting robot. Early evaluations of the system had shown a good result, with the system capable of harvesting 51.0% of 1,456 kiwifruits in four bays with an average cycle time of 5.5 s/fruit. However, the high cycle time of the harvester and high fruit loss rate (23.4%) hindered the commercial viability of the harvester. Williams *et al.* [72] improved the kiwifruit harvester designed by Williams *et al.* [64], with a capacity of successfully harvesting 86.0% of reachable fruit and 55.8% of all kiwifruit with a cycle time of 2.78 s/fruit.

CONCLUSION

In this chapter, a broad overview of the development of technology applied in apple and kiwifruit harvesting robots is given. Technology for the fruit harvesting robot includes two key elements, fruit detection and end-effector design. For detecting the target fruit, different types of features and image analysis algorithms are applied in the fruit harvesting robots. Color, shape, texture, and depth information of the target fruit have been widely used for fruit detection. Three-dimensional surface reconstruction of the target fruit requires data acquisition from a binocular stereo vision system or an RGB-D based vision system. The scheme of fruit detection may depend on the species of harvesting fruit. For example, the detection method of kiwifruit depends on whether the image is taken from the both sides or the bottom view. Moreover, the performance of the fruit detection system is also influenced by many factors, such as variable light, occlusions, and many others.

In addition to the fruit detection, the development of an end-effector design for fruit harvesting robot is also described in this chapter. The methods of fruit picking for the robotic harvest depend on the planting pattern of harvesting fruit. The structure of the apple tree is suitable for both selective picking and bulk picking, but kiwifruit with a vine planting pattern can only be picked by the selective method. Apple harvesting end-effectors have experienced development from a rigid mechanical structure to soft robotic gripper, and there may be better mechanisms in the future. The shake-and-catch harvester of apple fruit focuses on the selection of shaking points on the branch. The kiwifruit hangs naturally on the vine, and its end-effector mainly separates the fruit and the stem by cutting, twisting, and bending methods. Therefore, the reliability of harvesting robots must consider the environment in which the robot is working.

The development of mechanization in fruit production reflects the development level of agricultural engineering in a country, despite the fact that the gap in the popularization rate of agricultural machinery varies between different countries, and progress in the development of agricultural robot needs to be researched further from the following aspects: (1) The success of fruit detection has been limited by various field conditions, and researchers should focus on seeking ways to break the limits. (2) Research on harvesting robots should emphasize the importance of orchard management. (3) The development of agricultural robots needs to focus on agronomic production to enable the production of high-yield fruit by advanced, practical, and economical agricultural equipment.

CONSENT FOR PUBLICATION

Not applicable.

CONFLICT OF INTEREST

The authors confirm that this chapter contents have no conflict of interest.

ACKNOWLEDGEMENTS

This work was supported by the China Postdoctoral Science Foundation funded project (2019M663832); Fundamental Research Funds for the Central Universities of China (2452020170); National Natural Science Foundation of China (grant number 31971805); International Scientific and Technological Cooperation Foundation of Northwest A&F University (grant number A213021803).

REFERENCES

[1]   C.J. Hohimer, H. Wang, S. Bhusal, J. Miller, C. Mo, and M. Kaekee, "Design and field evaluation of a robotic apple harvesing system with a 3D-printed soft-robotic end-effector", *Trans. ASABE,* vol. 62, no. 2, pp. 405-414, 2019.
[http://dx.doi.org/10.13031/trans.12986]

[2]   A. Silwal, J.R. Davidson, M. Karkee, C. Mo, Q. Zhang, and K. Lewis, "Design, integration, and field evaluation of a robotic apple harvester", *J. F. Robot.,* vol. 34, no. 6, pp. 1140-1159, 2017.
[http://dx.doi.org/10.1002/rob.21715]

[3]   Z. Zhang, P.H. Heinemann, J. Liu, T.A. Baugher, and J.R. Schupp, "The development of mechanical apple harvesting technology: A review", *Trans. ASABE,* vol. 59, no. 5, pp. 1165-1180, 2016.
[http://dx.doi.org/10.13031/trans.59.11737]

[4]   L. He, H. Fu, M. Karkee, and Q. Zhang, "Effect of fruit location on apple detachment with mechanical shaking", *Biosyst. Eng.,* vol. 157, pp. 63-71, 2017.
[http://dx.doi.org/10.1016/j.biosystemseng.2017.02.009]

[5]   L. Fu, E. Tola, A. Al-Mallahi, R. Li, and Y. Cui, "A novel image processing algorithm to separate linearly clustered kiwifruits", *Biosyst. Eng.,* vol. 183, pp. 184-195, 2019.
[http://dx.doi.org/10.1016/j.biosystemseng.2019.04.024]

[6]   Y. Zhao, L. Gong, Y. Huang, and C. Liu, "A review of key techniques of vision-based control for harvesting robot", *Comput. Electron. Agric.,* vol. 127, pp. 311-323, 2016.
[http://dx.doi.org/10.1016/j.compag.2016.06.022]

[7]   L. Fu, Y. Feng, T. Elkamil, Z. Liu, R. Li, and Y. Cui, "Image recognition method of multi-cluster kiwifruit in field based on convolutional neural networks", *Nongye Gongcheng Xuebao (Beijing),* vol. 34, no. 2, pp. 205-211, 2018.

[8]   S. Chaivivatrakul, and M.N. Dailey, "Texture-based fruit detection", *Precis. Agric.,* vol. 15, no. 6, pp. 662-683, 2014.
[http://dx.doi.org/10.1007/s11119-014-9361-x]

[9]   G. Lin, Y. Tang, X. Zou, J. Xiong, and Y. Fang, "Color-, depth-, and shape-based 3D fruit detection", *Precis. Agric.,* vol. 21, no. 1, pp. 1-17, 2020.
[http://dx.doi.org/10.1007/s11119-019-09654-w]

[10]   D.M. Bulanon, T. Kataoka, Y. Ota, and T. Hiroma, "A segmentation algorithm for the automatic recognition of Fuji apples at harvest", *Biosyst. Eng.,* vol. 83, no. 4, pp. 405-412, 2002.
[http://dx.doi.org/10.1006/bioe.2002.0132]

[11]   E. (Efim) Kelman and R. Linker, "Vision-based localisation of mature apples in tree images using convexity", *Biosyst. Eng.,* vol. 118, pp. 174-185, 2014.
[http://dx.doi.org/10.1016/j.biosystemseng.2013.11.007]

[12]   J. Lv, D. Zhao, W. Ji, and S. Ding, "Recognition of apple fruit in natural environment", *Optik (Stuttg.),* vol. 127, no. 3, pp. 1354-1362, 2016.
[http://dx.doi.org/10.1016/j.ijleo.2015.10.177]

[13]   Y. Tang, M. Chen, C. Wang, L. Luo, J. Li, G. Lian, and X. Zou, "Recognition and localization methods for vision-based fruit picking robots : A review", *Front. Plant Sci.,* vol. 11, p. 510, 2020.
[http://dx.doi.org/10.3389/fpls.2020.00510] [PMID: 32508853]

[14]   S. Sun, H. Song, D. He, and Y. Long, "An adaptive segmentation method combining MSRCR and mean shift algorithm with K-means correction of green apples in natural environment", *Inf. Process. Agric.,* vol. 6, no. 2, pp. 200-215, 2019.
[http://dx.doi.org/10.1016/j.inpa.2018.08.011]

[15]   X. Liu, D. Zhao, W. Jia, W. Ji, and Y. Sun, "A detection method for apple fruits based on color and shape features", *IEEE Access,* vol. 7, pp. 67923-67933, 2019.
[http://dx.doi.org/10.1109/ACCESS.2019.2918313]

[16]   S.S. Dahikar, and S.V. Rode, "Agricultural crop yield prediction using artificial neural network approach", *Int. J. Innov. Res. Electr. Electron. Instrum. Control Eng.,* vol. 2, no. 1, pp. 683-686, 2014.

[17]   A. Bhargava, and A. Barisal, "Automatic detection and grading of multiple fruits by machine learning", *Food Anal. Methods,* vol. 13, no. 3, pp. 751-761, 2020.
[http://dx.doi.org/10.1007/s12161-019-01690-6]

[18]   J.J. Wang, D.A. Zhao, W. Ji, J.J. Tu, and Y. Zhang, "Application of support vector machine to apple recognition using in apple harvesting robot", *2009 IEEE Int. Conf. Inf. Autom. ICIA,* pp. 1110-1115, 2009.
[http://dx.doi.org/10.1109/ICINFA.2009.5205083]

[19]   W. Ji, D. Zhao, F. Cheng, B. Xu, Y. Zhang, and J. Wang, "Automatic recognition vision system

guided for apple harvesting robot", *Comput. Electr. Eng.,* vol. 38, no. 5, pp. 1186-1195, 2012.
[http://dx.doi.org/10.1016/j.compeleceng.2011.11.005]

[20]　Y. Tao, and J. Zhou, "Automatic apple recognition based on the fusion of color and 3D feature for robotic fruit picking", *Comput. Electron. Agric.,* vol. 142, pp. 388-396, 2017.
[http://dx.doi.org/10.1016/j.compag.2017.09.019]

[21]　A. Kamilaris, and F.X. Prenafeta-Boldú, "Deep learning in agriculture: A survey", *Comput. Electron. Agric.,* vol. 147, pp. 70-90, 2018.
[http://dx.doi.org/10.1016/j.compag.2018.02.016]

[22]　H. Kang, and C. Chen, "Fast implementation of real-time fruit detection in apple orchards using deep learning", *Comput. Electron. Agric.,* vol. 168, p. 105108, 2020.
[http://dx.doi.org/10.1016/j.compag.2019.105108]

[23]　H. Kang, and C. Chen, "Fruit detection,segmentation and 3D visualisation of environments in apple orchards", *Comput. Electron. Agric.,* vol. 171, p. 105302, 2020.
[http://dx.doi.org/10.1016/j.compag.2020.105302]

[24]　T.Y. Lin, P. Goyal, R. Girshick, K. He, and P. Dollar, "Focal Loss for Dense Object Detection", *IEEE Trans. Pattern Anal. Mach. Intell.,* vol. 42, no. 2, pp. 318-327, 2020.
[http://dx.doi.org/10.1109/TPAMI.2018.2858826] [PMID: 30040631]

[25]　R. Girshick, J. Donahue, T. Darrell, and J. Malik, "Rich feature hierarchies for accurate object detection and semantic segmentation", *Proc. IEEE Comput. Soc. Conf. Comput. Vis. Pattern Recognit.,* no. 1, pp. 580-587, 2014.
[http://dx.doi.org/10.1109/CVPR.2014.81]

[26]　S. Ren, K. He, R. Girshick, and J. Sun, "Faster R-CNN: Towards real-time object detection with region proposal networks", *IEEE Trans. Pattern Anal. Mach. Intell.,* vol. 39, no. 6, pp. 1137-1149, 2017.
[http://dx.doi.org/10.1109/TPAMI.2016.2577031] [PMID: 27295650]

[27]　K. He, G. Gkioxari, P. Dollár, and R. Girshick, "Mask R-CNN", *IEEE Trans. Pattern Anal. Mach. Intell.,* vol. 42, no. 2, pp. 386-397, 2020.
[http://dx.doi.org/10.1109/TPAMI.2018.2844175] [PMID: 29994331]

[28]　J.R.R. Uijlings, K.E.A. Van De Sande, T. Gevers, and A.W.M. Smeulders, "Selective search for object recognition", *Int. J. Comput. Vis.,* vol. 104, no. 2, pp. 154-171, 2013.
[http://dx.doi.org/10.1007/s11263-013-0620-5]

[29]　J. Gené-Mola, V. Vilaplana, J.R. Rosell-Polo, J.R. Morros, J. Ruiz-Hidalgo, and E. Gregorio, "Multi-modal deep learning for Fuji apple detection using RGB-D cameras and their radiometric capabilities", *Comput. Electron. Agric.,* vol. 162, pp. 689-698, 2019.
[http://dx.doi.org/10.1016/j.compag.2019.05.016]

[30]　J. Zhang, "Multi-class object detection using faster R-CNN and estimation of shaking locations for automated shake-and-catch apple harvesting", *Comput. Electron. Agric.,* vol. 173, no. April, p. 105384, 2020.
[http://dx.doi.org/10.1016/j.compag.2020.105384]

[31]　D. Wang, and D. He, "Recognition of apple targets before fruits thinning by robot based on R-FCN deep convolution neural network", *Nongye Gongcheng Xuebao (Beijing),* vol. 35, no. 3, pp. 156-163, 2019.

[32]　D. Zhao, R. Wu, X. Liu, and Y. Zhao, "Apple positioning based on YOLO deep convolutional neural network for picking robot in complex background", *Nongye Gongcheng Xuebao (Beijing),* vol. 35, no. 3, pp. 164-173, 2019.

[33]　Y. Majeed, "Apple tree trunk and branch segmentation for automatic trellis training using convolutional neural network based semantic segmentation", *IFAC-PapersOnLine,* vol. 51, no. 17, pp. 75-80, 2018.

[http://dx.doi.org/10.1016/j.ifacol.2018.08.064]

[34]   C. Mai, L. Zheng, H. Sun, and W. Yang, "Research on 3D reconstruction of fruit tree and fruit recognition and location method based on RGB-D camera", *Nongye Jixie Xuebao,* vol. 46, pp. 35-40, 2015.

[35]   J. Gené-Mola, "Fruit detection and 3D location using instance segmentation neural networks and structure-from-motion photogrammetry", *Comput. Electron. Agric.,* vol. 169, p. 105165, 2020. [http://dx.doi.org/10.1016/j.compag.2019.105165]

[36]   Y. Majeed, J. Zhang, X. Zhang, L. Fu, M. Karkee, and Q. Zhang, "Deep learning based segmentation for automated training of apple trees on trellis wires", *Comput. Electron. Agric.,* vol. 170, p. 105277, 2020. [http://dx.doi.org/10.1016/j.compag.2020.105277]

[37]   Y. Si, G. Liu, and J. Feng, "Location of apples in trees using stereoscopic vision", *Comput. Electron. Agric.,* vol. 112, pp. 68-74, 2015. [http://dx.doi.org/10.1016/j.compag.2015.01.010]

[38]   C. Wang, X. Zou, Y. Tang, L. Luo, and W. Feng, "Localisation of litchi in an unstructured environment using binocular stereo vision", *Biosyst. Eng.,* vol. 145, pp. 39-51, 2016. [http://dx.doi.org/10.1016/j.biosystemseng.2016.02.004]

[39]   W.S. Qureshi, A. Payne, K.B. Walsh, R. Linker, O. Cohen, and M.N. Dailey, "Machine vision for counting fruit on mango tree canopies", *Precis. Agric.,* vol. 18, no. 2, pp. 224-244, 2017. [http://dx.doi.org/10.1007/s11119-016-9458-5]

[40]   I. Sa, "Peduncle detection of sweet pepper for autonomous crop harvesting-combined color and 3-D information", *IEEE Robot. Autom. Lett.,* vol. 2, no. 2, pp. 765-772, 2017. [http://dx.doi.org/10.1109/LRA.2017.2651952]

[41]   G. Lin, Y. Tang, X. Zou, J. Xiong, and J. Li, "Guava detection and pose estimation using a low-cost RGB-D sensor in the field", *Sensors (Basel),* vol. 19, no. 2, pp. 1-15, 2019. [http://dx.doi.org/10.3390/s19020428] [PMID: 30669645]

[42]   Y. Chiu, P. Yang, and S. Chen, "Development of the end-effector of a picking robot for greenhouse-grown tomatoes", *Appl. Eng. Agric.,* vol. 29, no. 6, pp. 1001-1009, 2013.

[43]   W. Ji, Z. Qian, B. Xu, G. Chen, and D. Zhao, "Apple viscoelastic complex model for bruise damage analysis in constant velocity grasping by gripper", *Comput. Electron. Agric.,* vol. 162, pp. 907-920, 2019. [http://dx.doi.org/10.1016/j.compag.2019.05.022]

[44]   D. Zhao, J. Lv, W. Ji, Y. Zhang, and Y. Chen, "Design and control of an apple harvesting robot", *Biosyst. Eng.,* vol. 110, no. 2, pp. 112-122, 2011. [http://dx.doi.org/10.1016/j.biosystemseng.2011.07.005]

[45]   H. Kang, H. Zhou, and C. Chen, "Visual perception and modeling for autonomous apple harvesting", *IEEE Access,* vol. 8, pp. 62151-62163, 2020. [http://dx.doi.org/10.1109/ACCESS.2020.2984556]

[46]   Y. Miao, and J. Zheng, "Optimization design of compliant constant-force mechanism for apple picking actuator", *Comput. Electron. Agric,* vol. 170, p. 105232, 2020. [http://dx.doi.org/10.1016/j.compag.2020.105232]

[47]   Y. Li, Y. Chen, Y. Yang, and Y. Wei, "Passive particle jamming and its stiffening of soft robotic grippers", *IEEE Trans. Robot.,* vol. 33, no. 2, pp. 446-455, 2017. [http://dx.doi.org/10.1109/TRO.2016.2636899]

[48]   J. Li, M. Karkee, Q. Zhang, K. Xiao, and T. Feng, "Characterizing apple picking patterns for robotic harvesting", *Comput. Electron. Agric.,* vol. 127, pp. 633-640, 2016. [http://dx.doi.org/10.1016/j.compag.2016.07.024]

[49]   J. Zhang, L. He, M. Karkee, Q. Zhang, X. Zhang, and Z. Gao, "Branch detection for apple trees trained in fruiting wall architecture using depth features and Regions-Convolutional Neural Network (R-CNN)", *Comput. Electron. Agric.,* vol. 155, pp. 386-393, 2018.
[http://dx.doi.org/10.1016/j.compag.2018.10.029]

[50]   M.E. De Kleine, and M. Karkee, "A semi-automated harvesting prototype for shaking fruit tree limbs", *Trans. ASABE,* vol. 58, no. 6, pp. 1461-1470, 2015.
[http://dx.doi.org/10.13031/trans.58.11011]

[51]   W.F. Millier, G.E. Rehkugler, R.A. Pellerin, J.A. Throop, and R.B. Bradley, "Tree fruit harvester with insertable multilevel catchina system", *Trans. ASAE,* vol. 16, no. 5, p. 2020, 1973.
[http://dx.doi.org/10.13031/2013.37641]

[52]   T. Fadiji, C. Coetzee, L. Chen, O. Chukwu, and U.L. Opara, "Susceptibility of apples to bruising inside ventilated corrugated paperboard packages during simulated transport damage", *Postharvest Biol. Technol.,* vol. 118, pp. 111-119, 2016.
[http://dx.doi.org/10.1016/j.postharvbio.2016.04.001]

[53]   P. Komarnicki, R. Stopa, Ł. Kuta, and D. Szyjewicz, "Determination of apple bruise resistance based on the surface pressure and contact area measurements under impact loads", *Comput. Electron. Agric.,* vol. 142, pp. 155-164, 2017.
[http://dx.doi.org/10.1016/j.compag.2017.08.028]

[54]   A. Gongal, S. Amatya, M. Karkee, Q. Zhang, and K. Lewis, "Sensors and systems for fruit detection and localization: A review", *Comput. Electron. Agric.,* vol. 116, pp. 8-19, 2015.
[http://dx.doi.org/10.1016/j.compag.2015.05.021]

[55]   Y. Cui, S. Su, X. Wang, Y. Tian, P. Li, and F. Zhang, "Recognition and feature extraction of kiwifruit in natural environment based on machine vision", *Nongye Jixie Xuebao,* vol. 44, no. 5, pp. 247-252, 2013.

[56]   W. Zhan, D. He, and S. Shi, "Recognition of kiwifruit in field based on Adaboost algorithm", *Nongye Gongcheng Xuebao (Beijing),* vol. 29, no. 23, pp. 140-146, 2013.

[57]   J. Mu, J. Chen, G. Sun, F. Liu, Y. Ma, and F. Wang, "Characteristic parameters extraction of kiwifruit based on machine vision", *Nong-ji-hua Yanjiu,* vol. 36, no. 6, pp. 138-142, 2014.

[58]   L. Fu, B. Wang, Y. Cui, S. Su, Y. Gejima, and T. Kobayashi, "Kiwifruit recognition at nighttime using artificial lighting based on machine vision", *Int. J. Agric. Biol. Eng.,* vol. 8, no. 4, pp. 52-59, 2015.

[59]   L. Fu, S. Sun, V-A. Manuel, S. Li, R. Li, and Y. Cui, "Kiwifruit recognition method at night based on fruit calyx image", *Nongye Gongcheng Xuebao (Beijing),* vol. 33, no. 2, pp. 199-204, 2017.

[60]   L. Fu, "Kiwifruit detection in field images using Faster R-CNN with ZFNet", *IFAC-PapersOnLine,* vol. 51, no. 17, pp. 45-50, 2018.
[http://dx.doi.org/10.1016/j.ifacol.2018.08.059]

[61]   L. Mu, Z. Gao, Y. Cui, K. Li, H. Liu, and L. Fu, "Kiwifruit detection of far-view and occluded fruit based on improved alexnet", *Nongye Jixie Xuebao,* vol. 50, no. 10, pp. 24-34, 2019.

[62]   Z. Liu, "Improved kiwifruit detection using pre-trained VGG16 with RGB and NIR information fusion", *IEEE Access,* vol. 8, pp. 2327-2336, 2020.
[http://dx.doi.org/10.1109/ACCESS.2019.2962513]

[63]   R. Polat, I. Gezer, M. Guner, E. Dursun, D. Erdogan, and H.C. Bilim, "Mechanical harvesting of pistachio nuts", *J. Food Eng.,* vol. 79, no. 4, pp. 1131-1135, 2007.
[http://dx.doi.org/10.1016/j.jfoodeng.2006.03.023]

[64]   H.A.M. Williams, "Robotic kiwifruit harvesting using machine vision, convolutional neural networks, and robotic arms", *Biosyst. Eng.,* vol. 181, pp. 140-156, 2019.
[http://dx.doi.org/10.1016/j.biosystemseng.2019.03.007]

[65]   S. Hayashi, "Evaluation of a strawberry-harvesting robot in a field test", *Biosyst. Eng.,* vol. 105, no. 2,

pp. 160-171, 2010.
[http://dx.doi.org/10.1016/j.biosystemseng.2009.09.011]

[66]   L. Fu, F. Zhang, G. Yoshinori, Z. Li, B. Wang, and Y. Cui, "Development and experiment of end-effector for kiwifruit harvesting robot", *Nongye Jixie Xuebao,* vol. 46, no. 3, pp. 1-8, 2015.

[67]   C.W. Bac, E.J. Van Henten, J. Hemming, and Y. Edan, "Harvesting robots for high-value crops: state-of-the-art review and challenges ahead", *J. F. Robot.,* vol. 31, no. 6, pp. 888-911, 2014.
[http://dx.doi.org/10.1002/rob.21525]

[68]   J. Chen, H. Wang, H. Jiang, H. Gao, W. Lei, and G. Dang, "Design of end-effector for kiwifruit harvesting robot", *Trans. Chinese Soc. Agric. Mach,* vol. 43, no. 10, pp. 151-154,199, 2012.

[69]   H. Gao, H. Wang, and J. Chen, "Research and design of kiwi fruit harvesting robot", *Nong-ji-hua Yanjiu,* vol. 2, pp. 73-76, 2013.

[70]   S.S. Graham, W. Zong, J. Feng, and S. Tang, "Design and testing of a kiwifruit harvester end-effector", *Trans. ASABE,* vol. 61, no. 1, pp. 45-51, 2018.
[http://dx.doi.org/10.13031/trans.12361]

[71]   L. Mu, G. Cui, Y. Liu, Y. Cui, L. Fu, and Y. Gejima, "Design and simulation of an integrated end-effector for picking kiwifruit by robot", *Inf. Process. Agric.,* vol. 7, no. 1, pp. 58-71, 2020.
[http://dx.doi.org/10.1016/j.inpa.2019.05.004]

[72]   H. Williams, "Improvements to and large-scale evaluation of a robotic kiwifruit harvester", *J. F. Robot.,* vol. 37, no. 2, pp. 187-201, 2020.
[http://dx.doi.org/10.1002/rob.21890]

CHAPTER 10

Detection of Wheat Lodging Plots using Indices Derived from Multi-spectral and Visible Images

Zhao Zhang and **Paulo Flores**[*]

Department of Agricultural and Biosystems Engineering, North Dakota State University, North Dakota, USA

Abstract: Lodging is a critical issue in wheat production, resulting in reduced yield, low crop quality, and increased difficulties in the harvest. Wheat lodging detection contributes greatly to crop management and yield estimation, as well as insurance claim issues. The current manual measurement is labor-intensive, inefficient, and subjective. Aiming to develop a more efficient and objective method to distinguish lodging from non-lodging areas, this study collected aerial color and multi-spectral images using drones attached to different cameras. The experimental field consisted of 372 wheat plots of three different sizes and three days' datasets were collected. Individual images were first stitched to obtain an orthomosaic map and then each plot was visually classified as lodging or non-lodging. Features (*i.e.*, color, texture, NDVI, and height) of each plot were extracted. For each day's dataset, 300 plots (~80% of the total plots) were randomly selected to train the Support Vector Machine (SVM) model, while the remaining 72 plots (~20% of the total plots) were used to test the trained model. After training and testing 10 times, the prediction accuracy was obtained by averaging 10 prediction accuracies. When only using one feature to train the model, prediction accuracies ranged from 66% to 86%. The accuracy increased with more features incorporated for model training. When incorporating all four features, the prediction accuracy was about 90%, indicating its desirable performance in distinguishing lodging from non-lodging plots. The model prediction accuracy of using all four features is not significantly different from that of using only two factors (*i.e.*, texture and NDVI). Since data collection and processing workload increased with more features, researchers in the future could specifically focus on extracting and using texture, and NDVI features to train an SVM model for wheat lodging detection, instead of using four features (*i.e.*, color, texture, NDVI, and height).

Keywords: Color, Features, Height, NDVI, Support vector machine, Texture, Wheat lodging.

[*] **Corresponding author Paulo Flores:** Department of Agricultural and Biosystems Engineering, North Dakota State University, North Dakota, USA; Tel: +1-701-231-5348; E-mail: paulo.flores@ndsu.edu

Jiangbo Li & Zhao Zhang (Eds.)
All rights reserved-© 2021 Bentham Science Publishers

1. INTRODUCTION

Following corn and soybean, wheat (*Triticum aestivum* L.) ranks as the third most important crop in the US in terms of production, growing areas, and gross farm receipts [1]. As one of the most important staple crops, wheat is not only a major source of starch and energy in daily foods but also provides several components that are essential and beneficial for health, such as vitamins, dietary fiber, protein, and phytochemicals [2 - 5]. Furthermore, wheat consumption has been demonstrated to be able to reduce the risk of diseases, such as diabetes (type II), cardiovascular disease, and certain types of cancers [6 - 10]. Following Kansas, North Dakota ranks second in wheat production throughout the US, with a yield of 6.5×10^6 MT and $\$1.4 \times 10^9$ economic value in 2017 [11]. However, starting from 2008, the US wheat planted areas and yield continued to decrease due to lower returns and increased competition in the global market [12]. Therefore, there is a need to develop and adopt new technologies to assist with wheat field management to benefit the US wheat industry economically.

Crop lodging, defined as the permanent displacement of stems from an upright position due to external or internal factors, is one of the most critical issues during wheat production in both developed and developing countries [13 - 15]. Wheat lodging can occur either at stem or root [16, 17]. Stem lodging is caused by the bending or breaking of the lower culm internode, while root lodging can be attributed to a failure in root-soil integrity [18, 19]. Lodging can lead to lower yield and poor grain quality, resulting from self-shading, lowered canopy photosynthesis, increased respiration, reduced translocation of nutrients and carbon for grain filling, and high susceptibility to pests and diseases [20, 21]. It has been reported that wheat lodging could reduce yield up to 50% [14, 22 - 25]. In addition, lodging makes the mechanical harvest more difficult, as the low-level wheat spikes are difficult to be pulled into the combine header [26]. Thus, wheat lodging monitoring will contribute significantly to yield prediction, loss evaluation, and harvest strategy planning [27].

A majority of countries have implemented compensatory policies for agricultural losses caused by natural disasters [28 - 31]. In the US, these policies follow the USDA Risk Management Agency, which insures farmers' crops to a certain value of production [32]. While wheat lodging occurs, farmers have to identify the damaged areas by walking into the field and evaluating visually, after which they would submit a written notice of damage within a certain time period (48~72 hrs. from the initial discovery) [33]. Then, the third party of insurance loss adjuster would come to the farm, manually assess the loss, record measurement, and submit a claim, which finally determines whether the farmers would get paid or not [27]. Manual wheat lodging evaluation is laborious, as workers need to walk across a large field area at a high temperature (*e.g.*, 38°C). In addition, the manual approach is so subjective that each individual inspector may come with different conclusions, which may cause disagreement between farmers and representatives of an insurance company. Furthermore, considering error accumulation occurred during the manual measurement using inaccurate tools (*e.g.*, tape and measuring wheel), the calculated results may be significantly different from the real conditions, leading to under or overpayment. Therefore, it is desirable to have an automatic and objective lodging detection method to replace the manual approach.

Remote sensing technology, with a quick development over the past years, provides a potential tool to obtain timely information on crop lodging over large fields [34]. To date, three major technologies have been explored for crop lodging detection, including spectral image-based satellite sensing, radar-based optical sensing, and Unmanned Aerial Vehicles (UAVs) multiple imagery-based sensing [35, 36]. Though the satellite remote sensing covers huge land plots, its performance on lodging evaluation is weak because of limited spatial and temporal resolutions [37]. In addition, spectral differences supposed to be caused by lodging and non-lodging may be contributed by other factors, such as crop stress (*e.g.*, fertilizer, salinity, and drought) and diseases. The radar-based optical sensing has been tested, but its accuracy on lodging monitoring has not been proven [38]. One possible explanation is that radar system-based optical sensing is more suitable for homogeneous and large areas, while lodging usually occurs in a relatively small area [39]. Due to its relatively small area while occurring and high-resolution image requirement for wheat lodging detection, UAVs are considered as a potential tool [34, 40]. Compared to satellite- and radar-based detection method, UAVs have several advantages. On one hand, UAVs can fly relatively low above ground level and instantly capture bird's eye view images with high resolution; on the other hand, a variety of cameras (*e.g.*, thermal, RGB, and multi-spectral) can be customized and attached to UAVs according to different requirements [38]. In addition, with technological advances in computer vision and digital photogrammetry, aerial images can be processed in different approaches, such as producing geo-referred orthomosaic maps and generating digital surface models (DSMs) [41 - 43].

UAVs have been preliminary tested recently in crop lodging detection, such as rice, corn, and canola. For example, Chu *et al.* (2017) [44] used drones attached with RGB and near-infrared cameras for corn lodging severity detection. The collected imagery data was loaded into a photogrammetric software to construct a 3D canopy structure and DSMs, after which the crop height information was used for lodging severity detection. This study confirmed that the 3D model height was significantly correlated with manually measured results ($R^2 = 0.88$). Li *et al.* (2014) [36] attached an RGB camera to a drone for corn lodging detection [36].

Thirty features were extracted from the images to develop a method for identifying lodging areas, with a model accuracy of 93%. Reviewing existing literature, few studies have been conducted to incorporate comprehensive features of color, NDVI, texture, and crop height (DSMs) for wheat lodging detection. Research using five spectral channels (*e.g.*, red, blue, green, near-infrared, and red-edge) for lodging detection overlooked the role of crop height features [15]. Research using a visible color image for lodging detection ignored crop height and other spectral channels (*e.g.*, near-infrared and red-edge) [35, 45]. To fill this gap, this study explores the use of color, NDVI, texture, and crop height for wheat lodging detection in North Dakota wheat fields. First, individual images collected by drones were stitched to obtain orthomosaic maps, and then individual crop plot was manually cropped and visually classified as lodging or non-lodging. Features (*i.e.*, color, NDVI, texture, and crop height) were then extracted from each plot. 80% of the data were randomly selected to train an SVM and the remaining 20% were used for model testing. Finally, model prediction accuracy was reported.

2. MATERIALS AND METHODS

2.1. Field Experiments

Experimental fields were located at Thompson Country (Fig. **1**), North Dakota. A Total of 372 wheat plots were planted, with three different sizes of plots, 1.5 m x 3.6 m (5 ft. x 12 ft., 204 plots), 1.5 m x 5.4 m (5 ft. x 18 ft., 120 plots), and 1.5 m x 14.6 m (5 ft. x 48 ft., 48 plots). Wheat seeds were sown in the middle of May 2019, and they germinated after approximately a week. After germination, UAVs flew the mission every or every other week for data collection, depending on the weather conditions. Lodging occurred around the middle of July 2019 due to heavy rain and wind (North Dakota Agricultural Weather Network).

2.2. Image Acquisition and Processing

Two drones were used for data collection in this study. A DJI Phantom 4D RTK drone, attached with a Phantom 4 Pro V2.0 camera, was used to obtain color images (DJI-Innovations, Inc., ShenZhen, China). A DJI Matrice 600 Pro (DJI-Innovations, Inc., ShenZhen, China), attached with a multi-spectral camera (MicaSense RedEdge-MX Professional Multispectral Sensor, Simi Valley, CA, US), was used to obtain multi-spectral images. The DJI Phantom 4 flight altitude was ~15 m (~50 ft.), and the DJI Matric 600 flight altitude was ~46 m (~150 ft.). Since the lodging issue did not occur until the middle of July 2019, drone imagery used in this study was from three different dates – July 23, July 30, and August 8, 2019.

Fig. (1). Wheat field information established at Thompson County, North Dakota, US.

After each drone mission, obtained images were stitched together using the Pix4D mapper (Pix4D S.A., Switzerland) to obtain an orthomosaic map, during which a digital surface model was generated. MicaSense Atlas was used to process 5 bands (blue, 475 nm; green, 560 nm; red, 668 nm; red edge, 717 nm; near-infrared 840 nm) file generated by Pix4D. Vegetable indices (*e.g.*, NDVI) were calculated using proper bands. Eight Ground Control Points (GCPs) were installed into the field after seeding but were used before germination as geo-references for orthomosaic maps overlapping. Each individual plot was visually categorized as lodging or non-lodging.

2.3. Feature Extraction

In this study, four features (*e.g.*, color, texture, NDVI, and height) of each plot were extracted by developing proper code using MATLAB (V2019a, MathWorks, Natick, MA, USA) according to proper math formula.

2.3.1. RGB Color Feature

Color information could help distinguish objects [46]. According to equations (1) - (3), red (r), green (g), and blue (b) channel characteristics were extracted for each plot [47].

$$r = \frac{R}{R+G+B} \tag{1}$$

$$g = \frac{G}{R+G+B} \tag{2}$$

$$b = \frac{B}{R+G+B} \tag{3}$$

In the orthomosaic map, a non-lodging plot has an image filled mainly by wheat leaves, but a lodging plot mainly consists of stems or stems with leaves. Stems and leaves have different content of chlorophyll, which has a large reflection rate in green and a large absorption rate in red and blue. Hence, Extra Green value (ExG; eq 4) is probably a good indicator to distinguish lodging from non-lodging plots.

$$ExG = 2 \times g - r - b \tag{4}$$

2.3.2. Texture Feature

Texture feature is a repeating pattern of local variations in image intensity and can be potentially used to differ lodging from non-lodging plots [34]. In this study, the wheat crop texture feature of each plot was described in terms of coarseness (*Fcrs*), contrast (*Fcon*), line-likeness (*Flin*), and directionality (*Fdir*) [47].

Coarseness (*Fcrs*) is obtained using the following four steps.

Step 1: For each pixel *p(x, y)*, we compute six averages for different windows around the 'pixel'. Take average for every point of the input image over the neighborhood, with sizes of 1x1, 2x2, 4x4, 8x8, 16x16, and 32x32 (2^k, k = 0, 1, 2, 3, 4, 5).

$$A_k(x,y) = \left.\sum_{i=x-2^{k-1}}^{x+2^{k-1}-1}\sum_{j=y-2^{k-1}}^{y+2^{k-1}-1} f(i,j)\right/2^{2k} \tag{5}$$

Where, *f(i, j)* is the gray-value at each point *p(x, y)*.

Step 2: For each point, we take differences between pairs of averages corresponding to pairs of non-overlapping neighborhoods on the opposite side of the point in both horizontal (eq. 6) and vertical directions (eq. 7).

$$E_{k.h}(x,y) = |A_k(x + 2^{k-1}, y) - A_k(x - 2^{k-1}, y)| \tag{6}$$

$$E_{k.v}(x,y) = |A_k(x, y + 2^{k-1}) - A_k(x, y - 2^{k-1})| \tag{7}$$

Step 3: For each point, we pick the best size that generates the highest output value.

$$S_{best}(x,y) = 2^k \tag{8}$$

where k maximizes E in either direction. $E_k = E_{max} = \max(E_1, E_2, \ldots E_L)$.

Step 4: Finally, take the average of S_{best} throughout the plot to obtain *Fcrs* (eq. 9).

$$F_{crs} = \frac{1}{mxn} \sum_i^m \sum_j^n S_{best}(i,j) \tag{9}$$

Where, m and n are the effective width and height (measured in pixel) of the individual plot image.

Contrast (*Fcon*) is calculated following equation 10.

$$F_{con} = \frac{\sigma}{\sigma_4^{1/4}} \tag{10}$$

where $\sigma_4^{1/4}$ is the fourth moment about the mean of the individual, and σ is the variance.

Linelikeness (*Flin*) is defined as the average coincidence of edge direction that co-occurs at pixels separated by a distance *d* along the direction α. *Flin* is calculated as follows (eq. 11).

$$F_{lin} = \frac{\sum_{i=1}^m \sum_{j=1}^m P_{Dd(i,j)}\left(\cos(i-j)\frac{2\pi}{n}\right)}{\sum_{i=1}^m \sum_{j=1}^m P_{Dd(i,j)}} \tag{11}$$

Where, *PDd(i, j)* is the *m×m* local direction co-occurrence matrix of points at distance *d*. *PDd(i, j)* is defined as the relative frequency with which two neighboring cells are separated by a distance d along the edge direction occurring on the image.

Directionality (*Fdir*) consists of the edge strength and the direction of angle. HD(ϕ) is the histogram of local edge probabilities, which is computed against their directional angle. The histogram represents sufficiently global features of the input image, such as long lines and simple curves. Since the gradient is considered a vector, it has both magnitude and direction. The *Fdir* is computed using equation (12).

$$F_{dir} = \sum_p^{n_p} \sum_\phi (\Phi + \Phi_p)^2 H_D(\Phi) \tag{12}$$

Where, *p* represents the number of peaks in the histogram, *np* represents the total number of peaks in the histogram, and ϕp represents the pth peak location.

2.3.3. NDVI Feature

Normalized Difference Vegetation Index (NDVI) uses two parameters (*i.e.*, near-infrared and red light) to quantify vegetation [48]. The NDVI is rooted in the theory that vegetation strongly reflects the near-infrared but significantly absorbs the red light [49]. The NDVI value can be calculated following equation (13).

$$NDVI = (NIR - Red)/(NIR + Red) \tag{13}$$

2.3.4. Plant Height Feature

Crop height has the potential to be used as an index for lodging detection [44]. The raw height information at the pixel level in a plot was extracted from ArcMap (ArcMap 10.6, Redlands, CA, USA), and then Stata (V14, College Station, TX) was used for further statistic process. After going through the data preliminarily, it was noticed that wheat leaves/spikes are relatively small, and the height information at pixel level varies significantly and cannot represent the real crop height conditions in a very accurate manner. Thus, all heights at pixel level were first ranked from maximum to minimum, with the average of top 10% height values calculated to represent the plot height. All features extracted from plots are normalized before running to the next step. The data processing procedure is shown in Fig. (2).

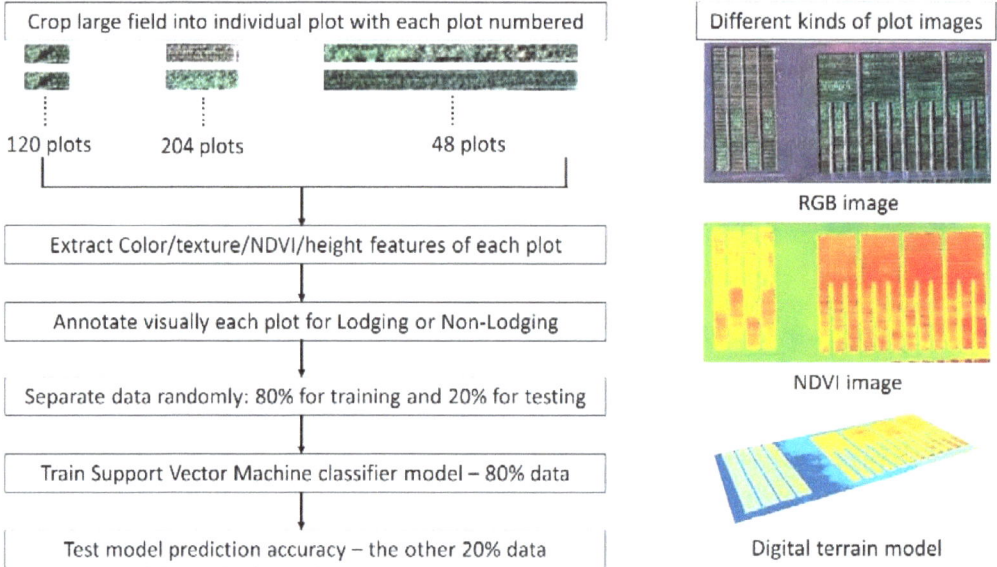

Fig. (2). The schematic diagram of data processing for wheat lodging detection.

2.4. Classifier and Datasets Separation

A Support Vector Machine (SVM), which is an effective binary linear classifier, was applied for data classification and prediction [50 - 55]. Features of red (eq. 1), green (eq. 2), and blue (eq. 3), extra green (eq. 4), coarseness (eq. 9), contrast (eq. 10), linelikeness (eq. 11), directionality (eq. 12), NDVI (eq. 13), and plant height were used to train the SVM model. The plot lodging and non-lodging were visually evaluated and used as labels during model training and testing. During the SVM model training, the Gaussian kernel is selected, and the KernelScale is set to auto. Each orthomosaic map contains 372 observations (plots), and the whole dataset is randomly divided into a training (80%) and testing (the left 20%) dataset.

3. RESULTS AND DISCUSSION

3.1.1. Color Feature Analysis

Fig. (3) shows the visual images of lodging and non-lodging individual plots, as well as the percentage distribution of r, g, b, and Extra Green for one lodging and non-lodging sample plot. For the index of r, the distributions of lodging and non-lodging samples are similar and significantly overlapped, indicating that red is probably not an ideal index to distinguish the two classes. The same scenario occurs to the b channel. Compared to r and b indices, the distribution of g value is little different between the lodging and non-lodging samples. For the g value below 120, the non-lodging area has a higher percentage than the lodging area, while it reverses for the g value above 120. The distribution of Extra Green index for non-lodging samples is concentrated at the left of the Figure (centered around 35), while the distribution for the lodging plot is concentrated at the right (centered around 110). By taking the advantages of amplifying g value and then subtracting by r and b value for each pixel, the Extra Green index appears more attractive in distinguishing the two classes. . A potential reason that the r and b value distributions overlapped significantly while g value distribution is a little different between the two classes is the chlorophyll, which absorbs red and blue light but reflects green light. For crops facing up vertically, the lodging-area is filled with a high percentage of leaves in the image. However, for the lodging plot, stems or even soil fill part of the image.

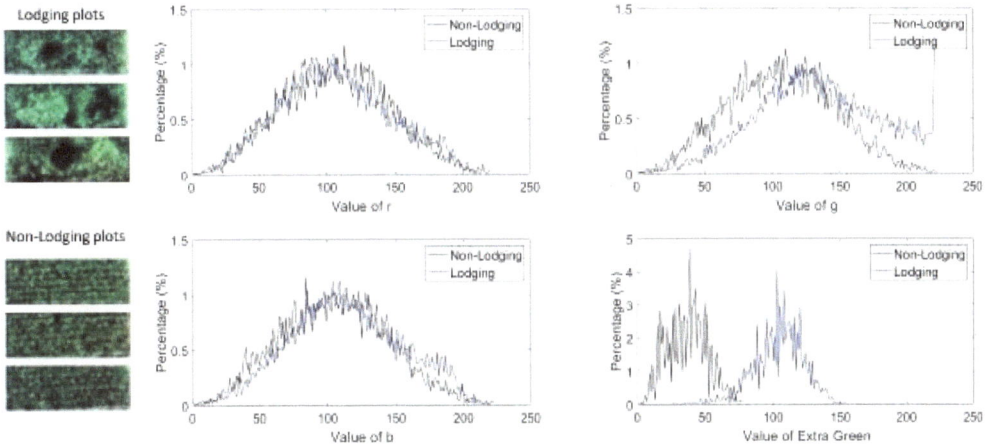

Fig. (3). Percentage distributions of different color characteristics (r, g, b, and extra green) for one randomly selected lodging and non-lodging plot in a wheat field (x-axis denotes color value range; y-axis denotes the ratio of a certain pixel value).

3.1.1. Texture Feature Analysis

Texture features of the individual plot were extracted and analyzed according to the formulas described in section 2.3.2. The black line in each plot (Fig. **4**) is the threshold value to distinguish lodging and non-lodging plots, which were calculated by maximizing intra-class (*i.e.*, lodging and non-lodging) variance. For the normalized coarseness parameter, the non-lodging plots have a majority of its values ranging from 0.4 to 0.6, while values for lodging plots spread more widely from 0.2 to 1. One possible explanation for the narrow range of coarseness values for the non-lodging plots is their uniformity – all leaves are in the images. However, the lodging plot consists of leaves and stems. The contrast distributions show that the values for non-lodging plots are generally smaller than 0.4, while a majority of the lodging values are above 0.4. Regarding the direction feature, lodging plots have larger values than that of non-lodging plots. For the linelikeness, the value range of lodging plots is from 0.6 to 1.0, while for non-lodging plots, it is from 0 to 1. Generally, the coarseness and contrast values are more desirable parameters to distinguish lodging from non-lodging plots over the parameters of direction and linelikeness. According to the preliminary analysis, using only one individual feature to differ lodging from non-lodging plots might be arbitrary, and more features incorporated may lead to desirable results.

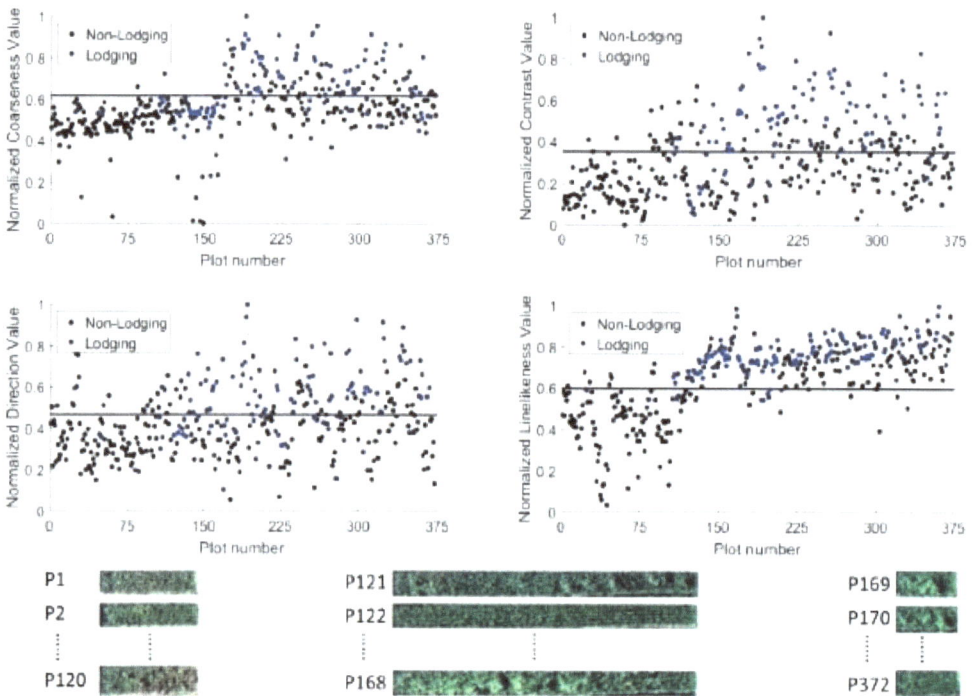

Fig. (4). Extracted texture feature values of each plot from an orthomosaic image; black color lines are the threshold values (calculated by maximizing the two classes' variance) to distinguish lodging and non-lodging crop plots (P represents plot).

3.1.2. NDVI Feature Analysis

The averaged NDVI value of all pixels in a plot was used to represent the plot's NDVI mean value. The standard deviation of NDVI values in a plot was also computed, and the results are shown in Fig. (5). The threshold line was calculated by maximizing the lodging and non-lodging class variance. For lodging plots, a majority of the normalized NDVI mean values are above 0.6, with only a few below this threshold. However, for non-lodging plots, the normalized NDVI mean values range from 0 to 1. The Normalized NDVI standard deviation data for lodging are mixed with those of non-lodging. On the right side of Fig. (**5**), 17 color images and their corresponding NDVI data are shown. In the NDVI map, the red and green/yellow color represent high and low values, respectively. The lodging plots (plots # 1, 2, 3, 4, 6, 7, 10, 13, and 14) have a larger NDVI value (more red) than the non-lodging areas (more green). However, using NDVI only may cause some errors, as some non-lodging areas (*e.g.*, plots 5 and 8) also have a reddish color NDVI map.

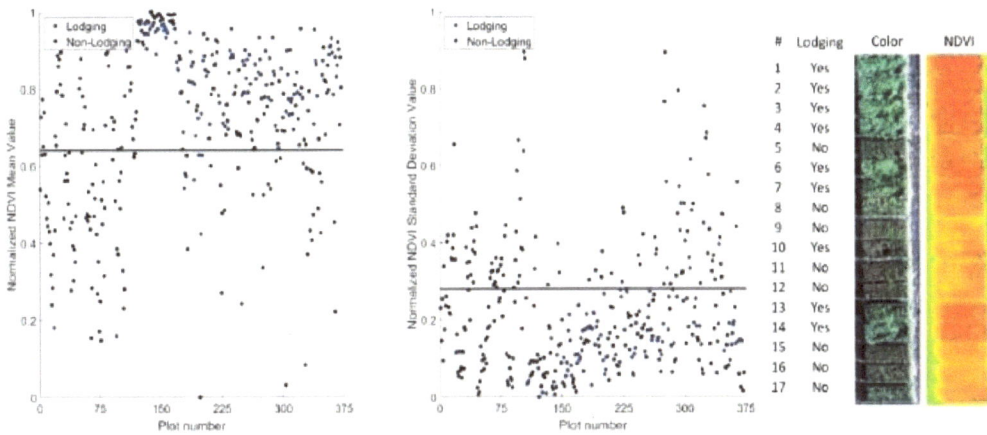

Fig. (5). RGB/NDVI images and extracted normalized NDVI mean and standard deviation values of each plot from an orthomosaic image (plot number refers to the Fig. (**4**); threshold line was calculated by maximizing the lodging and non-lodging intra-class variance).

3.1.3. Height Feature Analysis

The plot height information at the pixel level can be obtained from the digital terrain model top photo in Fig. (**6**). Theoretically, crops height feature could be a good indicator to differ lodging from non-lodging plots, as lodged crops' height is usually lower than that of the non-lodging crops. If crops are lodged partially, the plot height standard deviation would be larger for lodging plots than non-lodging plots. If all crops are lodged in a plot, the difference of height standard deviation between lodging and non-lodging areas would not be significant. Thus, the average of the top 10% height value was used as an additional parameter. The results are shown in Fig. (**6**), with a threshold line (calculated by maximizing the lodging and non-lodging class variance) to separate the two classes. For the standard deviation of normalized height, the data for lodging and non-lodging are mixed, and the same scenario occurs for the average of the top 10% height value.

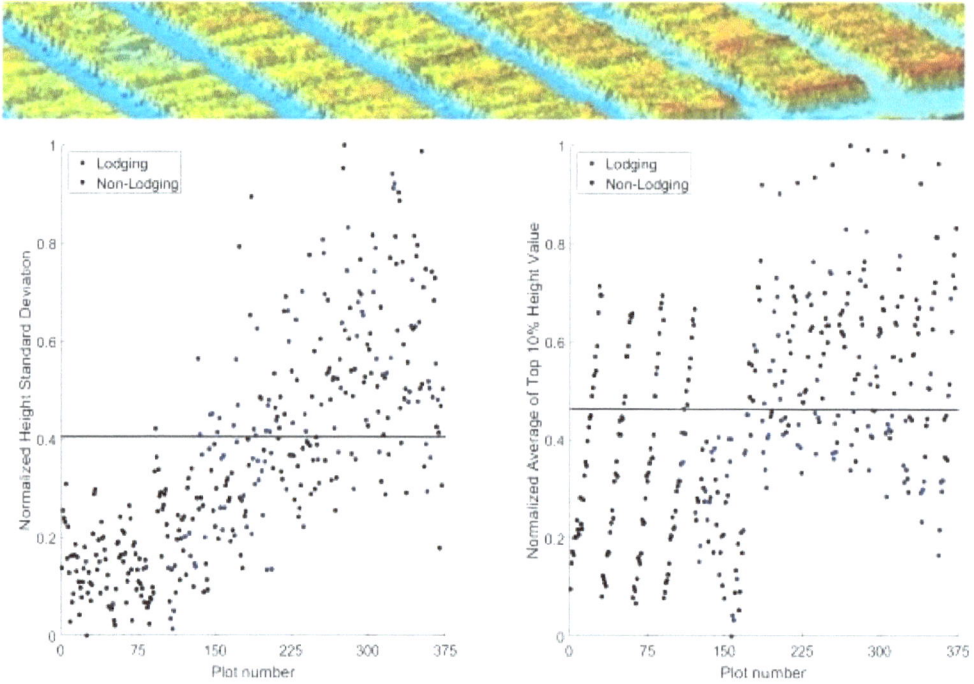

Fig. (6). Normalized plot height standard deviation and average of top 10% height value features extracted from orthomosaic images (plot number refers to the Fig. (**4**); threshold line was calculated by maximizing the lodging and non-lodging class variance).

3.2. SVM Training and Predicting

In this study, we used a single feature either color feature (CF), texture feature (TF), NDVI, or height feature (HF), to train the SVM model and test its accuracy. We also explored the performance of two features combination (CF+TF, CF+NDVI, CF+HF, TF+NDVI, TF+HF, and NDVI+HF), three features combination (CF+TF+NDVI, CF+TF+HF, TF+NDVI+HF, and CF+NDVI+HF), and four features combination (CT+TF+NDVI+HF) for model training and testing, with detailed results shown below.

For the July/23/2019 dataset, the more features used to train the model, the higher the model prediction accuracy is (Fig. **7**). Using four features individually to train the model, the prediction accuracies ranged from 71% to 81%. The prediction accuracy from the texture feature had a significantly higher accuracy (81%) than any of the other three individual factors, while the height feature has the lowest prediction accuracy (71%). The single color feature does not perform well which is mainly due to the significant overlapping for red, green, and blue (Fig. **3**). The NDVI does not perform well, which is probably because the lodging is incomplete, and crop leaves hide the stems in the top view image. When using two features, the prediction accuracies increase and range from 76% to 89%. Since the texture is the most significant individual feature, combinations between texture and any other feature have a relatively high prediction accuracy, ranging from 86% to 89%. A combination between color and NDVI generates the lowest prediction accuracy of 76%. When combining texture features with any other two or three features, the prediction accuracy ranges from 87% to 89%. However, the accuracy is only 79% for the combination of the other three features without texture (*i.e.*, C+N+H). The results show that the texture feature plays an important role in affecting the model's prediction accuracy.

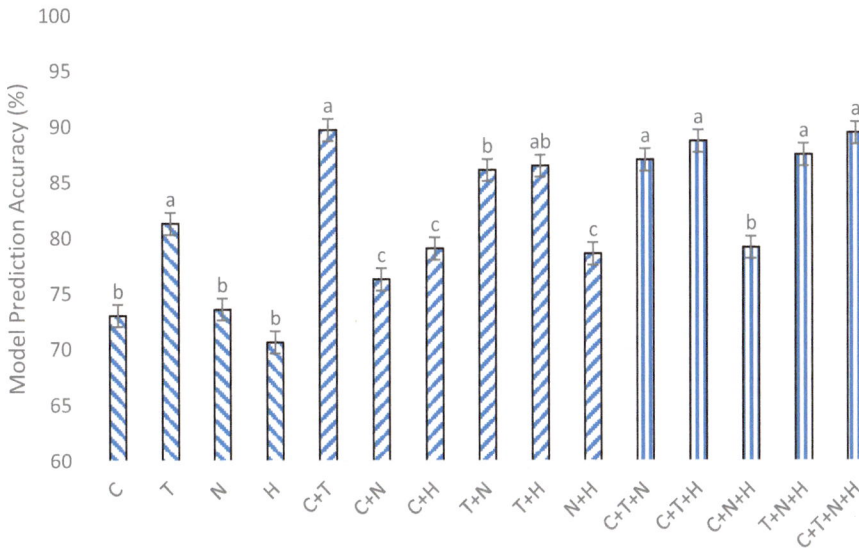

Fig. (7). Model prediction accuracy with different training features for the July/23/2019 dataset. C, T, N, and H stands for color, texture, NDVI, and height feature, respectively; statistic comparisons for the same number of training features (four factors combined with three factors) were conducted by Tukey T-Test at the level of 0.05 with different letters in each set showing significant difference.

Data processing results for the July/30/2019 dataset are similar to that of July/23/2019. Among all the single feature models, the model with only texture features results in the highest prediction accuracy of 84%. The reason that July/30 has a higher prediction accuracy than the dataset of July/23 when only using texture feature is probably that the lodging is more serious with time going on. Similar to the July/23 results, the crop height feature has the lowest model prediction accuracy of 66%, which is significantly lower than that from any other individual feature. The color and NDVI provide similar model prediction accuracy of 74%. For two-factor models, the inclusion of texture feature results in a relatively high prediction accuracy ranging from 87% to 88%. However, without texture features included, the prediction accuracies range from 76% to 86%. With texture feature included in the three or four-factor models, the prediction accuracies go from 86% to 90%. When the texture feature is excluded from the model, the prediction accuracy is only 83% (Fig. **8**).

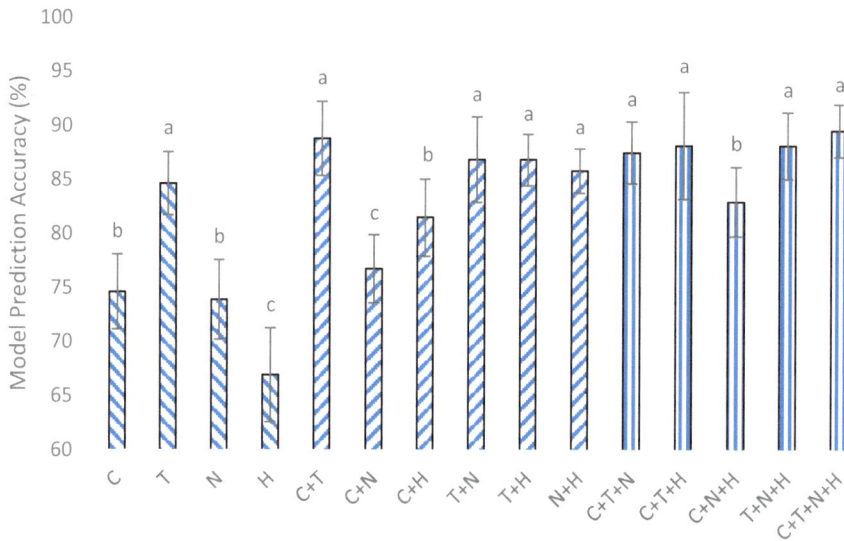

Fig. (8). Model prediction accuracy with different training features for July/30/2019 data. C, T, N, and H stands for color, texture, NDVI, and height feature, respectively; statistic comparisons for the same number of training features (four factors combined with three factors) were conducted by Tukey Test at the level of 0.05 with different letters in each set showing significant difference.

For the August/8/2019 dataset, the results are similar to the above analysis. With only the texture feature used for model training, the prediction accuracy is 88%, which is higher than that for the July/23 and July/30 datasets. Among the four features, the height information consistently has the lowest prediction accuracy of 73%. With two factors combined, the model prediction accuracies range from 83% to 93% with texture feature included, while the accuracies range from 80% to 87%, when excluding texture feature. When three or four factors are considered, the accuracy ranges from 90% to 92% with texture feature included; otherwise, it is only 87% (Fig. **9**).

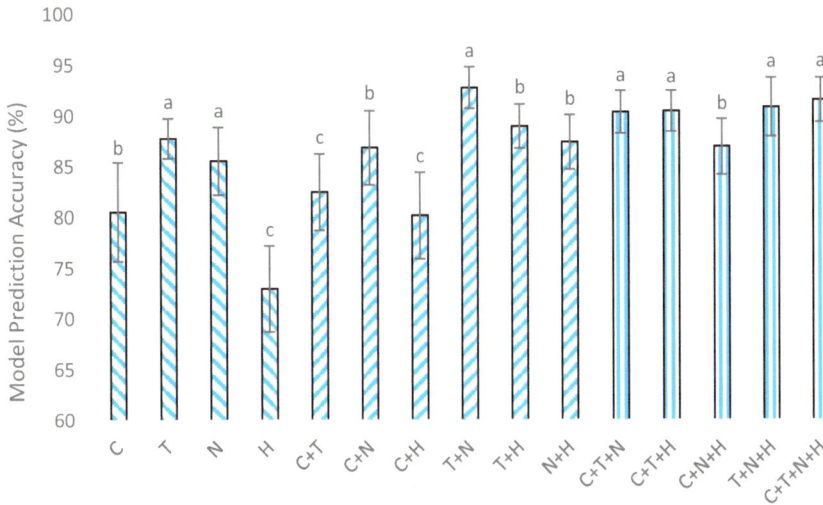

Fig. (9). Model prediction accuracy with different training features for August/8/2019 data. C, T, N, and H stands for color, texture, NDVI, and height feature, respectively; statistic comparisons for the same number of training features (four factors combined with three factors) were conducted by Tukey Test at the level of 0.05 with different letters in each setting showing significant difference.

3.3. Identifying Feature Combinations with Desirable Prediction Accuracy

From the above analysis of all three days' datasets, it has been identified that texture and combinations with texture included have a high model prediction accuracy. The combination of texture and NDVI gives a relatively high model prediction accuracy. Though the model with all four features included has a higher prediction accuracy, more features indicate more data collection and processing work. Hence, it is necessary to explore whether models with fewer features could lead to satisfactory prediction accuracy compared to all four features included. Fig. (**10**) shows the model prediction accuracy with different feature combinations (*e.g.*, texture, texture + NDVI, and all four features). For July/23/2019 dataset, models with only the texture feature have a prediction accuracy (81%), significantly lower than the other two combinations. The model prediction accuracy for the two-factor combination is not significantly different from that of the four-factor combination. For the July/30/2019 dataset, the model prediction accuracy with only texture feature is low, and the model prediction accuracy for two factors included is not significantly different from that with four factors integrated. A similar scenario occurs for the August/8/2019 dataset. Therefore, to simplify the data collection and process but not reduce the model prediction accuracy, the optimal model is to use the combination of texture and NDVI.

With time going on, the model prediction accuracy increases. For using texture, texture and NDVI, and four features, the model prediction accuracies increase from 81% to 88%, 86% to 93%, and 89% to 92%, respectively, from July/23/2019 to August/8/2019. This could probably be explained that lodging is a dynamic process, and with time going on, the angles between the crop stems and vertical line increase gradually.

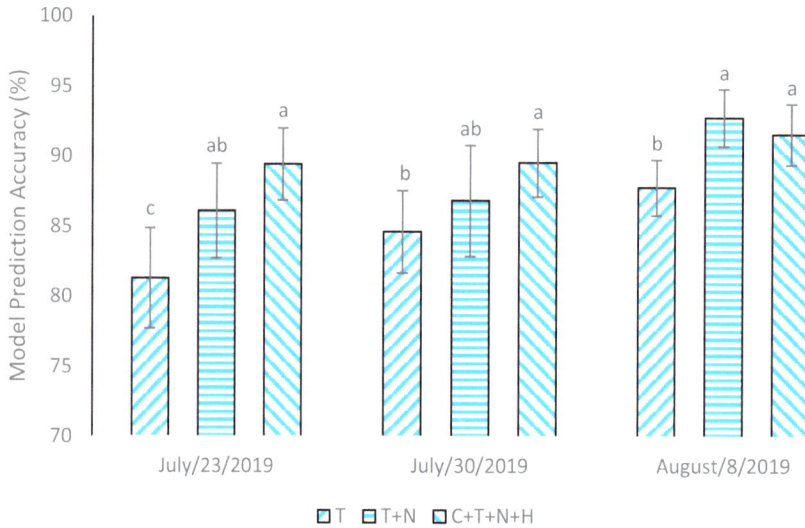

Fig. (10). Model prediction accuracies for three different days using one, two, and four features to train the Support Vector Machine model. (C, T, N, and H stands for color, texture, NDVI, and height feature, respectively; statistic comparisons for the three settings were conducted by Tukey Test at the level of 0.05 with different letters showing significant difference).

Since texture consists of four sub-features (*i.e.*, coarseness, contrast, direction, and linelikeness), we further explored prediction accuracies of using individual sub-features of texture, with results shown in Table **1**. Among the four sub-features, contrast and direction have higher model prediction accuracies than other sub-features. The model with only contrast has prediction accuracies of 75%±5% 69%±4%, and 81%±4% for the datasets of July/23, July/30, and August/8, respectively. When using direction sub-feature only, the model prediction accuracy is 71%±5%,81%±3%, 72%±5%, for the three datasets of July/23, July/30, and August/8, respectively. In addition, among all the three datasets, the sub-feature linelikeness consistently has the lowest model prediction accuracy, ranging from 63% to 67%.

Table 1. Model prediction accuracy using different sub-features of texture on data collected on different days.

Feature	July/23/2019	July/30/2019	August/8/2019
Coarseness	$72\pm3\%$	$70\pm5\%$	$67\pm4\%$
Contrast	$75\pm5\%$	$69\pm4\%$	$81\pm4\%$
Direction	$71\pm5\%$	$81\pm3\%$	$72\pm5\%$
Linelikeness	$67\pm5\%$	$63\pm5\%$	$64\pm5\%$

CONCLUSION

In this study, we first collected color and multi-spectral images by attaching two different cameras to drones for three days. Images were stitched together for an orthomosaic map and then individual plot images were cropped manually. Features of each plot were extracted, including color, texture, NDVI, and height. Each plot was classified into lodging or non-lodging visually. We randomly selected 80% of the data from each day to train the SVM and the remaining 20% for model testing. The trained model with four factors included had a prediction accuracy of ~91%, which demonstrated its desirable performance in lodging prediction. It was also found that the texture feature played an important role in improving the model prediction accuracy. The model prediction accuracy of using two features (texture and NDVI) is not significantly different from that of using four features (color, texture, NDVI, and height). Thus, this study validates a simple method of using fewer features (texture and NDVI) to replace four features to train an SVM model for lodging detection. This finding could significantly decrease further researchers' working load in data collection and processing. In addition, it was noticed that with time going on, the model prediction accuracy increased. A potential reason is that lodging is a dynamic process. Since the current method needs time for feature extraction and model training, future research should be directed towards real-time or near-real-time lodging detection. It would also be interesting to explore if this method could be used for other crops' lodging detection, such as corn, soybean, and canola. In addition, compared with other broadleaf crops (*e.g.*, corn), wheat has relatively small leaves. Thus, in this study, crop height information (*i.e.*, wheat digital surface model) does not perform well in predicting wheat lodging. However, further studies on correlating the proper height data percentile to the real wheat height could be conducted.

AUTHOR CONTRIBUTIONS

Z. Z. developed the main idea of this study and composed the manuscript, and PF helped with data processing.

FUNDING

This research was supported by the USDA.

CONSENT FOR PUBLICATION

Not applicable.

CONFLICT OF INTEREST

The authors confirm that this chapter contents have no conflict of interest.

ACKNOWLEDGEMENTS

Declared none.

REFERENCES

[1]   USDA, "Economic research service", Retrieved from: https://www.ers.usda.gov/topics/crops/wheat/

[2]   G.M. Borrelli, A. Troccoli, N. Di Fonzo, and C. Fares, "Durum wheat lipoxygenase activity and other quality parameters that affect pasta color", *Cereal Chem.,* vol. 76, no. 3, pp. 335-340, 1999.
        [http://dx.doi.org/10.1094/CCHEM.1999.76.3.335]

[3]   J.H. Cummings, L.M. Edmond, and E.A. Magee, "Dietary carbohydrates and health: do we still need the fibre concept?", *Clin. Nutr. Suppl.,* vol. 1, no. 2, pp. 5-17, 2004.
        [http://dx.doi.org/10.1016/j.clnu.2004.09.003]

[4]   A. Rodriguez-Mateos, C. Rendeiro, T. Bergillos-Meca, S. Tabatabaee, T.W. George, C. Heiss, and J.P. Spencer, "Intake and time dependence of blueberry flavonoid-induced improvements in vascular function: a randomized, controlled, double-blind, crossover intervention study with mechanistic insights into biological activity", *Am. J. Clin. Nutr.,* vol. 98, no. 5, pp. 1179-1191, 2013.
        [http://dx.doi.org/10.3945/ajcn.113.066639] [PMID: 24004888]

[5]   P.R. Shewry, and S.J. Hey, "The contribution of wheat to human diet and health", *Food Energy Secur.,* vol. 4, no. 3, pp. 178-202, 2015.
        [http://dx.doi.org/10.1002/fes3.64] [PMID: 27610232]

[6]   D. Vauzour, E.J. Houseman, T.W. George, G. Corona, R. Garnotel, K.G. Jackson, C. Sellier, P. Gillery, O.B. Kennedy, J.A. Lovegrove, and J.P. Spencer, "Moderate Champagne consumption promotes an acute improvement in acute endothelial-independent vascular function in healthy human volunteers", *Br. J. Nutr.,* vol. 103, no. 8, pp. 1168-1178, 2010.
        [http://dx.doi.org/10.1017/S0007114509992959] [PMID: 19943984]

[7]   M.J. Keenan, R.J. Martin, A.M. Raggio, K.L. McCutcheon, I.L. Brown, A. Birkett, S.S. Newman, J. Skaf, M. Hegsted, R.T. Tulley, E. Blair, and J. Zhou, "High-amylose resistant starch increases hormones and improves structure and function of the gastrointestinal tract: a microarray study", *J. Nutrigenet. Nutrigenomics,* vol. 5, no. 1, pp. 26-44, 2012.
        [http://dx.doi.org/10.1159/000335319] [PMID: 22516953]

[8]   G.E. Lobley, G. Holtrop, D.M. Bremner, A.G. Calder, E. Milne, and A.M. Johnstone, "Impact of short term consumption of diets high in either non-starch polysaccharides or resistant starch in comparison with moderate weight loss on indices of insulin sensitivity in subjects with metabolic syndrome", *Nutrients,* vol. 5, no. 6, pp. 2144-2172, 2013.
        [http://dx.doi.org/10.3390/nu5062144] [PMID: 23752495]

[9]   D. Badawy, H.M. El-Bassossy, A. Fahmy, and A. Azhar, "Aldose reductase inhibitors zopolrestat and

ferulic acid alleviate hypertension associated with diabetes: effect on vascular reactivity", *Can. J. Physiol. Pharmacol.,* vol. 91, no. 2, pp. 101-107, 2013.
[http://dx.doi.org/10.1139/cjpp-2012-0232] [PMID: 23458193]

[10]   K.J. Humphreys, M.A. Conlon, G.P. Young, D.L. Topping, Y. Hu, J.M. Winter, A.R. Bird, L. Cobiac, N.A. Kennedy, M.Z. Michael, and R.K. Le Leu, "Dietary manipulation of oncogenic microRNA expression in human rectal mucosa: a randomized trial", *Cancer Prev. Res. (Phila.),* vol. 7, no. 8, pp. 786-795, 2014.
[http://dx.doi.org/10.1158/1940-6207.CAPR-14-0053] [PMID: 25092886]

[11]   USDA, *North Dakota Agriculture.,* 2018. https://www.nass.usda.gov/Statistics_by_State/ North_Dakota/Publications/Annual_Statistical_Bulletin/2018/ND-Annual-Bulletin18.pdf

[12]   USDA, *National Agricultural Statistics Service,* 2019. https://www.nass.usda.gov/

[13]   M.J. Pinthus, "Lodging in wheat, barley, and oats: the phenomenon, its causes, and preventive measures", *Adv. Agron.,* vol. 25, pp. 209-263, 1974.
[http://dx.doi.org/10.1016/S0065-2113(08)60782-8]

[14]   P.M. Berry, and J. Spink, "Predicting yield losses caused by lodging in wheat", *Field Crops Res.,* vol. 137, pp. 19-26, 2012.
[http://dx.doi.org/10.1016/j.fcr.2012.07.019]

[15]   S. Mardanisamani, F. Maleki, S. Hosseinzadeh Kassani, S. Rajapaksa, H. Duddu, and M. Wang, "Crop lodging prediction from uav-acquired images of wheat and canola using a DCNN augmented with handcrafted texture features", *Proceedings of the IEEE Conference on Computer Vision and Pattern Recognition Workshops.,* 2019.

[16]   M. Neenan, and J.L. Spencer-Smith, "An analysis of the problem of lodging with particular reference to wheat and barley", *J. Agric. Sci.,* vol. 85, no. 3, pp. 495-507, 1975.
[http://dx.doi.org/10.1017/S0021859600062377]

[17]   M.J. Crook, and A.R. Ennos, "The mechanics of root lodging in winter wheat, *Triticum aestivum* L"., *J. Exp. Bot.,* vol. 44, no. 7, pp. 1219-1224, 1993.
[http://dx.doi.org/10.1093/jxb/44.7.1219]

[18]   M. Sterling, C.J. Baker, P.M. Berry, and A. Wade, "An experimental investigation of the lodging of wheat", *Agric. For. Meteorol.,* vol. 119, no. 3-4, pp. 149-165, 2003.
[http://dx.doi.org/10.1016/S0168-1923(03)00140-0]

[19]   D. Peng, X. Chen, Y. Yin, K. Lu, W. Yang, Y. Tang, and Z. Wang, "Lodging resistance of winter wheat (*Triticum aestivum* L.): Lignin accumulation and its related enzymes activities due to the application of paclobutrazol or gibberellin acid", *Field Crops Res.,* vol. 157, pp. 1-7, 2014.
[http://dx.doi.org/10.1016/j.fcr.2013.11.015]

[20]   H. Hitaka, "Studies on the lodging of rice plants", *Jpn. Agric. Res. Q.,* vol. 4, no. 3, pp. 1-6, 1969.

[21]   T.L. Setter, E.V. Laureles, and A.M. Mazaredo, "Lodging reduces yield of rice by self-shading and reductions in canopy photosynthesis", *Field Crops Res.,* vol. 49, no. 2-3, pp. 95-106, 1997.
[http://dx.doi.org/10.1016/S0378-4290(96)01058-1]

[22]   R.O. Weibel, and J.W. Pendleton, "Effect of artificial lodging on winter wheat grain yield and quality", *Agron. J.,* vol. 56, no. 5, pp. 487-488, 1964.
[http://dx.doi.org/10.2134/agronj1964.00021962005600050013x]

[23]   M. Stapper, and R.A. Fischer, "Genotype, sowing date and plant spacing influence on high-yielding irrigated wheat in southern New South Wales. II. Growth, yield and nitrogen use", *Aust. J. Agric. Res.,* vol. 41, no. 6, pp. 1021-1041, 1990.
[http://dx.doi.org/10.1071/AR9901021]

[24]   P.M. Berry, R. Sylvester-Bradley, and S. Berry, "Ideotype design for lodging-resistant wheat", *Euphytica,* vol. 154, no. 1-2, pp. 165-179, 2007.
[http://dx.doi.org/10.1007/s10681-006-9284-3]

[25]   P.M. Berry, M. Sterling, C.J. Baker, J. Spink, and D.L. Sparkes, "A calibrated model of wheat lodging compared with field measurements", *Agric. For. Meteorol.,* vol. 119, no. 3-4, pp. 167-180, 2003. [http://dx.doi.org/10.1016/S0168-1923(03)00139-4]

[26]   A.S. Peake, N.I. Huth, P.S. Carberry, S.R. Raine, and R.J. Smith, "Quantifying potential yield and lodging-related yield gaps for irrigated spring wheat in sub-tropical Australia", *Field Crops Res.,* vol. 158, pp. 1-14, 2014. [http://dx.doi.org/10.1016/j.fcr.2013.12.001]

[27]   M.D. Yang, K.S. Huang, Y.H. Kuo, H. Tsai, and L.M. Lin, "Spatial and spectral hybrid image classification for rice lodging assessment through UAV imagery", *Remote Sens.,* vol. 9, no. 6, p. 583, 2017. [http://dx.doi.org/10.3390/rs9060583]

[28]   B.J. Barnett, and O. Mahul, "Weather index insurance for agriculture and rural areas in lower-income countries", *Am. J. Agric. Econ.,* vol. 89, no. 5, pp. 1241-1247, 2007. [http://dx.doi.org/10.1111/j.1467-8276.2007.01091.x]

[29]   G. Enjolras, and P. Sentis, "Crop insurance policies and purchases in France", *Agric. Econ.,* vol. 42, no. 4, pp. 475-486, 2011. [http://dx.doi.org/10.1111/j.1574-0862.2011.00535.x]

[30]   H.H. Chang, and D. Zilberman, "On the political economy of allocation of agricultural disaster relief payments: application to Taiwan", *Eur. Rev. Agric. Econ.,* vol. 41, no. 4, pp. 657-680, 2013. [http://dx.doi.org/10.1093/erae/jbt037]

[31]   O. Okhrin, M. Odening, and W. Xu, "Systemic weather risk and crop insurance: the case of China", *J. Risk Insur.,* vol. 80, no. 2, pp. 351-372, 2013. [http://dx.doi.org/10.1111/j.1539-6975.2012.01476.x]

[32]   USDA, *Risk Management,* 2019. Retrieved from: https://www.rma.usda.gov/

[33]   B. Horvatic, *Using Drone Mapping for Crop Insurance.,* 2019. Retrieved from: https://www.precisionag.com/in-field-technologies/drones-uavs/using-d-one-mapping-for-crop-insurance/

[34]   T. Liu, R. Li, X. Zhong, M. Jiang, X. Jin, and P. Zhou, "Estimates of rice lodging using indices derived from UAV visible and thermal infrared images", *Agric. For. Meteorol.,* vol. 252, pp. 144-154, 2018. [http://dx.doi.org/10.1016/j.agrformet.2018.01.021]

[35]   X. Li, K. Wang, Z. Ma, and H. Wang, "Early detection of wheat disease based on thermal infrared imaging", *Nongye Gongcheng Xuebao (Beijing),* vol. 30, no. 18, pp. 183-189, 2014. [http://dx.doi.org/10.3901/JME.2014.10.183]

[36]   Z. Li, Z. Chen, L. Wang, J. Liu, and Q. Zhou, "Area extraction of maize lodging based on remote sensing by small unmanned aerial vehicle", *Nongye Gongcheng Xuebao (Beijing),* vol. 30, no. 19, pp. 207-213, 2014.

[37]   L.Y. Liu, J.H. Wang, X.Y. Song, C.J. Li, W.J. Huang, and C.J. Zhao, "The canopy spectral features and remote sensing of wheat lodging", *Journal of Remote Sensing.,* vol. 9, Beijing, no. 3, p. 323, 2005.

[38]   H. Yang, E. Chen, Z. Li, C. Zhao, G. Yang, and S. Pignatti, "Wheat lodging monitoring using polarimetric index from RADARSAT-2 data", *Int. J. Appl. Earth Obs. Geoinf.,* vol. 34, pp. 157-166, 2015. [http://dx.doi.org/10.1016/j.jag.2014.08.010]

[39]   B. Somers, G.P. Asner, L. Tits, and P. Coppin, "Endmember variability in spectral mixture analysis: A review", *Remote Sens. Environ.,* vol. 115, no. 7, pp. 1603-1616, 2011. [http://dx.doi.org/10.1016/j.rse.2011.03.003]

[40]   A.S. Laliberte, and A. Rango, "Texture and scale in object-based analysis of subdecimeter resolution

unmanned aerial vehicle (UAV) imagery", *IEEE Trans. Geosci. Remote Sens.,* vol. 47, no. 3, pp. 761-770, 2009.
[http://dx.doi.org/10.1109/TGRS.2008.2009355]

[41]   D. Turner, A. Lucieer, and C. Watson, "An automated technique for generating georectified mosaics from ultra-high resolution unmanned aerial vehicle (UAV) imagery, based on structure from motion (SfM) point clouds", *Remote Sens.,* vol. 4, no. 5, pp. 1392-1410, 2012.
[http://dx.doi.org/10.3390/rs4051392]

[42]   J.P. Dandois, and E.C. Ellis, "High spatial resolution three-dimensional mapping of vegetation spectral dynamics using computer vision", *Remote Sens. Environ.,* vol. 136, pp. 259-276, 2013.
[http://dx.doi.org/10.1016/j.rse.2013.04.005]

[43]   S. Yahyanejad, and B. Rinner, "A fast and mobile system for registration of low-altitude visual and thermal aerial images using multiple small-scale UAVs", *ISPRS J. Photogramm. Remote Sens.,* vol. 104, pp. 189-202, 2015.
[http://dx.doi.org/10.1016/j.isprsjprs.2014.07.015]

[44]   T. Chu, M. Starek, M. Brewer, S. Murray, and L. Pruter, "Assessing lodging severity over an experimental maize (*Zea mays* L.) field using UAS images", *Remote Sens.,* vol. 9, no. 9, p. 923, 2017.
[http://dx.doi.org/10.3390/rs9090923]

[45]   M. Du, and N. Noguchi, "Monitoring of wheat growth status and mapping of wheat yield's within-field spatial variations using color images acquired from UAV-camera system", *Remote Sens.,* vol. 9, no. 3, p. 289, 2017.
[http://dx.doi.org/10.3390/rs9030289]

[46]   G.E. Meyer, and J.C. Neto, "Verification of color vegetation indices for automated crop imaging applications", *Comput. Electron. Agric.,* vol. 63, no. 2, pp. 282-293, 2008.
[http://dx.doi.org/10.1016/j.compag.2008.03.009]

[47]   H. Tamura, S. Mori, and T. Yamawaki, "Textural features corresponding to visual perception", *IEEE Trans. Syst. Man Cybern.,* vol. 8, no. 6, pp. 460-473, 1978.
[http://dx.doi.org/10.1109/TSMC.1978.4309999]

[48]   T.N. Carlson, and D.A. Ripley, "On the relation between NDVI, fractional vegetation cover, and leaf area index", *Remote Sens. Environ.,* vol. 62, no. 3, pp. 241-252, 1997.
[http://dx.doi.org/10.1016/S0034-4257(97)00104-1]

[49]   H. Cen, L. Wan, J. Zhu, Y. Li, X. Li, Y. Zhu, H. Weng, W. Wu, W. Yin, C. Xu, Y. Bao, L. Feng, J. Shou, and Y. He, "Dynamic monitoring of biomass of rice under different nitrogen treatments using a lightweight UAV with dual image-frame snapshot cameras", *Plant Methods,* vol. 15, no. 1, p. 32, 2019.
[http://dx.doi.org/10.1186/s13007-019-0418-8] [PMID: 30972143]

[50]   V. Vapnik, *The nature of statistical learning theory.* Springer science & business media, 2013.

[51]   Y. Lu, and R. Lu, *Detection of surface and subsurface defects of apples using structured-illumination reflectance imaging with machine learning algorithms.,* 2018.
[http://dx.doi.org/10.13031/trans.12930]

[52]   Z. Zhang, P.H. Heinemann, J. Liu, T.A. Baugher, and J.R. Schupp, "The development of mechanical apple harvesting technology: A review", *Trans. ASABE,* vol. 59, no. 5, pp. 1165-1180, 2016.
[http://dx.doi.org/10.13031/trans.59.11737]

[53]   Z. Zhang, A.K. Pothula, and R. Lu, "A review of bin filling technologies for apple harvest and postharvest handling", *Appl. Eng. Agric.,* vol. 34, no. 4, pp. 687-703, 2018.
[http://dx.doi.org/10.13031/aea.12827]

[54]   Z. Zhang, C. Igathinathane, J. Li, H. Cen, Y. Lu, and P. Flores, "Technology progress in mechanical harvest of fresh market apples", *Comput. Electron. Agric.,* vol. 175, p. 105606, 2020.
[http://dx.doi.org/10.1016/j.compag.2020.105606]

[55]   Z. Zhang, P. Flores, C.L. Igathinathane, D. Naik, R. Kiran, and J.K. Ransom, "Wheat lodging detection from uas imagery using machine learning algorithms", *Remote Sens.,* vol. 12, no. 11, p. 1838, 2020.
[http://dx.doi.org/10.3390/rs12111838]

SUBJECT INDEX

A

Absorption 13, 18, 19, 29, 80, 142, 170, 174, 177, 181
  estimated 18
  resonant magnetic energy 13
  spectroscopy 29
AChE immobilization 53
Acouso-optic tunable filters (AOTF) 87, 141
Acoustic(s) 1, 2, 19, 22, 102, 168, 170, 182, 183, 185, 190, 192, 222
  collision 190
  energy 183
  resonance 185
  sensing technology 182, 185, 192
Acoustic signal 183, 190, 191
  device 191
Acoustic vibration 19, 20, 21, 102, 103, 104, 105, 106, 107
  method 20, 21, 102, 103, 104, 105, 106, 107
  technique 21
Acoustic wave 19, 187
  attenuation 187
  transmitted 19
Adaptability 55, 100, 192, 200, 255
  high environmental 255
Agricultural intelligent sensing technologies 169
Agricultural product quality 6, 18, 21, 168, 170, 172, 181, 196, 216
  application of optical sensor technology in 172, 181
  assessment 21
Agricultural products 2, 139, 168, 169, 173, 181, 185, 192, 197, 199, 222
  analyzing 192
  commercial 185
  contaminated 139
  dry 199
  granular 197
  industrial 192

irradiated 2
  quality detection 222
  safety detection 173
Agricultural product tissues 13, 180
Agricultural products processing 168, 169, 170, 172, 180, 181, 182, 193, 197, 200, 202, 205, 222
  industry 168, 169
  technology 169, 181
Agro-products 22, 51, 55, 58, 68, 216
  animal-derived 55
  contaminated 58
  evaluation 51, 68
  industry 22
  processing 216
Algorithms 17, 19, 90, 131, 143, 144, 146, 151, 157, 251, 254, 257
  advanced chemometric 151
  genetic 143
Allergens 49, 52, 61, 62, 66, 67
Allergic reactions, life-threatening 61
Analysis 4, 6, 11, 19, 52, 53, 103, 138, 143, 146, 147, 161, 222, 251
  agro-product 52
  chemical 143, 147, 222
  cluster 4, 19
  faecal contamination 11
  fluorescence data 146
  high-throughput 161
  high-throughput screening 138
  non-linear regression 251
  parallel factor 6
  sensitive 53
  theoretical 103
Anthocyanin 79, 207
Antigen, integrated exogenous 61
Apparent diffusion coefficient (ADC) 34, 208
Apple 244, 250
  grading system 244
  robot harvesting systems 250

www.ingramcontent.com/pod-product-compliance
Lightning Source LLC
Chambersburg PA
CBHW050811220326
41598CB00006B/173